Bioelectrochemistry of Biomembranes and Biomimetic Membranes

Bioelectrochemistry of Biomembranes and Biomimetic Membranes

Rolando Guidelli

WILEY

Published by John Wiley & Sons, Inc., Hoboken, New Jersey

Published simultaneously in Canada

Library of Congress Cataloging-in-Publication Data:

Names: Guidelli, R. (Rolando)
Title: Bioelectrochemistry of biomembranes and biomimetic membranes / Rolando Guidelli.
Description: Hoboken, New Jersey : John Wiley & Sons, Inc., [2017] | Includes bibliographical references and index.
Identifiers: LCCN 2016026090| ISBN 9781119271055 (cloth) | ISBN 9781119278405 (epub)
Subjects: LCSH: Biomimetic materials. | Bioelectrochemistry. | Membranes (Biology)
Classification: LCC R857.M3 G86 2017 | DDC 610.28–dc23 LC record available at https://lccn.loc.gov/2016026090

Set in 10/12pt Warnock by SPi Global, Chennai, India

Printed in the United States of America

10 9 8 7 6 5 4 3 2 1

To my grandchildren Neri and Petra

Contents

Preface

This book is largely based on the lecture notes for the course of Bioelectrochemistry, which I have been holding for students of the "biennio di laurea specialistica" (roughly equivalent to the last year of a UK college) at the Department of Chemistry of Florence University, from 2005 to 2010. It is accessible to graduate students and final year undergraduate students in chemistry and biology, as well as researchers in related disciplines including biology, physics, physiology, and pharmacology. It is particularly suitable for students attracted by the fascinating area of biological membranes and their functions and interested in reading a supplemental text that takes practically no argument for granted and leads the reader, step by step, through gradually more fundamental aspects of this area. To this end, the manuscript describes the essential electrochemical basics required to understand why and how electrochemical and electrophysiological tools are fundamental in elucidating the mode of ion transport across biomembranes. In this respect, it is also of interest to many electrochemists attracted by the biological realm (as it happened to me 20 years ago) and wishing to get a better view of the potentialities of the their own background and tools to move more closely into this area. On the other hand, it is of interest to biophysicists and biochemists willing to get an exhaustive overview of the potential of biomimetic membranes for the investigation of the function of membrane peptides and proteins.

The book deals with the bioelectrochemistry of biological membranes and their mimics in a homogeneous and thematically unified way. It is not a collection of selected, separate topics on the bioelectrochemistry of membranes. To understand in depth the structure and function of biological membranes, it is also essential to understand and apply principles of physical chemistry. In particular, the fundamental role played by the transmembrane potential in modulating the function of biomolecules incorporated in the membrane allows us to regard and treat the membrane as an outright electrified interface. Hence, to understand the function of biological membranes and the properties of their experimental models, called biomimetic membranes, a knowledge of some basic principles of electrochemistry and of the most significant electrochemical techniques is required. The purpose of this manuscript is to construct a coherent thermodynamic and electrochemical framework to achieve this goal.

The book is composed of seven chapters. In Chapter 1, some basic concepts of membrane biochemistry and the relative terminology are briefly outlined. Chapter 2 deals with the electrostatics of biomembranes; here, the relation between electric field and electric potential is briefly touched upon, in order to introduce a simplified derivation of the Poisson equation and, subsequently, the derivation of the Poisson–Boltzmann equation and the Gouy–Chapman theory, which are required to determine the profile of the electric potential across a bilayer lipid membrane. In this chapter, thermodynamic concepts are still not adopted, except for Boltzmann's factor, which is used to illustrate the competition between the ordering effect of the electric field and the disordering effect of temperature. Chapter 3 is devoted to thermodynamics; after a brief introduction to some basic concepts of chemical thermodynamics (including the concept of electrochemical potential, which is not clear to many students), the thermodynamics of irreversible processes is explained and its importance in providing the basis for the function of all ion pumps is pointed out. This chapter ends with a number of examples of coupling of chemical reactions to the vectorial transport of molecules across different types of membranes. The first three chapters serve to introduce the reader to the heart of the problem of passive and active transport across membranes, as carried out by embedded proteins. Passive transport by ion channels is dealt with in Chapter 4, whereas primary active transport by ion pumps and secondary active transport by transporters are discussed in Chapter 5. Particular emphasis is placed on the role of the transmembrane potential in modulating the function of ion channels and pumps. In Chapter 4, the channel current is discussed at two levels: at the first level, it is considered to be proportional to the transmembrane potential according to Ohm's law, with the necessary inclusion of the open probability; at the second level, it is obtained by derivation of the Goldman–Hodgkin–Katz equation, and the outcomes of the two approaches are compared. Two levels are also adopted for explaining the transmission of nerve impulses; the second level leads to the derivation of the equation for the action potential. Chapter 5 deals with the energy levels of the enzymatic cycle of ion pumps and describes in detail how this cycle can be investigated under both steady-state and pre-steady-state conditions by means of electrochemical/electrophysiological techniques. Chapter 6 describes the various biomimetic membranes used to incorporate peptides and proteins in order to investigate their functional activity by electrochemical means, starting from the traditional "bilayer lipid membrane" (BLM) and considering the evolution toward more robust and sophisticated experimental models of biomembranes. Before classifying them in detail on the basis of their structural features and before scrutinizing both their advantages and drawbacks, a number of electrochemical techniques used for their investigation is described and illustrated by selected examples. Possible future developments and applications are foreseen. Finally, Chapter 7 describes the salient features of a number of nonelectrochemical, auxiliary techniques that are frequently employed to characterize solid-supported biomimetic membranes, either at a macroscopic or a molecular level; they are complemented by a few illustrative applications.

I am very grateful to those publishers who have allowed me to reproduce figures that appeared in their own publications, and to Wikipedia for the nice images that I downloaded from Wikimedia Commons. The sources of each of these are indicated in the legends. I would also like to acknowledge the constant collaboration of Lucia Becucci, who started working with me on mercury-supported biomimetic membranes as a student in 1992 and has been conducting research in this area until now, demonstrating notable skillfulness in organizing and carrying out research work and in devising stimulating experiments. It is a pleasure to thank my colleague Jacek Lipkowski for his invaluable suggestions and helpful criticism on Chapter 7. Special thanks go to Ms. Anita Lekhwani for her cheerful assistance in the preparation of this book proposal and to Ms. Sumathi Elangovan, the Project Editor, for her helpful advice and excellent editorial suggestions. Recognition should also be accorded to my daughter Ilaria and her husband, Niccolò Amerini, for managing my computer's performance.

1

Biological Membranes

1.1 Introduction

Biological membranes maintain the spatial organization of life. The first living cells felt the need to be enveloped by a selectively permeable barrier to protect and shield the set of their life-sustaining chemical transformations from changes in the environment. This envelope, which encloses all living cells, is the plasma membrane. The plasma membrane prevents undesirable agents from entering the cell, while keeping needed molecules on its inside. To function effectively, it must also selectively pass molecules, ions, and signals from one side to the other. Its permeability properties ensure that essential molecules such as glucose, amino acids, and lipids readily enter the cell, metabolic intermediates remain in the cell, and waste compounds leave the cell. The organization of the metabolic activity of the cell requires an additional compartmentalization in its inside, via specialized subunits called organelles, which are again enveloped by a membrane and carry out specific functions. The aqueous solution inside the organelles often contains solutes that are different from those of the solution directly contained in the cell (the *cytosol*), and the organelle membranes maintain such a difference.

Since all membranes are interposed between two aqueous media, the strategy underlying their function involves creating a hydrophobic barrier, which is formed by a thin lipid layer. The water-soluble compounds present both within cells and organelles as well as outside of them are not soluble in the lipid medium of the membrane, and pass slowly or not at all through it. The lipid material is particularly convenient for a number of reasons. Thus, it assembles spontaneously into two juxtaposed lipid monolayers that generate a highly hydrophobic region in their interior, while exposing a relatively hydrophilic surface to the aqueous solutions bathing the two sides of the membrane. Moreover, the lipid bilayer so formed is highly fluid and allows an easy incorporation of different biomolecules capable of spanning it and of transferring molecules, energy, or information across it in a selective way. Finally, the hydrophobic interior of the lipid bilayer, thanks to its low dielectric constant, concentrates by far the majority of the electric potential difference between the two sides of the membrane, that is, the *transmembrane potential*. The change of the transmembrane potential

Bioelectrochemistry of Biomembranes and Biomimetic Membranes, First Edition. Rolando Guidelli.

over time, often induced by some external stimulus, modulates the function of many biomolecules incorporated in the membrane.

Life, similar to all other processes in our universe, obeys the laws of physics and chemistry. Consequently, all biological processes may occur only if they are accompanied by a decrease in the corresponding Gibbs energy, in accordance with the second law of thermodynamics. Some biological processes, when taken separately and out of their context, may seem to proceed with an increase in Gibbs energy. However, an attentive examination of their context reveals unavoidably that they are intimately coupled to some other process that proceeds with a decrease in Gibbs energy, such that the combination of the two is still characterized by a decrease in Gibbs energy. This coupling finds its justification in the thermodynamics of irreversible processes.

The thickness of the plasma membrane is much smaller than the radius of a cell, allowing us to treat the membrane surface as if it were planar; this simplifies the mathematics to a significant extent. To understand in depth the structure and function of biological membranes, it is also essential to understand and apply the principles of physical chemistry. In particular, the fundamental role played by the transmembrane potential in modulating the function of the biomolecules incorporated in the membrane allows us to consider and treat the membrane as a proper *electrified interface*. Hence, to understand the function of biological membranes and the properties of their experimental models, called *biomimetic membranes*, a knowledge of some basic principles of electrochemistry and of the most significant electrochemical techniques is required. The purpose of this work is to construct a coherent thermodynamic and electrochemical framework to achieve this goal. The role of the electrochemical foundations and techniques for the investigation of processes of biological relevance was first recognized in the 1980s, when Gutmann and Keyzer (1986) coined the name of *bioelectrochemistry* to denote this area of science.

In addition to membrane processes, bioelectrochemistry deals with the investigation of the properties and functions of water-soluble biomolecules, usually by adsorbing them on surface-modified and derivatized electrodes and by studying electron transfer reactions between them and the electrode (Bartlett, 2008; Alkire et al., 2011). However, the electrochemical behavior of biomolecules at electrodes does not necessarily pertain to bioelectrochemistry: this is true only if it provides some useful piece of information on the role played by these molecules in biological processes. These aspects of bioelectrochemistry are beyond the scope of this book. Bioelectrochemistry has also many applications in practical devices such as biosensors and biofuel cells.

1.2 The Biological Membranes

To stay alive, all living things need biological membranes (briefly, *biomembranes*). Biomembranes are thin layers that form the outer boundary of living cells, separating their inside (the *cytoplasm*) from their outside (the *extracellular fluid*). The cytoplasm comprises the *cytosol* (a gel-like substance enclosed within the cell membrane) and a number of substructures called *organelles*, which are also enclosed by a membrane. The membrane enclosing a cell is called *plasma membrane*. One important component of biological membranes consists of two monolayers (*leaflets*) of lipid molecules (Fig. 1.1). Lipid molecules are *amphiphilic*, that is, they have a hydrophobic section (the *hydrocarbon tail*) and a hydrophilic section (the *polar head* or *headgroup*). In biomembranes, the two lipid monolayers are oriented with the hydrocarbon tails directed toward each other and the polar heads turned toward the aqueous solutions that bath the two sides of the membrane. The resulting *lipid bilayer*, about 6 nm thick, is a matrix that incorporates different proteins performing a variety of functions.

Biomembranes form a highly selective barrier between the inside and the outside of living cells. They are highly insulating to inorganic ions, and large ion concentration gradients can be maintained across them. The permeability and structural properties of biological membranes are sensitive to the chemical nature of the membrane components and to events that occur at the interface or within the bilayer. For example, biomembranes provide the environmental matrix for proteins that specifically transport certain ions and other molecules, for receptor proteins and for signal transduction molecules. The lipid and protein portions of biomembranes are also sensitive to the presence of *lipophilic* perturbants, that is, molecules with a high affinity for lipids. Anesthetics, for example, readily partition into lipid membranes, altering their electrical and permeability characteristics. The various responses observed in biomembranes are concentration dependent, usually very rapid and reversible, and frequently dependent upon the transmembrane potential.

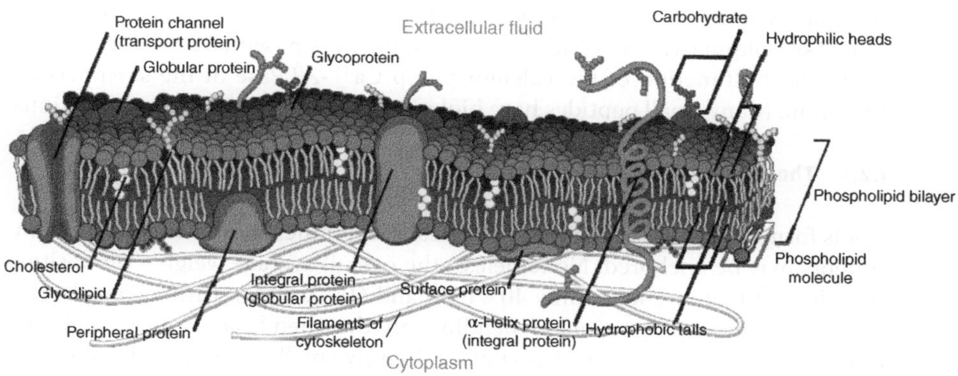

Figure 1.1 Schematic picture of a plasma membrane, showing the bimolecular layer of lipid molecules (including cholesterol), integral proteins spanning the lipid bilayer, peripheral proteins, filaments of cytoskeleton (the cellular "scaffolding" present in the cytoplasm), as well as glycoproteins, which expose their covalently attached oligosaccharide chains (glycans) to the extracellular fluid. Source: https://commons.wikimedia.org/wiki/File:Cell_membrane_detailed_diagram_de.svg.

Proteins are composed of linear chains of polymers of amino acids (*polypeptide chains*) linked by amide bonds, called *peptide bonds*. Some of the proteins (the *structural proteins*) simply support the texture of the membrane. A more important group of proteins (the *functional proteins*) participates directly in membrane processes such as flow of matter, energy, or information. Some proteins (the *integral proteins*) are embedded in the lipid bilayer with the hydrophobic section of their polypeptide chains, and protrude from the bilayer surface into one or both the adjacent aqueous solutions with the extrinsic, more hydrophilic section of the chains. Other proteins (the *peripheral proteins*) are weakly bound to the surface of the bilayer by electrostatic interactions or by hydrogen bonds, and interact with the polar heads of the lipid or with the integral membrane proteins. These are *globular proteins*, namely proteins in which the polypeptide chain folds spontaneously in a way that removes the hydrophobic sections from contact with the aqueous solution, burying them in the interior of the protein; conversely, the hydrophilic sections remain on the surface of the protein, where they form hydrogen bonds with water and between themselves. These proteins are water soluble. Some of them (e.g., cytochrome *c*, plastocyanin, ferredoxin) contain electrophilic metal ions and exchange electrons with the integral proteins. The majority of redox proteins, namely proteins containing one or more redox sites, have no biological function when taken alone; rather, they are associated with other redox proteins.

Short linear chains of polymers of amino acids linked by peptide bonds are called *peptides*. The distinction between peptides and proteins is not well defined. This distinction is often based on the number of amino acid residues. Peptides are considered to have less that 50 residues, polypeptides from 50 to 100 residues, and proteins more than 100 residues. However, according to a different classification, even polypeptide chains of less than 50 residues are referred to as proteins, provided they have a well-defined *secondary structure*, consisting of regularly repeating conformations of their polypeptide backbone. This is the case of sarcolipin (with 31 residues) or phospholamban (with 51 residues), which modulate the function of the calcium pump Ca^{2+}-ATPase of the sarcoplasmic reticulum. Many small peptides have biological activity and important functions.

1.2.1 The Lipids and the Lipid Bilayer

Lipids form the basic structure of biological membranes in/on which other components may be anchored. Nonetheless, the overall average weight composition of biological membranes is only 40% lipid and 60% protein, with a considerable variation in this proportion, depending on the membrane type. Lipids are amphiphilic compounds consisting of a relatively small hydrophilic head bound to two long hydrocarbon tails. They can be classified into derivatives of glycerol (*glycerolipids*), derivatives of sphingosine (*sphingolipids*), and *sterols*.

In glycerolipids, the glycerol is usually esterified by two fatty acids and by phosphoric acid. Without further esterification, one obtains *phosphatidic acid* (PA). However, the phosphate group of PA almost always undergoes a further esterification with ethanolamine, choline, or serine, giving rise to

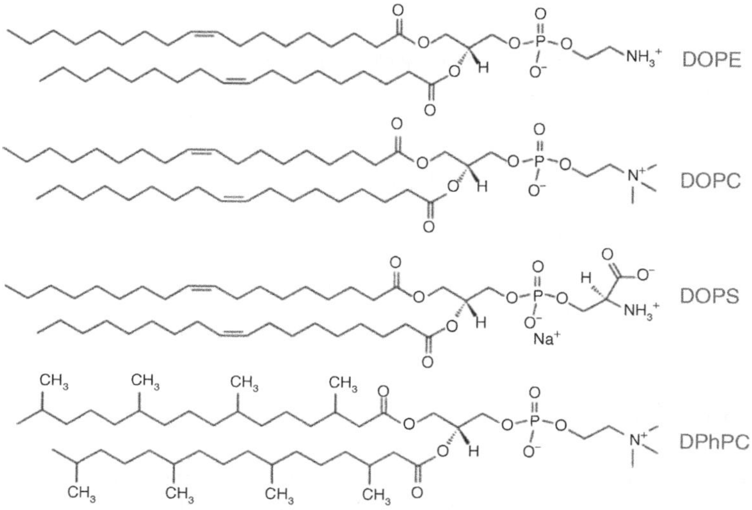

Figure 1.2 Structures of dioleoylphosphatidylethanolamine (DOPE), dioleoylphosphatidylcholine (DOPC), dioleoylphosphatidylserine (DOPS), and diphytanoylphosphatidylcholine (DPhPC).

phosphatidylethanolamine (PE), *phosphatidylcholine* (PC), and *phosphatidylserine* (PS) (Fig. 1.2). At physiological pH, both PE and PC are neutral, albeit zwitterionic, while PS carries a negative charge. The variously esterified phosphate group constitutes the hydrophilic polar head of the glycerolipid.

Sphingolipids are based on the backbone of *sphingosine,* a long-chain alcohol containing a hydrocarbon tail, an amino group that may attach a fatty acyl group yielding *ceramide,* and a C-1 hydroxyl group (Fig. 1.3). Esterification of the C-1 hydroxyl group with phosphoric acid and esterification of the latter with choline yields *sphingomyelin,* one of the most important lipids of this class. On the other hand, if the C-1 hydroxyl group of sphingosine binds to a sugar residue by a β-glycosidic linkage, it gives rise to a *glycosphingolipid.* Sphingolipids are particularly abundant in the tissues of the nervous system. Sterols are rigid, compact molecules derived from a system of four fused saturated rings (Fig. 1.4). The major sterol in the plasma membrane of all mammalian cells is *cholesterol.* Plant membranes do not contain cholesterol, but contain other sterols, such as *ergosterol.*

Glycerophospholipids and sphingolipids dispersed in an aqueous phase tend to spontaneously form lipid bilayers. In the absence of a solid support, these bilayers have a spherical shape and consist of multilamellar aggregates in which stacked lipid bilayers are separated by water layers (*multilamellar vesicles*). Excess water and rapid stirring may break these aggregates giving rise to *unilamellar vesicles,* which consist of a single spherical lipid bilayer (Fig. 1.5). The term *liposome* is used to denote both multi- and unilamellar vesicles. *Micelles* are a different type of lipid aggregates in which the hydrocarbon tails of the lipid molecules converge

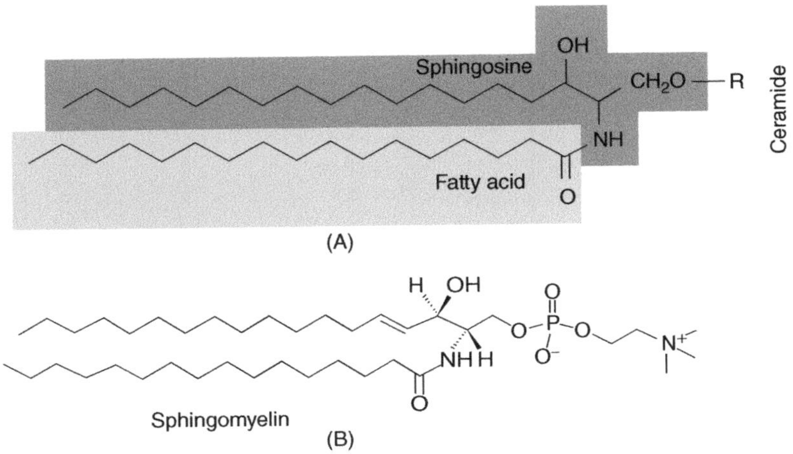

Figure 1.3 (A) Basic building blocks of sphingolipids; ceramide consists of the combination of the two differently highlighted portions, with R = H. (B) Structure of the predominant form of egg sphingomyelin.

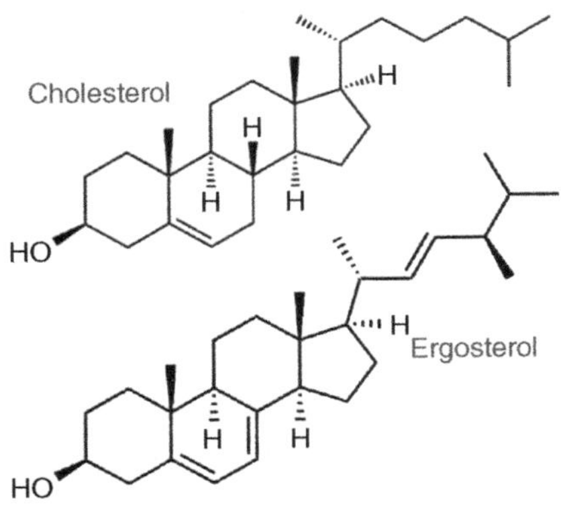

Figure 1.4 Structures of cholesterol and ergosterol.

on the center of the aggregate, while their polar heads are exposed to the aqueous solution.

Lipid bilayers are formed when the overall cross-sectional area of the two hydrocarbon tails of the lipid molecule is almost equal to that of the polar heads. Thus, *lysophospholipids*, which have only one hydrocarbon tail, cannot form lipid bilayers, because the head is appreciably larger than the tail. Instead, they tend to form micelles in aqueous solution. Lipid bilayer formation is accompanied by a decrease in the Gibbs energy, $G = H - TS$, and, hence, occurs spontaneously. The decrease in the entropy S, due to the increased order of lipid molecules in the bilayer, is more than compensated for by the increased translational entropy of the water molecules escaped from direct contact with the lipid. Moreover, the attractive van der Waals interactions between adjacent lipid molecules within the bilayer and the increase in the number of hydrogen bonds of the water molecules

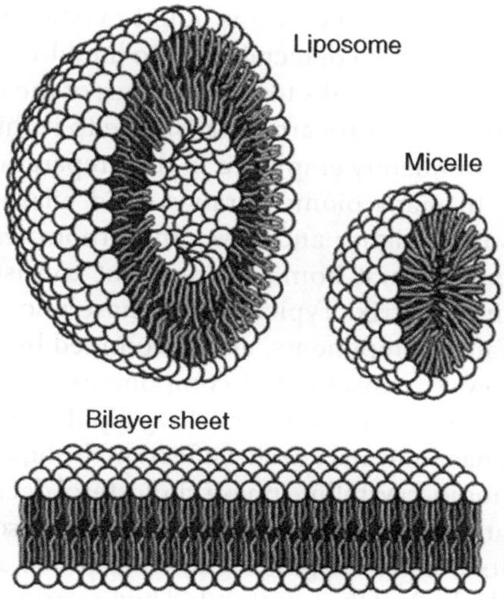

Figure 1.5 Sections of a unilamellar vesicle, a micelle, and a lipid bilayer. Source: https://commons.wikimedia.org/wiki/File:Phospholipids_aqueous_solution_structures.svg.

escaped from contact with the lipid contribute in decreasing the enthalpy H of the system.

A lipid bilayer can exist in a *solid-ordered* (s_o) *state,* called *gel state,* if the temperature is low enough. The gel state is anisotropic, tightly packed, and has limited lateral mobility and axial rotation of the hydrocarbon tails. As temperature is increased, a *melting temperature* T_m is reached at which a phase transition from the s_o state to a *liquid-crystalline state* takes place. The latter state is isotropic, loosely packed, and has a high degree of lateral mobility and axial rotation. However, even in this state, the lipid molecules do not easily pass from one leaflet of the bilayer to the other, due to the high resistance opposed by the hydrophobic core of the bilayer to the penetration of the hydrophilic polar head, as required by this *flip-flop* movement.

Lipids with one or more carbon–carbon double bonds (*unsaturated lipids*) have low T_m values, because each double bond produces a kink in the acyl chain, creating extra free space within the bilayer, disrupting its regular packing, and imparting additional flexibility to the hydrocarbon tails. On the other hand, lipids without carbon–carbon double bonds (*saturated lipids*) have melting temperatures that may exceed room temperature, if the length of their hydrocarbon tails is long enough. Longer hydrocarbon tails increase the van der Waals interactions between adjacent lipid molecules, decreasing the lipid mobility. Increasing the length of a saturated hydrocarbon tail by one carbon usually increases the melting temperature of a saturated lipid by 10 °C or less. In particular, saturated phospholipids with hydrocarbon tails longer than 14 carbons are in the s_o state at room temperature, while those with fewer than 14 are in the liquid-crystalline state. Thus, dipalmitoylphosphatidylcholine and palmitoylsphingomyelin, with saturated, 16-carbon hydrocarbon tails, have a T_m value of 41 °C, while dioleoylphosphatidylcholine, with two 18-carbon monounsaturated tails, has a T_m value of −20 °C. However, DPhPC (Fig. 1.2), albeit saturated, forms liquid-crystalline bilayers at room temperature and does not exhibit a detectable gel to liquid-crystalline phase transition from −120 °C to +120 °C, thanks to the presence of the methyl groups regularly distributed along the two hydrocarbon tails. Due to its high fluidity and resistance to oxidation, it is frequently employed in the preparation of biomimetic membranes.

In many biomembranes, lipid bilayers consist of mixtures of lipids with T_m values higher and lower than room temperature. In this case, the lipid bilayer is often not homogeneous, but consists of lipid microdomains immersed in a lipid matrix. Typically, gel-phase microdomains, enriched in the high-T_m saturated components, are surrounded by a liquid-crystalline phase enriched in the low-T_m unsaturated components (Brown and London, 1998; 2000). The presence of cholesterol in these lipid mixtures has a disordering effect on the gel phase and an ordering effect on the liquid-crystalline phase. More precisely, in the liquid-crystalline phase cholesterol decreases the axial rotation of acyl chains and, at concentrations above 30 mol%, it also decreases lateral diffusion. This is due to its nonpolar rigid ring system, which allows the cholesterol molecule to fit into the kink of the unsaturated hydrocarbon tails; this increases van der Waals contact and decreases the membrane fluidity. Conversely, cholesterol increases lateral diffusion and axial rotation in the gel-phase microdomains, causing them to become isotropic (Parasassi et al., 1995). This is revealed by a morphological investigation of these lipid mixtures by two-photon excitation fluorescence microscopy (Bagatolli, 2006). Thus, gel-phase microdomains have irregular shapes due to their anisotropic structure. Conversely, the microdomains in the presence of cholesterol have a roundish shape, because they are isotropic similar to the surrounding liquid-crystalline phase. In other words, cholesterol causes the microdomains to pass from the gel phase to the so-called *liquid-ordered* (l_o) phase. This phase is considered as liquid, thanks to its sufficient lateral diffusion, and ordered because it is more tightly packed than the liquid-crystalline phase, due to the still modest axial rotation of its acyl chains. These l_o microdomains are called *lipid rafts* (Pike, 2004) (Fig. 1.6). In contraposition with the l_o phase, the liquid-crystalline phase is often referred to as the *liquid-disordered* (l_d) phase.

The effect of cholesterol on gel-phase microdomains is again ascribable to its nonpolar rigid ring system, which prevents close contact between the saturated hydrocarbon tails, acting as an impurity that lowers the T_m of the microdomain. The dual role of cholesterol in animal plasma membranes serves to maintain the fluidity of the membrane constant through fluctuations in temperature. Lipid rafts are considered to regulate the membrane function in *eukaryotic* cells, namely cells containing a nucleus, as distinct for *prokaryotic* cells, which do not contain it (Simons and Ikonen, 1997). They are involved in cell signaling and molecular trafficking. A number of peripheral proteins tend to associate with lipid rafts, thus communicating more easily between themselves.

The lipid composition of biomembranes, albeit constant for a given cell type, varies notably in passing from one cell to another. Membranes that do not contain cholesterol, such as the inner and outer membranes of mitochondrion, do not form lipid rafts and are often in the l_d state. Lipids are distributed asymmetrically

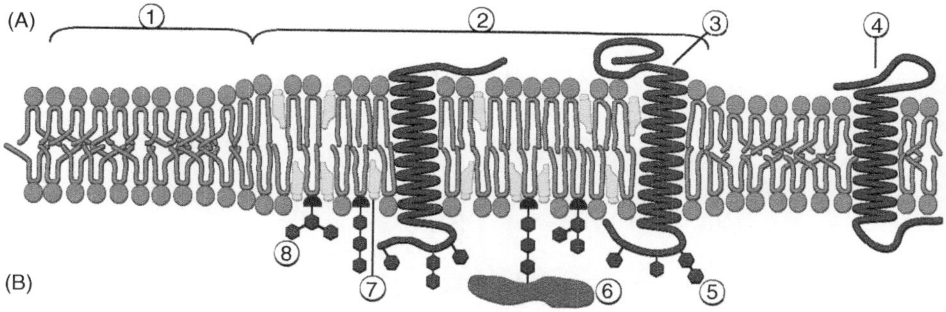

Figure 1.6 Schematic picture of a lipid raft (2), with a predominance of sphingolipids (8) and cholesterol (7), surrounded by the liquid-disordered matrix (1). (A) denotes the cytosolic side, (B) the extracellular side. Source: https://commons.wikimedia.org/wiki/File:Lipid_raft_organisation_scheme.svg.

between the inner and outer leaflets of the bilayer, thanks to the very slow flip-flop movement. For example, the plasma membrane of human red blood cells (the *erythrocytes*) contains sphingomyelin and PC mainly in the outer leaflet, and PE and the negatively charged PS exclusively in the inner leaflet.

1.2.2 The Membranes of Cells and Organelles

The *cytoplasm*, enclosed within the plasma membrane, is a highly organized system concerned with the activity of the cell (Fig. 1.7). The plasma membrane separates this highly organized system from the relative chaos that exists outside the cell. To maintain and increase its high degree of organization and to respond to its environment, the cell requires a continuous exchange of matter, energy, and information with its surroundings, which takes place across the plasma membrane. Analogous exchanges also take place across other important membranes that envelope several *organelles*; these subunits reside inside eukaryotic cells and are suspended in the *cytosol*, the aqueous medium of the cytoplasm (Rawn, 1989). The *nucleus* is an organelle that contains more than 95% of the cell genetic material, organized as multiple long linear *deoxyribonucleic acid* (DNA) molecules, and is the control center of eukaryotic cells. The nucleus is enclosed by a double membrane called *nucleus envelope*, which allows the movement of proteins and *ribonucleic acid* (RNA) from the nucleus to the cytoplasm. DNA replication and RNA transcription, the first steps in the expression of genetic information, take place in the nucleus.

Another important organelle, the *mitochondrion*, is the site of the major part of the energy production of *aerobic* cells, that is, cells that require oxygen for their function. It is enveloped by an unfolded outer membrane and by an inner membrane folded into membrane curtains called *cristae*. The aqueous solution enclosed by the inner membrane is called *matrix*, whereas the region enclosed between the inner and outer membranes is called *intermembrane space*.

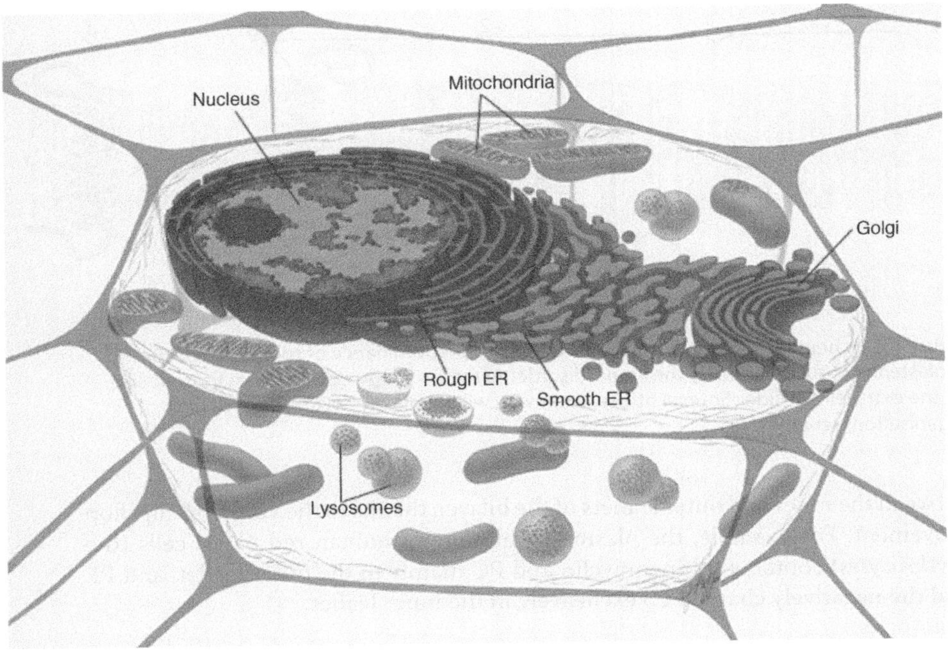

Figure 1.7 Cross section of an animal cell. Source: https://commons.wikimedia.org/wiki/File:Animal_cell_NIH.jpg.

The matrix is the site of *glycolysis* and the *citric acid cycle*, which consist of numerous enzymatic oxidation reactions of carbohydrates, amino acids, and fatty acids. The final product of many of these reactions is the reduced form of *nicotinamide adenosine dinucleotide* (NADH). The inner mitochondrial membrane incorporates a system of three multiprotein electron-transport chain complexes that carry electrons from NADH to molecular oxygen across the membrane (cf. Section 3.4.2), according to the net reaction:

$$\mathrm{NADH} + 1/2\mathrm{O_2} + \mathrm{H^+} \rightarrow \mathrm{H_2O} + \mathrm{NAD^+} \tag{1.1}$$

The net effect of this *respiratory electron-transport chain* is the generation of an electrochemical potential gradient of protons across the inner mitochondrial membrane; the driving force that moves protons from the matrix to the intermembrane space is sometimes referred to as the *proton-motive force*. The three multiprotein complexes are also referred to as the *proton-pumping oxidoreductases*. The protons pumped into the intermembrane space move back into the matrix spontaneously along an integral protein embedded in the inner mitochondrial membrane, called *ATP synthase*, which exploits part of the Gibbs energy involved in this spontaneous proton flow to synthesize *adenosine 5′-triphosphate* (ATP). This highly energetic molecule powers several cellular processes, such as H^+, Na^+, K^+, and Ca^{2+} pumping across membranes, by dissociating into *adenosine*

5′-diphosphate (ADP) and inorganic phosphate, briefly denoted by P_i. A respiratory electron-transport chain similar to that present in the inner mitochondrial membrane is also found in the plasma membrane of bacteria. The respiratory electron-transport chain is a typical example of energy flow across membranes.

Other membranes present in eukaryotic cells are those enveloping the *rough endoplasmic reticulum*, which form a network through the cell and merge with the outer membrane of the nucleus envelope (Fig. 1.7). They are coated with *ribosomes,* large complexes of RNA and proteins that catalyze the formation of proteins from individual amino acids, using messenger RNA as a template. The *smooth endoplasmic reticulum* has no attached ribosomes and is in charge of lipid production. It is associated with a membrane system consisting of sets of flattened sacs called the *Golgi complexes,* where proteins are provided with carbohydrate or lipid moieties, yielding glycoproteins or lipoproteins.

Muscle fiber cells contain a specialized form of endoplasmic reticulum called *sarcoplasmic reticulum*, a network of pockets that acts as a calcium reservoir, where the calcium concentration is from three to four orders of magnitude higher than in the cytoplasm. Excitation by a nerve impulse changes the electric potential across the membrane of the muscle fiber cell, inducing the sarcoplasmic reticulum to release Ca^{2+} ions into the cytoplasm (called the *sarcoplasm* in muscle cells) and promoting muscle contraction. The calcium pump *Ca^{2+}-ATPase*, an integral protein embedded in the membrane of the sarcoplasmic reticulum and powered by ATP, drives Ca^{2+} ions from the sarcoplasm back into the lumen of the sarcoplasmic reticulum, promoting muscle relaxation. Cells also contain vesicles filled with several enzymes, called *lysosomes*. They digest excess of worn-out organelles and food particles, and engulf viruses or bacteria. When a cell dies, its lysosomes rupture, and the enzymes so released hydrolyze the components of the dead cell.

The shape of cells is maintained by a protein scaffolding consisting of a network of different structures, called *cytoskeleton,* contained within the cytoplasm of all cells; it is made up of three types of protein filaments: *actin filaments* (also called microfilaments), *intermediate filaments,* and *microtubules*. This ordered network of filaments forms a band just beneath the plasma membrane; besides providing mechanical support to the plasma membrane, it has additional functions, such as moving organelles within the cell or linking transmembrane proteins (especially receptors) to cytoplasmic proteins.

The structure of plant cells is in many ways similar to that of animal cells. However, plant cells also contain specialized organelles called *chloroplasts,* which are the site of green-plant photosynthesis (Fig. 1.8). The chloroplast is enveloped by a double membrane that surrounds an aqueous solution called *stroma*. This contains a continuous membrane called *thylakoid membrane,* which is highly folded into flattened, sac-like vesicles, called *grana* (singular: *granum*), and encloses an internal space (the *lumen*). An electron-transport chain, consisting of a sequence of redox proteins embedded in the thylakoid membrane, absorbs light and couples the photon energy with a redox reaction. This process transfers electrons from water to the oxidized form of *nicotinamide adenine dinucleotide phosphate* ($NADP^+$) yielding its reduced form, NADPH, and molecular oxygen

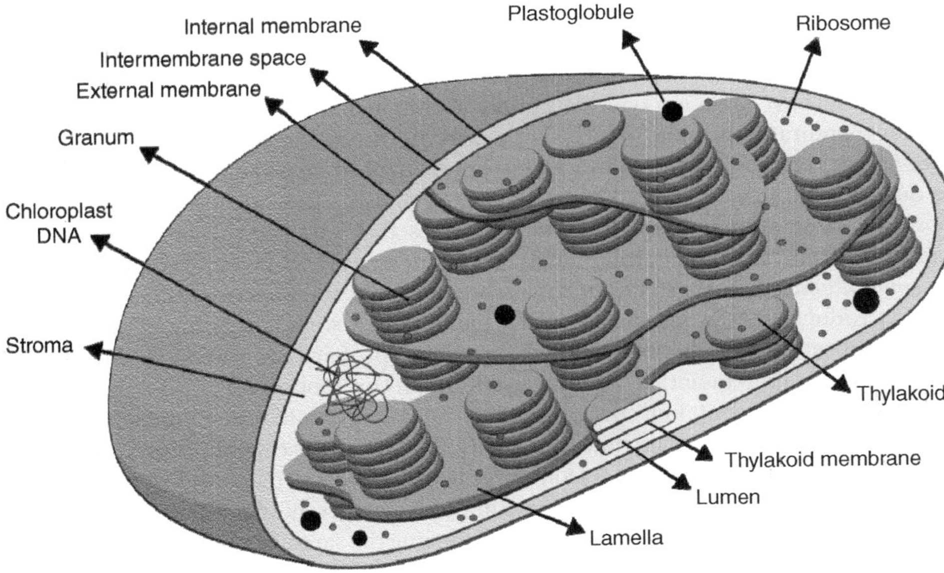

Figure 1.8 Cross section of a chloroplast. Source: Miguelsierra, https://commons.wikimedia.org/wiki/File:Scheme_Chloroplast-en.svg. Used under CC BY-SA 4.0 https://creativecommons.org/licenses/by-sa/4.0/.

(cf. Section 3.4.3). Such a reaction is accompanied by an increase in Gibbs energy and, therefore, it could not occur spontaneously without coupling with light absorption. Even this electron-transport chain generates an electrochemical potential gradient of protons across the thylakoid membrane. The spontaneous proton flow from the stroma to the lumen along an *ATP synthase* embedded in the thylakoid membrane is exploited by this protein to synthesize ATP from ADP, similar to that in mitochondrial respiration. This is a further example of energy flow across membranes.

The *vacuoles* of plant and fungal cells are analogous to animal lysosomes in their relatively low pH and in their content of enzymes (called *proteinases* or *peptidases*), which hydrolyze the peptide bonds of the polypeptide chains of proteins. The smooth flat membrane that surrounds them is called *tonoplast*.

1.3 The Proteins

A membrane incorporates a number of proteins, which are constituted by linear chains of polymers of amino acids. If a protein is hydrolyzed under accurately controlled conditions, it releases up to a maximum of 20 different amino acids. Since a genetic code is responsible for the formation of the amino acid chain, and since all organisms use the same genetic code, it is probable that evolution of life stared from a common ancestral cell. The conformation and function of a protein are determined by its amino acid composition and by the sequence in which they are bound to form the polypeptide chain. This sequence constitutes the *primary structure* of the protein. Each protein amino acid has a central carbon, referred to as the α-carbon, which is attached to four different groups: a basic amino group ($-NH_2$), an acidic carboxyl group ($-COOH$), a hydrogen atom, and a group called *side chain*, which characterizes one amino acid with respect to all others (Fig. 1.9).

Of the 20 protein amino acids, 11 have hydrophobic side chains (alkyl chains or aromatic rings, with the exclusion of cysteine and methionine, which also have a sulfur atom); 4 have hydrophilic side chains, due to the presence of a hydroxyl or carbamidic group ($H_2N-C=O$); 3 have positively charged side chains, due to the presence a charged nitrogen atom (lysine, arginine, and histidine); and 2 have a negatively charged chain due to the presence of a carboxyl group (aspartate and glutamate).

At physiological pH, amino acids are present in the *zwitterionic form*, in which the proton of the carboxyl group is donated to the amino group, giving rise to a neutral dipolar molecule. When amino acids join up in a chain, the α-carboxylate group of one amino acid condenses with the α-amino group of another, with loss of a water molecule and formation of a particular amide bond called *peptide bond* (Fig. 1.9). Formation of a sequence of peptide bonds generates a polypeptide, a long, continuous, and unbranched peptide, which is the structural element of proteins. The linked amino acids are referred to as the *amino acid residues*. The free amino group and the free carboxyl group at opposite ends of a polypeptide chain are called *termini*: amino terminus or N-terminus on one side, and carboxyl terminus or C-terminus on the opposite side. Both groups are ionized at pH 7. The side chains, denoted by R_1 and R_2 in Fig. 1.9, confer different properties to amino acids, and hence to the corresponding amino acid residues. Proteins consist of one or more polypeptides arranged in a biologically functional way and are often bound to *cofactors*, nonprotein chemical compounds that assist the protein in its transformations.

Integral proteins consist of polypeptide chains alternating, on the average, more hydrophobic and more hydrophilic sections. They are organized in the membrane with the hydrophobic sections embedded in the lipid bilayer, while the hydrophilic sections protrude from the lipid bilayer and are exposed to the two aqueous solutions that bath the membrane. The membrane organization is a

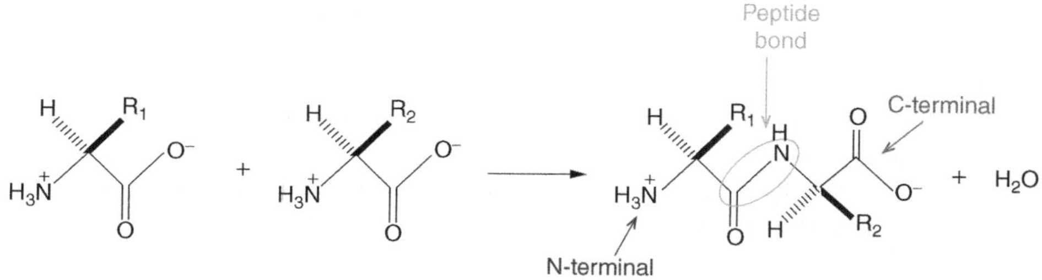

Figure 1.9 Formation of a peptide bond.

direct consequence of this partitioning between its two components: lipids and proteins. Since these components are not held together by chemical bonds, they are free to diffuse and move independently within the membrane plane. Clear experimental evidence in favor of protein mobility in membranes is obtained by labeling the protein with a fluorophore, in order to follow its movement. A short burst of intense excitation light is projected onto the membrane, destroying the fluorescence of the fluorophore molecules in a well-defined spot, a photochemical process called *photobleaching*. The gradual fluorescence recovery within the given spot is followed as a function of time, thus permitting an estimate of the diffusion coefficient of the protein. The model assuming free independent mobility of the various membrane components is termed *fluid mosaic model*, after Singer and Nicolson (1972).

Nowadays, the acknowledgment of the existence of ordered microdomains, called lipid rafts, within the more disordered lipid matrix, and of noncovalent bonds between plasma membrane proteins and the components of the cytoskeleton, has partly modified some concepts of the fluid mosaic model. In fact, not all components of a membrane are necessarily distributed in a totally homogeneous and casual way.

1.4 The Membrane Functions

Biological membranes mediate the interaction of the cells and organelles that they enclose with their environment. This mediation may give rise to three types of flows across membranes, namely flow of matter, flow of energy, and flow of information. The *flow of matter* involves the transport of molecules from one side of the membrane to the other (called *translocation*), and the movement of reactants and products of reactions catalyzed by membrane-bound enzymes. The *flow of energy* takes place across specialized membranes, that is, the inner mitochondrial membranes, the bacterial membranes, and the thylakoid membranes; here, the Gibbs energy released by the oxidation of nutrients or the energy of electromagnetic radiation can be exploited and stored in the form of an electrochemical potential gradient of protons and, exceptionally, of sodium ions, and ultimately, in the form of highly energetic molecules, such as ATP. The *flow of information* is restricted to some specialized membranes. They must be equipped with appropriate receptors for tactic stimuli, hormones, antigens, and any other piece of information that may arrive from the environment. They must also contain an enzymatic machinery capable of processing the information received, via the binding to a form recognizable by the intracellular components, where the final acceptor of the external stimulus is located.

The different types of flow across membranes are due to the various proteins embedded in the membrane. On the average, biomembranes contain more than one-half their weight of proteins, and it is this membrane component that endows them with a broad range of functional capabilities. A lipid bilayer lacking proteins is a mere diaphragm opposing an almost insurmountable potential energy barrier to any type of flow. It is, therefore, convenient to classify membrane proteins on the basis of their specific role. The proteins that span membranes may perform several diversified functions, and a single membrane protein may have multiple functions. These functions can be classified into several categories, which are discussed in the following sections.

1.4.1 Transport

The flow of molecules and ions across membranes is governed by sophisticated transport systems, usually integral membrane proteins, which play fundamental roles for the correct functioning of the cell. Thus, they regulate the cell volume and maintain the intracellular pH and ionic composition within a very narrow range to provide a favorable medium to the enzymatic activity; they allow the passage of molecules that are employed for energy production, or else as building blocks, and expel toxic substances; they generate electrochemical potential gradients essential for nerve and muscle excitability and for the transport of various substances.

The flow of substances across biological membranes, called *membrane transport*, may be categorized as either *active* or *passive*. In active transport, the transported species moves from a place where its electrochemical potential is lower to one where it is higher, namely against its electrochemical potential gradient. The active transport is an *endergonic* process, that is, a process characterized by an increase in Gibbs energy; as such, when taken separately, it cannot proceed spontaneously, as opposed to *exergonic* processes, that is, processes accompanied by a decrease in Gibbs energy. In fact, thermodynamics states that chemical processes may proceed only if they are accompanied by a decrease in their Gibbs energy. Thus, in order to proceed, an endergonic process must be coupled to an exergonic process, in such a way that the combined process is accompanied by a decrease in Gibbs energy. In other words, active transport requires the coupling to a primary energy source, such as light or chemical energy. In passive transport, the transported species moves from a place where its electrochemical potential is higher to one where it is lower, namely down its electrochemical potential gradient. Consequently, this type of transport is exergonic.

Membrane transport may also be classified on the basis of the number of different transported species and of the relative directions of flow. Transport is defined as *uniport* if it involves a single molecular species. The concerted and simultaneous movement of two species is distinguished between *symport* and *antiport*. The symport (also called *cotransport*) is the transport of two different molecular species in the same direction, whereas the antiport (also called *countertransport*) is the transport of two different molecular species in opposite directions. The proteins in charge of cotransport are called *cotransporters*, while those in charge of countertransport are called *countertransporters*. Ion transport across biological membranes may assume an *electroneutral* or *electrogenic* character. It is denoted as electroneutral or *electrically silent* if it does not give rise to a net flow of charge, as may be the case with the symport of ions of equal charge magnitude but opposite sign, or the antiport of ions of equal charge magnitude and sign. Conversely, electrogenic ion transport is characterized by a net flow of charge and, therefore, generates a charge separation across the membrane through which it occurs.

Examples of passive transport are offered by a particular category of organic molecules, the *ionophores*. They are lipid-soluble molecules usually synthesized by microorganisms to transport ions across cell membranes. These molecules are frequently used to investigate transport phenomena, besides being important toxins or drugs. There are two broad classifications of ionophores: *ion carriers* and *channel formers* (also called *pore formers*). An ion carrier is a molecule that complexes a particular ion via hydrophilic groups lining the interior of its molecular structure (often carbonyl groups of its peptide chain), while it has a hydrophobic exterior. In this way, it shields the charge of the ion from the surrounding environment, thus facilitating its translocation across the hydrophobic interior of the membrane. A typical example is represented by *valinomycin*, consisting of a sequence of two amino acids and two carboxylic acids bound via amide and ester bridges, which is repeated three times to form a cyclic structure; it specifically cages K^+ ions in the extracellular fluid and shuttles them into a cell across its plasma membrane.

Channel formers are molecules that form a pore whose interior is lined with a number of hydrophilic groups, while its exterior is hydrophobic and interacts attractively with the lipid molecules of the membrane. Channel formers span the whole membrane, allowing the ions that move within the pore to pass through at a rate several orders of magnitude higher than that of ions shuttled by ion carriers. They are usually formed by aggregation of a number of α-helical peptide monomers, which turn their hydrophilic side toward the lumen of the peptide bundle, giving rise to a hydrophilic *ion channel*. Incidentally, the α-helix is a common secondary structure of proteins with a right-hand-coiled conformation in which every backbone $N-H$ group donates a hydrogen to the backbone $C=O$ group of the amino acid four residues earlier ($i+4 \rightarrow i$ hydrogen bonding). The lumen of α-helixes is too narrow to allow the passage of ions, even if desolvated. This is the reason why α-helical monomers must cluster to form an ion channel. A quite atypical, but extensively investigated, channel former is *gramicidin*, a linear peptide that transports monovalent cations. The diameter of the lumen of its helix is larger than that of α-helixes, such that, differently than these latter, it allows the passage of ions through it. Since its length matches that of a single lipid monolayer, the gramicidin channel that spans a lipid bilayer is a dimer formed by two aligned monomeric units.

All known processes of active transport are mediated by asymmetrically oriented integral proteins. These proteins are characterized by reaction cycles during which they undergo conformational transitions that modify the orientation and the affinity of the binding site toward the transported species. Active transport is classified into two main categories: *primary* and *secondary*. Primary active transport is powered by primary energy sources, such as ATP hydrolysis (*transport ATPases*), electron transport (e.g., *proton-pumping oxidoreductases* in the inner mitochondrial membrane), and light (e.g., *bacteriorhodopsin* and *alorhodopsin*).

These systems of primary active transport are called *ion pumps* and transport ions selectively, generating electrochemical potential gradients. In turn, these gradients provide the Gibbs energy required to activate several physiological processes that, taken separately, are endergonic. Secondary active transport refers to systems that exploit the Gibbs energy of electrochemical potential gradients to transport ions and neutral molecules against their own electrochemical potential gradient. Typical examples of secondary active transport are offered by cotransporters and countertransporters.

1.4.2 Signal Transduction

A membrane protein, called *receptor*, has a binding site with specific shape that fits the shape of a *chemical messenger*, for example, a hormone. The external messenger (the signaling molecule) induces a conformational change in the protein, known as receptor activation, allowing it to relay the message to the inside of the cell. Receptor activation is always the initial step, leading to the cell's ultimate responses to the messenger. Signal transduction covers both signaling from the receptor located in the plasma membrane and signaling via molecules located within the cell. The movement of signals can be simple, similar to that associated with the class of acetylcholine receptor molecules. These receptors constitute channels that, upon ligand interaction, allow signals to be passed in the form of small ion movements, either into or out of the cell. These ion movements result in changes in the transmembrane potential of the cell that, in turn, propagates the signal along the cell. More complex signal transductions involve the coupling of ligand–receptor interactions to many intracellular events. These events include addition of a phosphate group to a protein or another organic molecule (*phosphorylation*) by a protein in charge of this task (*kinase*). Protein phosphorylations change enzyme activities and protein conformations. The eventual outcome is an alteration in cellular activity and changes in the program of genes expressed within the responding cell, by regulating the initiation of transcription in the messenger RNA.

1.4.3 Cell–Cell Recognition

It is the ability of cells to distinguish one type of neighboring cell from another. This is important, for example, in the sorting of cells into tissues or in the rejection of a foreign cell by the immune system. Some proteins of the plasma membrane are *glycoproteins*, namely proteins having a short, branched chain of usually less than 15 sugar units, exposed to the extracellular side of the membrane. These chains are also covalently bound to lipids, forming molecules called *glycolipids*. The carbohydrates on the extracellular side of plasma membranes vary from one cell type to another. Quite often, a cell recognizes another cell by binding to the carbohydrate moieties of glycolipids or glycoproteins on the extracellular surface of the plasma membrane of the other cell.

1.4.4 Enzymatic Activity

A protein built into the membrane may act as an enzyme, catalyzing various reactions related to the plasma membrane, with its active site exposed to substances in the external solution. It can be considered as a transmembrane receptor that binds an extracellular ligand, causing enzymatic activity on the intracellular side of the membrane. It has an extracellular ligand-binding domain, a transmembrane helix, and an intracellular domain, which has a catalytic function. The signaling molecule binds to the receptor outside the cell and causes a conformational change in the catalytic functional group of the receptor, which is located inside the cell. In some cases, several enzymes in a membrane are organized as a team that carries out sequential steps of a metabolic pathway.

1.4.5 Intercellular Joining

Membrane proteins of adjacent cells may hook together in various kinds of junctions, such as *gap junctions* or *tight junctions*. A gap junction is a specialized intercellular connection between animal cells. It directly links the cytoplasm of two cells, allowing various molecules to pass freely between them. The first gap junction protein was isolated from the liver and is called CX32. A gap junction channel is composed of two *connexons* (or *hemichannels*), one per cell, which keep the two interacting plasma membranes at a distance of about 2–4 nm apart. When the connexons in the plasma membranes of two cells in contact are aligned, they form a continuous aqueous channel that joins the two cell interiors. Each connexon is formed from six four-pass transmembrane proteins called *connexins*. Dye-injection experiments suggest a functional pore diameter of about 1.5 nm, implying that coupled cells share their small molecules, such as inorganic ions, sugars, amino acids, subunits of nucleic acids such as DNA or RNA (the *nucleotides*), and vitamins, but not their macromolecules. This cell coupling has important functional implications.

A tight junction is composed of a branching network of strands that seals the plasma membranes of two adjacent cells. Each sealing strand consists of a long row of transmembrane adhesion proteins embedded in each of the two interacting plasma membranes. The extracellular domains of these proteins join directly to one another to occlude the intercellular space. The major transmembrane adhesion proteins are the *claudins* and *occludins*, which differ from one tight junction to another. Strands act independent from each other, and hence the sealing efficiency of tight junctions increases with an increase in the number of strands. Tight junctions hold the cells together. Moreover, by pinning adjacent membranes together, they prevent the lateral diffusion of integral membrane proteins, confining them to the particular portion of the plasma membrane (either the apical or the basal portion) where they are required to exert a specialized function. The epithelial tissue, which lines all internal and external body surfaces, is made up of cells closely packed and ranged in one or more layers, often held together by tight junctions.

References

Alkire, R. C., D. M. Kolb, J. Lipkowski, *Advances in Electrochemical Science and Engineering*, Vol. 13, *Bioelectrochemistry*, Wiley-VCH, Weinheim, Germany, 2011.

Bagatolli, L. A., *Biochim. Biophys. Acta* 1758 (2006) 1541–1556.

Bartlett, P. N., *Bioelectrochemistry. Fundamentals, Experimental Techniques and Applications*, John Wiley & Sons, Chichester, England, 2008.

Brown, D. A., E. London, *J. Membrane Biol.* 164 (1998) 103–114.

Brown, D. A., E. London, *J. Biol. Chem.* 275 (2000) 17221–17224.

Gutman, F., H. Keyzer, *Modern Bioelectrochemistry*, Springer, 1986.

Parasassi, T., A. M. Giusti, M. Raimondi, E. Gatton, *Biophys. J.* 68 (1995) 1895–1902.

Pike, L. J., *Biochem. J.* 378 (2004) 281–292.

Rawn, J. D., *Biochemistry*, N. Patterson Publishers, Burlington, U.S.A., 1989.

Simons, K., E. Ikonen, *Nature* 387 (1997) 569–572.

Singer, S. J., G. L. Nicolson, *Science* 175 (1972) 720–731.

2

Electrostatics of Biomembranes

In this chapter, we briefly touch upon some basic elements of electrostatics, which are required to understand how the electric potential varies across biological membranes.

2.1 The Electric Field

Coulomb's law states that the electrostatic force $\boldsymbol{F}$ of interaction between two point charges q_1 and q_2 in vacuum is directed along their conjunction line; moreover, its magnitude is directly proportional to the scalar product of the magnitudes of the charges and inversely proportional to the square of the distance r between them. Therefore, its magnitude is given by

$$F = Kq_1q_2/r^2 \ \text{ with } \ K = 1 \text{ (CGS electrostatic) or } = 1/(4\pi\varepsilon_0)\text{(MKSQ)}, \tag{2.1}$$

where K is a proportionality constant equal to 1 in the centimeter-gram-second (CGS) electrostatic system, and equal to $1/(4\pi\varepsilon_0)$ in the meter-kilogram-second-coulomb (MKSQ) system; $\varepsilon_0 = 8.854 \times 10^{-12}$ (C)^2/(J m) is the *vacuum permittivity*. It follows that the magnitude of the infinitesimal force dF exerted by a point charge q on a charge dq, so infinitesimally small as not to perturb the distribution of other possible surrounding charges, is given by $\mathrm{d}F = Kq\,\mathrm{d}q/r^2$. The ratio $\mathrm{d}F/\mathrm{d}q \equiv E = Kq/r^2$ is the magnitude of the force exerted by a point charge q on a positive unit *test* charge, that is, a unit charge whose presence is arbitrarily assumed to exert no effect on the surrounding charge distribution. The corresponding vector is called the *electric field* created by the point charge q and is denoted by $\boldsymbol{E}$. As all *vector fields* in space, the electric field assigns a vector to each point in space. A vector field in a plane can be viewed as an arrow, with a given magnitude and direction, attached to every point in the plane. A vector in a three-dimensional space is defined when its three components along the axes of a Cartesian coordinate system are specified, each expressed by a number. As distinct from a vector field, a scalar field in space associates a single number to each point in space.

Bioelectrochemistry of Biomembranes and Biomimetic Membranes, First Edition. Rolando Guidelli.

The electric field satisfies an important property according to which, at each point in space, it is equal to the opposite of the gradient, $\nabla\phi$, of a scalar field ϕ, termed *electric potential* (Shadowitz, 1975). In general terms, by gradient of a scalar quantity at a point **P** in space we mean a vector whose direction is that along which the scalar quantity increases more rapidly; its magnitude is given by the rate of increase of the scalar quantity along that direction. For example, let us show that the electric potential ϕ created by a point charge q at a distance r from its position in space is given by $\phi = Kq/r$. The electric field created by the point charge q can be represented by its *lines of force*, that is, ideal lines in space, such that the tangent at any point of these lines gives the direction of the vector field at that point. Moreover, the density of the lines of force at any point is proportional to the magnitude of the vector field at that point.

The lines of force generated by a point charge are radially outward from the point charge if it is positive (Fig. 2.1), and radially inward to it if it is negative. This is because a positive charge q repels an infinitesimal positive test charge, while a negative charge attracts it. The density of the lines of force in Fig. 2.1 decreases progressively as we depart from the charge q, similar to the electric field, of magnitude q/r^2. Let us consider a point **P**, at a distance r from the charge q, and let us move away from it by an infinitesimal vector d$\boldsymbol{r}$ along the line of force passing by **P**, so as to reach the point **P** + d$\boldsymbol{r}$. We stated without proof that the electric potential at a point **P** is given by the scalar quantity $\phi = Kq/r$. The gradient of ϕ, denoted by $\nabla\phi$, is the rate of increase of ϕ along the line of force passing by **P**. Therefore, its magnitude is given by the ϕ value at **P** + d$\boldsymbol{r}$ minus that at **P**, divided by dr:

$$\nabla\phi = \frac{K\frac{q}{r+\mathrm{d}r} - K\frac{q}{r}}{\mathrm{d}r}\hat{\boldsymbol{r}} = \frac{-K\frac{q\,\mathrm{d}r}{r^2+r\,\mathrm{d}r}}{\mathrm{d}r}\hat{\boldsymbol{r}} \cong \frac{-K\frac{q\,\mathrm{d}r}{r^2}}{\mathrm{d}r}\hat{\boldsymbol{r}} = -K\frac{q}{r^2}\hat{\boldsymbol{r}}. \tag{2.2}$$

Here, $\hat{\boldsymbol{r}}$ is the unit vector (*versor*) along the direction of the vector d$\boldsymbol{r}$. We then see that the electric field at a point **P**, $K(q/r^2)\,\hat{\boldsymbol{r}}$, is equal to the opposite of the gradient of the electric potential at that point. In this case, as well as in general,

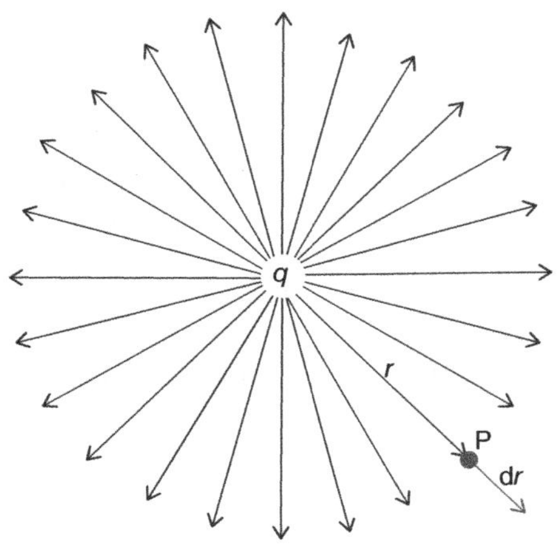

Figure 2.1 Lines of force generated by a positive point charge q.

a positive test charge moves from points where the electric potential is higher to those where it is lower; the negative sign in front of the gradient of the electric potential accounts for this behavior.

Let us now consider the particularly simple case in which the electric field has the same direction at all points of space and let us denote this direction by the x-axis. In this case, the electric potential varies exclusively along this axis and the problem passes from tridimensional to *unidimensional* (or *unidirectional*). The only nonzero component, E_{x_1}, of the electric field at a point x_1 along the x-axis is given again by the opposite of the difference between the electric potential value at a point $(x_1 + \mathrm{d}x)$ and its value at point x_1, divided by dx:

$$E_{x_1} = -\frac{\phi(x_1 + \mathrm{d}x) - \phi(x_1)}{\mathrm{d}x} = -\left(\frac{\partial \phi}{\partial x}\right)_{x_1}. \qquad (2.3)$$

This is just the opposite of the derivative of the electric potential with respect to the distance along the x-axis.

2.2 Electric Field Created by a Uniform, Infinitely Extended Planar Charge Distribution

The electric field in the immediate vicinity of a biomembrane can be considered as arising from a uniform, infinitely extended planar charge distribution. In fact, it vanishes at a much smaller distance from the membrane surface than the radius of the cell or organelle enveloped by the membrane. Thus, in an aqueous solution of 0.1 M KCl, the electric field practically vanishes at distances greater than 10 Å from the membrane surface. In other words, a hypothetical infinitesimally sized "observer," placed at a distance from the membrane surface at which the electric field is about to vanish, sees the membrane as a practically infinitely extended planar surface, in spite of its curvature.

To determine this electric field, let us consider a continuous charge distribution in a tridimensional space, characterized by a local *volume charge density* $\rho = \mathrm{d}q/\mathrm{d}V$, where d$V$ is an infinitesimal volume element and dq is the infinitesimal charge enclosed in it. Let us now imagine enclosing the whole charge distribution into an ideal closed surface of arbitrary shape and, then, gathering it into a single point **P** inside the closed surface. The resulting point charge Q is given by the volume integral of ρ over the volume enclosed by the ideal closed surface:

$$Q = \iiint_V \rho \, \mathrm{d}V. \qquad (2.4)$$

Let us now further simplify the geometry of our system with a thought experiment involving the replacement of the closed surface of arbitrary shape by an ideal spherical surface of radius r that contains it completely (without including further charges) and has its center at point **P**, as shown in Fig. 2.2. Each volume element of the original charge distribution is a *source* of lines of force if positively charged, or a *sink* of them if negatively charged. It is quite intuitive that the lines of

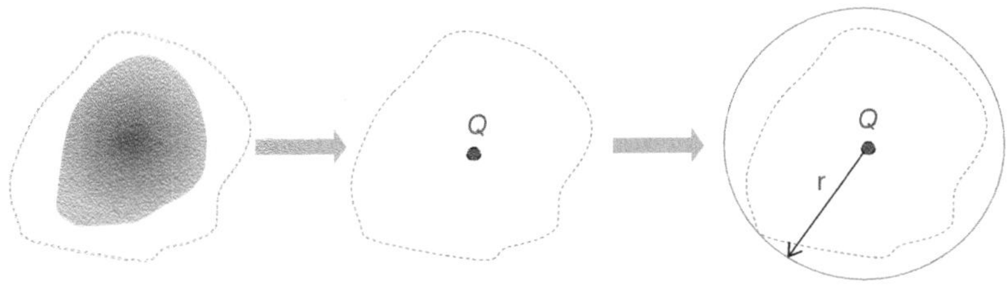

Figure 2.2 The three steps of the thought experiment; the dark area is the original charge distribution and the irregularly shaped curve is the original closed surface enclosing it.

force that stem from the original charge distribution and cross the original closed surface are equal in number to those that stem from the point charge Q and cross the spherical surface. The electric field generated by the point charge Q on the ideal spherical surface of radius r is equal to KQ/r^2 and is normal to this surface, for obvious symmetry reasons.

By definition, the flux of a vector field through a closed surface is obtained by integrating over its surface area the scalar product of the vector by the versor $\hat{n}$, normal to the surface and directed outward (Shadowitz, 1975). The flux of the electric field is, therefore, given by

$$\iint_S \boldsymbol{E} \cdot \hat{n} \, \mathrm{d}S, \tag{2.5}$$

where dS is an infinitesimal surface element. Clearly, the flux is a maximum on the surface elements where $\boldsymbol{E}$ is normal to the surface, and zero where $\boldsymbol{E}$ is parallel to it. In general, the flux of the electric field through a closed surface measures the number of lines of force that cross it. In the present case, we have

$$\begin{aligned}\iint_S \boldsymbol{E} \cdot \hat{\boldsymbol{n}} \, \mathrm{d}S &= K \iint_S \frac{Q}{r^2} \, \mathrm{d}S = K\frac{Q}{r^2} \iint_S \mathrm{d}S \\ &= 4\pi r^2 K \frac{Q}{r^2} = 4\pi KQ = \frac{Q}{\varepsilon_0} \text{(in MKSQ system)}.\end{aligned} \tag{2.6}$$

In what follows, we always adopt the MKSQ system, setting $K = (4\pi\varepsilon_0)^{-1}$. Note that the flux is independent of the radius r of the sphere, since the higher its radius, the weaker the electric field created by the point charge Q on the sphere surface is. Equation 2.6 was obtained via an approximate thought experiment, but it could be demonstrated by rigorous arguments for any charge distribution enclosed by any closed surface (Shadowitz, 1975).

Let us now apply to the present case the mathematical *divergence theorem*, which states that the outward flux of a vector field through a closed surface is equal to the volume integral of the *divergence* of the vector field over the region enclosed by this surface. Thus, we have

$$\iint_S \boldsymbol{E} \cdot \hat{\boldsymbol{n}} \, \mathrm{d}S = \iiint_V \mathrm{div} \boldsymbol{E} \, \mathrm{d}V. \tag{2.7}$$

The divergence of a vector $\boldsymbol{F}$ is defined as

$$\operatorname{div} \boldsymbol{F} = \frac{\partial}{\partial x}F_x + \frac{\partial}{\partial y}F_y + \frac{\partial}{\partial z}F_z. \tag{2.8}$$

F_x, F_y, and F_z are the components of the $\boldsymbol{F}$ vector along the x, y, and z axes of a Cartesian orthogonal coordinate system. The divergence of a vector field at a point **P** measures the number of lines of force of the given vector field that leave an infinitesimal volume element $\mathrm{d}V$ enclosing **P** minus those that enter it, divided by $\mathrm{d}V$. When extended to all the points inside a finite volume V, the lines of force that leave the volume element enclosing the given point **P** are equal in number to those that enter the volume elements enclosing all the nearest neighboring points, thus canceling each other out, except for the points located on the surface enclosing the whole volume V. This justifies why the volume integral of the divergence of the electric field over a volume V equals its flux out of the surface enclosing that volume.

On combining Eqs. 2.4, 2.6, and 2.7, we obtain

$$\iint_S \boldsymbol{E} \cdot \hat{\boldsymbol{n}} S = \iiint_V \operatorname{div} \boldsymbol{E}\, \mathrm{d}V = Q/\varepsilon_0 = \varepsilon_0^{-1} \iiint_V \rho\, \mathrm{d}V. \tag{2.9}$$

The equality of the two different volume integrals yields, after differentiation,

$$\operatorname{div} \boldsymbol{E} = \rho/\varepsilon_0. \tag{2.10}$$

The rationale behind this equation can be explained as follows. If an infinitesimal volume element enclosing a point **P** contains a net charge, this is a source of lines of force (if positive) or a sink (if negative); hence, the lines of force leaving the given volume element are different in number from those entering it, and the same is true for the divergence of the electric field. Conversely, if no charges are present in the volume element, then the lines of force simply pass through it without being generated or extinguished within it, and the divergence equals zero.

Eq. 2.10 is *Poisson's equation* of electrostatics. In the unidirectional case defined by the x-axis, the only nonzero component of the electric field is that along the x-axis, $E_x = -\mathrm{d}\phi/\mathrm{d}x$, and, from the definition of the divergence (Eq. 2.8), Poisson's equation takes the form

$$\mathrm{d}E_x/\mathrm{d}x = -\mathrm{d}^2\phi/\mathrm{d}x^2 = \rho/\varepsilon_0. \tag{2.11}$$

Let us now consider a uniform, infinitely extended planar charge distribution of *surface charge density* $\sigma = \mathrm{d}q/\mathrm{d}S$, where $\mathrm{d}S$ is the area of an infinitesimal surface element and $\mathrm{d}q$ is the infinitesimal charge on it. Let us now consider an ideal cylinder of unit cross-sectional area, with its axis normal to the plane of the uniform charge distribution. Clearly, the charge enclosed in the cylinder equals σ (Fig. 2.3). For obvious symmetry reasons, the electric field $\boldsymbol{E}$ created by the planar charge distribution is normal to the distribution and, hence, parallel to the lateral surface of the cylinder. Its flux through this surface is, therefore, equal to zero. In other words, no line of force passes through the lateral surface. Conversely, $\boldsymbol{E}$ is normal to both bases of the cylinder. Consequently, its outward flux equals

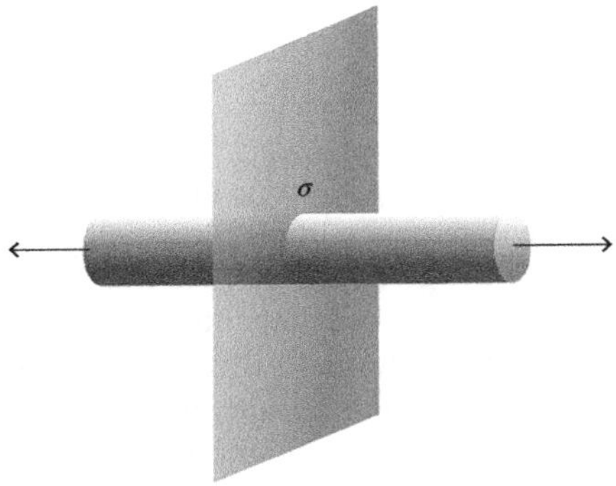

Figure 2.3 Portion of a uniform, infinitely extended planar charge distribution of surface charge density σ, with an ideal cylinder of unit cross-sectional area normal to it. The arrows are unit vectors directed outward.

$E_x \cdot \hat{x} \times$ (unit area) through the right base in Fig. 2.3, and $-E_x \cdot (-\hat{x}) \times$ (unit area) through the left base, where $\hat{x}$ is the versor along the x-axis. Setting the flux of $\boldsymbol{E}$ through the whole surface of the cylinder equal to $Q/\varepsilon_0 = \sigma/\varepsilon_0$ in view of Eq. 2.6, we obtain

$$2E_x \times (\text{unit area}) = \sigma/\varepsilon_0 \times (\text{unit area}) \rightarrow E_x = \sigma/(2\varepsilon_0). \qquad (2.12)$$

This equation shows that the electric field created by an infinitely extended planar charge distribution is independent of the distance x from the distribution. Clearly, if the charge distribution were not infinitely extended and a hypothetical observer could see its edge, then this equation would not hold, because edge effects would be felt.

By definition, the electric field is the force that acts on the unit positive charge. Hence, the product of the electric field E_x at point x_1 by an infinitesimal displacement dx expresses the work done by the electric field to shift the unit positive charge from point x_1 to point $x_1 + \mathrm{d}x$. In view of Eq. 2.3, this work is given by

$$E_x\,\mathrm{d}x = -(\mathrm{d}\phi/\mathrm{d}x)\,\mathrm{d}x = -\mathrm{d}\phi = \phi(x_1) - \phi(x_1 + \mathrm{d}x), \qquad (2.13)$$

and is equal to the electric potential value at point x_1 minus that at point $x_1 + \mathrm{d}x$. It might be demonstrated by rigorous arguments that this result holds for any distance between the two points and in the three-dimensional space as well. In other words, the work done by the electric field to bring the unit positive charge from point **P** to point **Q** in three-dimensional space equals the electric potential value at point **P** minus that at point **Q**. Consequently, this work does not depend on the particular path followed to bring the unit positive charge from point **P** to point **Q**. This property of the electric field is expressed by stating that the electric field is *conservative*.

For the electrical work done *by* the electric field (not that done *against* it) on the unit positive charge to be positive, the electric potential at the point of departure must be more positive than that at the point of arrival. In other words, in the absence of forces different from the electric force, the unit positive charge moves from points where the electric potential is higher to those where it is lower. In general, the electric force acting on a given charge q is given by the product of the charge by the electric field. Hence, a negative charge will move from points where the electric potential is lower to those where it is higher, in the absence of forces different from the electric force.

2.3 The Parallel-plate Capacitor

A *parallel-plate capacitor* consists of two parallel uniform planar charge distributions of equal magnitude but opposite sign. To exclude edge effects, let us assume that the two charge distributions are infinitely extended. In Fig. 2.4, the charge distribution on the left is positive and that on the right is negative. The two planar surfaces of the capacitor are termed *plates*. The electric field created by the plate on the left, being the force acting on the unit positive charge, repels this charge away from the plate. On the contrary, the electric field created by the negatively charged plate on the right is directed toward this plate. Since the absolute value of the surface charge density is the same on the two plates, the absolute value of the electric field created by each of the plates is also the same. Having assumed that the two plates are infinitely extended, these electric fields are independent of the distance. The total electric field is the vectorial sum of the two partial electric fields created by the two plates. Consequently, the two partial electric fields add up between the plates and cancel out outside them. In view of Eq. 2.12, the magnitude, $|\boldsymbol{E}|$, of the total electric field between the plates is constant and equal to

$$|\boldsymbol{E}| = |\sigma|/\varepsilon_0. \tag{2.14}$$

The independence of the electric field, $E_x = -(d\phi/dx)$, from distance implies that the electric potential decreases linearly with distance from the positive to the negative plate, as shown in Fig. 2.4.

Outside the plates, the electric field vanishes and the electric potential is, therefore, constant. In view of the linear dependence of the electric potential on distance, the electric potential difference, $\phi(x_2) - \phi(x_1)$, between the negative and the positive plate is obtained from Eqs. 2.3 and 2.14:

$$E_x = \frac{\sigma_1}{\varepsilon_0} = -\frac{\phi(x_2) - \phi(x_1)}{x_2 - x_1} \rightarrow \phi(x_2) - \phi(x_1) = -\frac{\sigma_1 d}{\varepsilon_0} \quad \text{with } d \equiv x_2 - x_1. \tag{2.15}$$

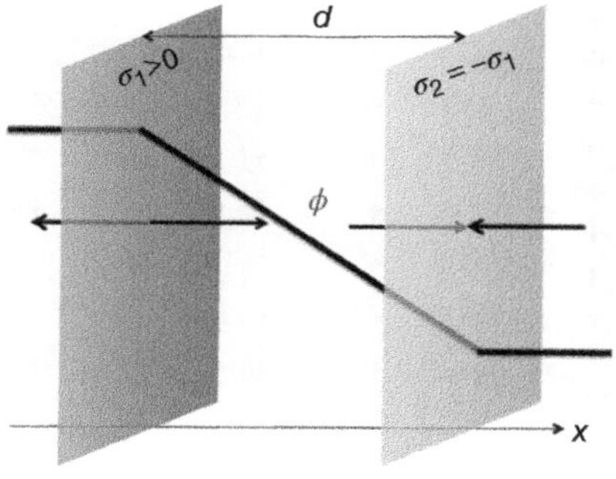

Figure 2.4 A parallel-plate capacitor. The thick line is the electric potential profile and the arrows are the partial electric fields created by the corresponding uniform charge distributions.

Here, σ_1 is the surface charge density on the positive plate on the left. The ratio of the absolute value of the surface charge density on any of the two plates to the absolute value of the potential difference between the plates is termed *capacitance* and is denoted by C:

$$C = |\sigma_1|/|\phi(x_2) - \phi(x_1)| = \varepsilon_0/d. \tag{2.16}$$

The term "capacitance" has the significance of spaciousness, in that it measures how much charge can be stored on the two plates of the capacitor, for a given value of the electric potential difference between them.

2.4 The Dielectric Constant

The considerations made so far refer to the case in which the electric field created by the charged plates of the capacitor is in vacuum or, at most, in air. In a condensed phase, such as a liquid or a solid, many molecules are *dipolar*, namely have the center of positive charge displaced with respect to the center of negative charge. These molecules are schematically represented by a positive point charge q (*the positive pole*) and a negative point charge $-q$ (*the negative pole*), separated by a distance d. The *permanent dipole moment* of a dipolar molecule is a vector directed from the negative to the positive pole, and its magnitude is given by the product qd. A good example of dipolar molecule is the water molecule, whose asymmetry leads to a permanent dipole moment directed from the oxygen atom to the midpoint between the two hydrogen atoms. An external electric field tends to orient dipolar molecules with their dipole moment aligned along the direction of the field, by exerting a *force couple*, as shown in Fig. 2.5.

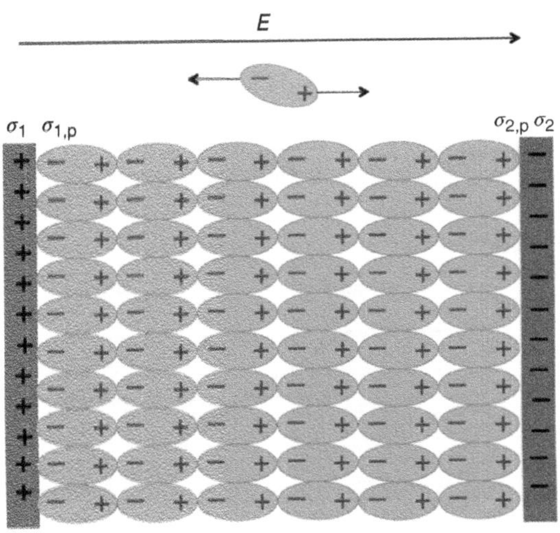

Figure 2.5 Scheme of molecular dipoles aligned along the direction of the electric field created by the true surface charge densities, σ_1 and $\sigma_2 = -\sigma_1$, on the plates of a parallel-plate capacitor. The corresponding surface polarization charge densities are denoted by $\sigma_{1,p}$ and $\sigma_{2,p}$.

Electric fields applied to a bulk condensed phase, such as an aqueous solution, orient dipolar molecules only very slightly. In other words, the component of the dipole moment of a dipolar molecule along the direction of the external electric field is only a small fraction of its total dipole moment. Molecules with mirror symmetry such as oxygen, nitrogen, carbon dioxide, and carbon tetrachloride are such that the centers of positive and negative charges coincide; they are termed *apolar* because they have no permanent dipole moment. However, an external electric field tends to separate the center of positive charge from that of negative charge by moving them in opposite directions, with this tendency being opposed by the electrostatic attraction between the two centers. The balance between these two opposing forces generates an *induced dipole moment* totally aligned along the direction of the external field. This field exerts the same effect on dipolar molecules, which are, therefore, subject to both *orientation polarization*, due to their permanent dipole moment, and *distortion polarization*, which is responsible for their induced dipole moment.

If the space between the plates of a parallel-plate capacitor is filled with a condensed phase, the dipole moments induced by the electric field created by the surface charge densities, σ_1 and $\sigma_2 = -\sigma_1$, on the two plates are aligned along the direction of the field. As already stated, the dipole moments induced by the electric field are much less than their permanent dipole moments, qd, if any. In this way, the positive (negative) pole of each induced dipole is in contact with the negative (positive) pole of the nearest neighboring induced dipole, as shown in Fig. 2.5. In practice, the opposite poles of adjacent induced dipoles are mutually neutralized, with the only exception of the negative poles in direct contact with the positively charged plate and the positive poles in contact with the negatively charged plate. The surface charge density, $\sigma_{1,p}$, of the terminal negative poles adjacent to the positive plate and the surface charge density, $\sigma_{2,p} = -\sigma_{1,p}$, of the terminal positive poles adjacent to the negative plate are called *surface polarization charge densities*. A surface polarization charge density differs from a *true surface charge density*, such as σ_1, because it exists only in the presence of a surface polarization charge density, equal in magnitude but opposite in sign, located somewhere on the closed surface enclosing the condensed phase. The surface polarization charge density $\sigma_{1,p}$ counterbalances to a large extent the true surface charge density σ_1, and the same is obviously true for $\sigma_{2,p}$ with respect to σ_2. In practice, the electric field within the plates and the potential difference across them are determined by the overall surface charge densities $(\sigma_1 + \sigma_{1,p})$ and $(\sigma_2 + \sigma_{2,p}) = -(\sigma_1 + \sigma_{1,p})$; hence, they are both decreased by the factor $(\sigma_1 + \sigma_{1,p})/\sigma_1$, whose reciprocal is called *dielectric constant* and denoted by the symbol ε. Hence, in a condensed phase, the expressions in Eqs. 2.11 and 2.15 become

$$-(\mathrm{d}^2\phi/\mathrm{d}x^2) = \rho/(\varepsilon_0\varepsilon), \tag{2.17}$$

$$E_x = \frac{\sigma_1}{\varepsilon_0\varepsilon} = -\frac{\phi(x_2) - \phi(x_1)}{x_2 - x_1} \rightarrow \phi(x_2) - \phi(x_1) = -\frac{\sigma_1 d}{\varepsilon_0\varepsilon} \quad \text{with } d \equiv x_2 - x_1. \tag{2.18}$$

Conversely, the capacitance C of a parallel-plate capacitor in Eq. 2.16 is increased by the factor ε:

$$C = |\sigma_1|/|\phi(x_2) - \phi(x_1)| = \varepsilon_0\varepsilon/d. \tag{2.19}$$

This is because, to generate a given potential difference across the plates, it is necessary to increase the true surface charge density $|\sigma_1|$ by a factor equal to ε with respect to vacuum, with a resulting increase in the "spaciousness" of the plates by the same factor.

Clearly, the dielectric constant of a condensed phase increases with an increase in the permanent dipole moment of the molecules that compose it. Thus, the aqueous phase, whose molecules have a very high permanent dipole moment as compared with their size, has a dielectric constant of about 78. Conversely, hydrocarbons such as pentane or decane, whose molecules are practically apolar, have a dielectric constant of about 2.

2.5 The Electric Potential Across a Membrane

Let us consider a very rough membrane model consisting of a phospholipid bilayer interposed between two aqueous solutions, with the hydrocarbon tails of the two lipid leaflets directed toward each other and the polar heads exposed to the two bathing solutions. The polar heads of phosphoglycerides are the site of a potential difference of about 200–250 mV, positive toward the hydrocarbon tails (Clarke, 1997; Becucci et al., 2000). This is due to a permanent dipole moment that is believed to be located at the ester linkages between the glycerol backbone and the carboxyl groups of the fatty acids of the phospholipid. These linkages are present in all phosphoglycerides, both neutral, such as phosphatidylcholine, and charged, such as phosphatidylserine. The aforementioned permanent dipole is rigidly blocked in the lipid bilayer and does not reorient under an external electric field. Consequently, the potential difference created by the array of these dipoles in each of the two leaflets of the lipid bilayer (the *surface dipole potential*) will be considered as constant and denoted by the Greek letter χ.

Let us first consider the case in which the polar heads of the two lipid leaflets are electrically neutral. Under these conditions, a potential difference across the membrane (*transmembrane potential*) induced by an external stimulus can only be produced by a predominance of positive ions over negative ones in the aqueous solution bathing one side of the membrane and by a predominance of negative ions over positive ones in the aqueous solution bathing the other side. This imbalance between positive and negative ions decreases as we move away from any of the two membrane surfaces, fading away at a distance from the membrane that depends on the electrolyte concentration bathing its two sides. For a 0.1 M concentration of a 1,1-valent electrolyte, such as KCl or NaCl, the solution layer adjacent to the membrane surface beyond which this imbalance practically vanishes is about 10 Å thick and is termed *diffuse layer* (Mohilner, 1966). Beyond the diffuse layer, in any infinitesimal volume element of the solution, the charge of positive

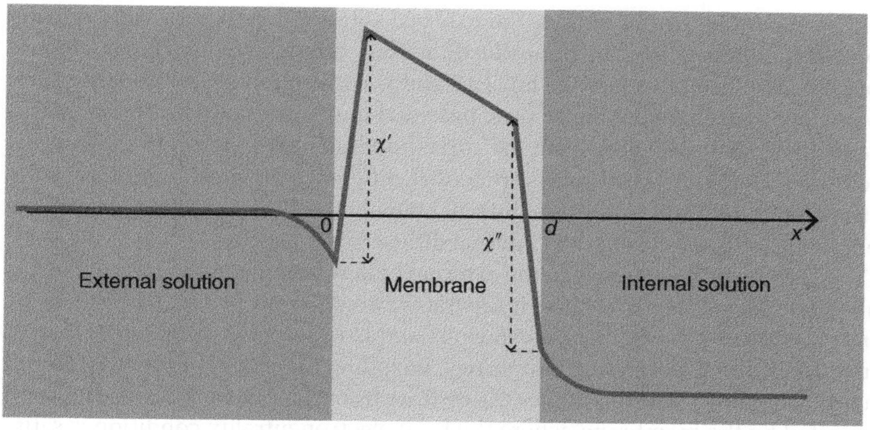

Figure 2.6 Electric potential profile across a neutral bilayer lipid membrane, for a negative transmembrane potential $\Delta\phi$.

ions matches exactly that of negative ions (*local electroneutrality principle*). By *bulk solution* we mean the solution beyond the diffuse layer.

Let x denote an axis normal to the membrane, with its origin, $x = 0$, placed on the external membrane surface. Denoting by d the membrane thickness, $x = d$ is the plane of the internal membrane surface. If both leaflets of the lipid bilayer consist of phosphoglycerides, two surface dipole potentials reside in the two opposite polar heads: χ' on the external surface and $-\chi''$ on the internal one (Fig. 2.6). The difference in sign accounts for the usage according to which the transmembrane potential, $\Delta\phi$, of a plasma membrane is the electric potential on the intracellular side relative to that on the extracellular side, considered as zero. Recalling that the surface dipole potential χ is directed toward the hydrocarbon tails, χ' makes a positive contribution to $\Delta\phi$, while χ'' makes a negative one. Figure 2.6 shows the electric potential profile across the membrane.

If an external perturbation generates a negative transmembrane potential across an electrically neutral membrane, this is necessarily determined by an excess of positive ions over negative ones on the external side of the membrane (external diffuse layer), and by an equal and opposite excess of negative ions over positive ones on its internal side (internal diffuse layer). The resulting electric potential profile is schematically depicted in Fig. 2.6. The electric potential in the external solution decreases gradually when proceeding from the bulk solution to the membrane external surface. This is because the cations, which are free to move in the solution under the external electric field, accumulate on the membrane external surface, since they are attracted toward it by an electric potential negative with respect to that in the external bulk solution. For the same reason, anions move away from the external membrane surface, contributing to generate an excess of positive charges in the external diffuse layer. Note that the presence of a net positive charge should by no means be considered as the "source" of an electric field, but rather as the effect of a favorable external applied potential. In the bulk of the external solution, no electric field moving oppositely charged ions in opposite directions exists, and hence the electric potential is constant in view of Eq. 2.18 and is conventionally set equal to zero. By analogous arguments, the electric potential in the internal solution increases gradually when proceeding from the corresponding bulk solution to the internal membrane surface. The resulting potential difference attracts anions from the internal bulk solution toward the internal membrane surface and repels cations, generating a negatively charged internal diffuse layer.

The region enclosed between the external bulk solution and the internal one is called *interfacial region* because it separates two different phases. In general, by *phase* we mean a perfectly homogeneous and isotropic region of space, that is, a region in which all physical properties, including the electric potential, do not change from one point to another. As distinct from the two bulk aqueous phases separated by the membrane, where the local electroneutrality condition is satisfied and the electric potential is constant, the interfacial region is characterized by a separation of charges, which generates a potential difference between the two aqueous phases, as in the case of Fig. 2.6. The sum of all charges enclosed in the interfacial region is necessarily equal to zero, that is, the interfacial region is electroneutral as a whole. In fact, if *ab absurdo* it were charged, it would create in the two adjacent aqueous phases an electric field that would move the ions therein, causing them to arrange themselves at the boundary of the phases, ultimately annihilating the overall charge of the interfacial region. This is the reason why the excess of positive charges on one side of an electroneutral membrane is necessarily equal in magnitude and opposite in sign to the excess of negative charges on the other side.

The excess of charges in the external diffuse layer can be quantified if we imagine enclosing it in an ideal cylinder of unit cross-sectional area, with its axis normal to the membrane plane, and with one base located in the external bulk phase and the other on the external membrane surface (see Fig. 2.7). Since the imbalance between negative and positive ions vanishes in the two bulk phases, the total charge enclosed in the cylinder does not depend on its length. The charge enclosed

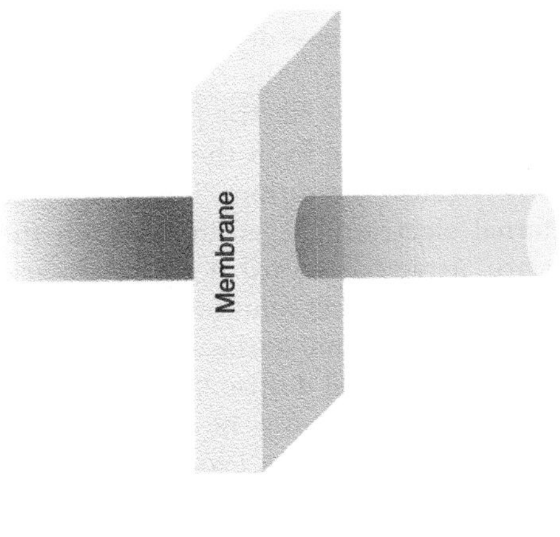

Figure 2.7 Fragment of a bilayer lipid membrane, with two ideal cylinders of unit cross-sectional area protruding from its two surfaces toward the respective bulk phases. The fading halftones mimic the fading local charge densities as we move away from the membrane surface.

in this ideal cylinder, albeit distributed in space, can be referred to the unit surface of the membrane and expressed as a surface charge density σ'_d, where the subscript stands for diffuse layer. Henceforth, quantities relative to the external side of the membrane will be denoted by a prime, those relative to the internal side by a double prime.

Analogous considerations hold for the internal diffuse layer. Thus, the overall charge enclosed in an ideal cylinder of unit cross-sectional area, with one base located in the internal bulk phase and the other on the internal membrane surface, is referred to the unit surface and denoted by σ''_d. In view of the electroneutrality of the whole interface, we can write

$$\sigma'_\mathrm{d} + \sigma''_\mathrm{d} = 0. \tag{2.20}$$

Within the diffuse layer, ions are subject both to electrostatic forces and to the randomizing forces of thermal agitation. The former tend to attract ions to the membrane surface or to repel them from it, depending on their sign; the latter tend to mix ions so as to induce local electroneutrality conditions. Under these two contrasting forces, ions assume a compromised spatial distribution, which determines the electric potential profile within the diffuse layer. This profile can be calculated by estimating the ion distribution according to *Boltzmann statistics* (Hill, 1960), which accounts for the competition between electric field strength and thermal agitation, and by relating the Boltzmann distribution of ionic charges to the electric potential through Poisson's equation of electrostatics (Eq. 2.17).

2.6 Poisson–Boltzmann Equation and Gouy–Chapman Theory

The following considerations refer to the solution adjacent to a single side of the membrane, regardless of whether internal or external. Therefore, the membrane side is not specified. According to Boltzmann statistics, the concentration $c_h(x)$ of a hth ionic species of charge number z_h at a distance x from the membrane is given by the following equation:

$$c_h(x) = c_h \exp\left[-\frac{z_h F\phi(x)}{RT}\right]. \tag{2.21}$$

Here, c_h is the bulk concentration of the hth species and $\phi(x)$ is the electric potential at the given distance x, as measured with respect to the electric potential in the bulk phase, conventionally set equal to zero. The concentration distribution expressed by Eq. 2.21 is governed by the Boltzmann exponential factor, whose exponent is equal to the opposite of the ratio of the potential energy of the particle under consideration to the unidirectional energy of thermal agitation (kT for a single particle or RT for a mole of particles). In the present case, the potential energy of the ionic particle stems exclusively from its charge $z_h e$ and is, therefore, given by its electrostatic potential energy; this is expressed by the product of the charge of one mole of the particle ($z_h F$) by the electrostatic potential $\phi(x)$ at the distance x at which the concentration $c_h(\mathrm{x})$ is estimated.

The Boltzmann exponential factor is < 1 when its exponent is negative, that is, when the charge number z_h of the ion and the electric potential $\phi(x)$ measured with respect to the bulk solution have the same sign; in this case, the electrostatic force repels the ion away from the membrane, thus decreasing the local concentration $c_h(x)$ with respect to its bulk value c_h. Naturally, the higher the temperature T, the higher the energy RT of thermal agitation that tends to oppose the order imposed by the electric potential $\phi(x)$. This opposition is adequately accounted for by the Boltzmann exponential factor, which approaches the unit value the more, the higher the energy RT of thermal agitation is with respect to the electrostatic potential energy $z_hF\phi(x)$. Analogous considerations hold when the electrostatic potential energy $z_hF\phi(x)$ is negative, that is, when z_h and $\phi(x)$ have opposite sign; in this case, the electric potential difference, $\phi(x) - \phi(\infty) \equiv \phi(x)$, between the position $x = x$ and the bulk solution tends to attract the charged particles toward the membrane surface, and $c_h(x)$ is now greater than c_h. It must be stressed that ions, being free to move, accumulate where the electric potential has sign opposite to that of their charge. In this respect, ions *are subjected* to the electric field and, hence, differ from fixed charges, such as those of the negative polar heads of phosphatidylserine, which *create* an electric field by their presence.

Any change in the local concentrations of the different ionic species with respect to their bulk values, at a given distance x from the membrane surface, generates a *volume charge density* $\rho(x)$ different from zero. By $\rho(x)$ we mean the ratio of the infinitesimal charge dq contained in a volume element located at a distance x from the membrane surface, divided by its infinitesimal volume dV. This is given by the sum of the local concentrations of all ionic species at x, each multiplied by the corresponding molar charge, z_hF:

$$\rho(x) = \sum_h z_h F c_h(x). \tag{2.22}$$

For simplicity, let us assume that the solution contains only a single 1,1-valent electrolyte, that is, an electrolyte consisting of a monovalent cation and a monovalent anion. In this case, Eq. 2.22 becomes

$$\rho(x) = Fc_+(x) - Fc_-(x). \tag{2.23}$$

To proceed further, we must now apply Poisson's equation (Eq. 2.17):

$$-(\mathrm{d}^2\phi/\mathrm{d}x^2) = \rho(x)/(\varepsilon_0\varepsilon). \tag{2.24}$$

In this case, ε is the dielectric constant of the solvent. From a qualitative point of view, this equation states that the presence of a local charge density ρ at x causes a change in the electric field $E_x = -\mathrm{d}\phi/\mathrm{d}x$ and hence is the source or sink of a line of force. On the other hand, if $\rho(x)$ equals zero, the electric field is constant at x, and the electric potential varies linearly with distance, as is the case between the plates of a parallel-plate capacitor. By replacing $\rho(x)$ from Eq. 2.23 into Eq. 2.24 and by then replacing the local concentrations $c_+(x)$ and $c_-(x)$, as expressed by

Eq. 2.21, into the resulting equation, we obtain the so-called *Poisson–Boltzmann equation*:

$$\frac{d^2\phi}{dx^2} = -\frac{Fc}{\varepsilon_0\varepsilon}(e^{-f\phi} - e^{f\phi}) \text{ with } f \equiv \frac{F}{RT}. \tag{2.25}$$

Here, c is the common value of the bulk concentrations of the monovalent cation and anion.

This second-order differential equation can be solved analytically, so as to obtain the expression of the electric potential $\phi(x)$ as a function of the distance x. Its reduction to a first-order differential equation can be easily carried out upon multiplying both members of Eq. 2.25 by $2(d\phi/dx)$ and noting that $2(d\phi/dx)(d^2\phi/dx^2)$ equals $(d/dx)(d\phi/dx)^2$:

$$\frac{d}{dx}\left(\frac{d\phi}{dx}\right)^2 = -\frac{2Fc}{\varepsilon_0\varepsilon}(e^{-f\phi} - e^{f\phi})\frac{d\phi}{dx}. \tag{2.26}$$

By integrating this equation between $x = x$ and $x = \infty$, we obtain

$$\left(\frac{d\phi}{dx}\right)_\infty^2 - \left(\frac{d\phi}{dx}\right)_x^2 = +\frac{2RTc}{\varepsilon_0\varepsilon}\left|e^{-f\phi} + e^{f\phi}\right|_x^\infty. \tag{2.27}$$

Here, $x = \infty$ denotes any point in the bulk solution, where the electric potential is conventionally set equal to zero. When proceeding toward the bulk solution (i.e., for $x \to \infty$), both the electric potential and the electric field, $-d\phi/dx$, tend to zero. Setting $\phi(\infty) = 0 = (d\phi(x)/dx)_\infty$ in Eq. 2.27 and taking the square root of both members, we have

$$\left(\frac{d\phi}{dx}\right)_x = +\left(\frac{2RTc}{\varepsilon_0\varepsilon}\right)^{1/2}(e^{f\phi} + e^{-f\phi} - 2)^{1/2} = -\left(\frac{2RTc}{\varepsilon_0\varepsilon}\right)^{1/2}(e^{f\phi/2} - e^{-f\phi/2}). \tag{2.28}$$

The choice of the negative sign in front of the third member of Eq. 2.28 is based on the following considerations. The difference between the two exponentials in the third member of this equation is positive or negative depending on whether $\phi(x)$ is positive or negative, respectively. Now, if $\phi(x)$ is negative within the diffuse layer, $d\phi(x)/dx$ is certainly positive, because the electric potential tends to increase gradually when proceeding toward the bulk solution, where its value, $\phi(\infty)$, is conventionally set equal to zero; if, on the contrary, $\phi(x)$ is positive, $d\phi(x)/dx$ is certainly negative by analogous considerations. In both cases, for the first member in Eq. 2.28 to have the same sign as the third one, the minus sign in front of the third member is required.

Equation 2.28 may be written in a more compact form by considering the definition of the hyperbolic sine of an independent variable y:

$$\sinh(y) = (e^y - e^{-y})/2. \tag{2.29}$$

On the basis of this definition, Eq. 2.28 takes the form

$$\frac{d\phi}{dx} = -\left(\frac{8RTc}{\varepsilon_0\varepsilon}\right)^{1/2}\sinh\left(\frac{f\phi}{2}\right). \tag{2.30}$$

This equation can be further integrated over x, thus providing an analytical expression for $\phi(x)$. Since this expression is relatively involved for our purposes, suffice here to know that it exists. The Poisson–Boltzmann equation at a planar charged surface was solved analytically by Gouy (1910) and, independently, by Chapman (1913). For this reason, the resulting theory is commonly called the *Gouy–Chapman theory*.

Here, we confine ourselves to integrating Eq. 2.30 in the simple case in which the absolute value of the argument of $\sinh(f\phi/2)$ is much less than unity, which amounts to assuming that the energy of thermal agitation is much greater than the electrostatic potential energy. In this case, the $\sinh(y)$ function simplifies as follows:

$$\sinh(y) = (e^{y} - e^{-y})/2 \cong [(1+y) - (1-y)]/2 = y \quad \text{for } y \ll 1, \tag{2.31}$$

where the series expansion of the two exponentials is interrupted after the first two terms ($e^{y} = 1 + y + \cdots$). Consequently, Eq. 2.30 becomes

$$\frac{d\phi}{dx} = -\kappa\phi \text{ with } \kappa \equiv \left(\frac{2F^2c}{\varepsilon_0\varepsilon RT}\right)^{1/2}. \tag{2.32}$$

The quantity κ is called the *reciprocal Debye length*. Equation 2.32 can be readily integrated by separation of variables between the minimum distance, $x = 0$, from the membrane surface at which the Poisson–Boltzmann equation is applicable, and a generic distance $x > 0$:

$$\int_{\phi_0}^{\phi(x)} \frac{d\phi}{\phi} = -\kappa\int_0^x dx = -\kappa x = |\ln\phi|_{\phi_0}^{\phi(x)}, \tag{2.33}$$

where ϕ_0 is the electric potential at $x = 0$. Writing Eq. 2.33 in exponential form, we obtain

$$\phi(x) = \phi_0 \exp(-\kappa x) \quad \text{for } x > 0. \tag{2.34}$$

According to this equation, the electric potential decays exponentially toward its limiting bulk value, $\phi(\infty) = 0$, with a decay constant κ. The reciprocal, $1/\kappa$, of this constant has the dimension of length, and expresses the distance from the $x = 0$ plane at which the electric potential $\phi(x)$ is reduced to a fraction $1/e = 0.3678$ of its value ϕ_0 at $x = 0$. In this respect, κ^{-1} measures the *diffuse layer thickness*, namely the thickness of the solution layer where ionic concentrations are altered to an appreciable extent with respect to their bulk values. For an aqueous solution of a 1,1-valent electrolyte at 25 °C, κ^{-1} is approximately equal to 100 Å for $c = 10^{-3}$ M, and to 10 Å for $c = 0.1$ M.

To determine the relationship between the electric potential ϕ_0 on the membrane surface plane and the charge density σ_d in the adjacent diffuse layer, consider an ideal cylinder of unit cross-sectional area, identical with the right-hand cylinder in Fig. 2.7. The flux of the electric field through this cylinder is given by the following equation:

$$\iint_S \mathbf{E}\cdot\hat{\mathbf{n}}\, dS = (\varepsilon_0\varepsilon)^{-1}\iiint_V \rho\, dV, \tag{2.35}$$

which differs from Eq. 2.9, relative to vacuum, by the presence of the dielectric constant ε of the solvent, since we are now dealing with a condensed phase. The flux of $\boldsymbol{E}$ through the unit base located in the bulk solution equals zero since, by definition of a bulk phase, the electric potential does not vary with distance. Consequently, the electric field component along the x-axis, $E_x = -\mathrm{d}\phi/\mathrm{d}x$, equals zero in the bulk solution. For obvious symmetry reasons, the electric field $\boldsymbol{E}$ created by a planar, infinitely extended charged membrane is normal to the membrane plane and, hence, E_x is the only nonzero electric field component. Since the lateral surface of the ideal cylinder is normal to the membrane plane and, hence, parallel to E_x, the flux of the electric field through the lateral surface equals zero. The only nonzero contribution to the flux of $\boldsymbol{E}$ through the cylinder surface is that through its unit base positioned on the membrane surface plane, $x = 0$. Application of Eq. 2.35 to our cylinder yields

$$\iint_S \boldsymbol{E} \cdot \hat{\boldsymbol{n}}\, \mathrm{d}S = -E_x(x=0) \iint_S \mathrm{d}S = -E_x(x=0) \times (\text{unit area})$$

$$= (\mathrm{d}\phi/\mathrm{d}x)_{x=0} = (\varepsilon_0\varepsilon)^{-1} \iiint_V \rho\, \mathrm{d}V = (\varepsilon_0\varepsilon)^{-1}\sigma_\mathrm{d}. \tag{2.36}$$

In writing this equation, we considered that E_x is parallel to the versor $\hat{\boldsymbol{n}}$ normal to the unit base at $x = 0$, which, however, has an opposite direction with respect to E_x since it is directed outward with respect to the cylinder closed surface, by definition of flux. Consequently, the scalar product $\boldsymbol{E} \cdot \hat{\boldsymbol{n}}$ is equal to the opposite of $E_x(x=0) = -(\mathrm{d}\phi/\mathrm{d}x)_{x=0}$. Moreover, the cylinder encloses the whole charge density within the diffuse layer; hence, the triple integral of the volume charge density ρ over the cylinder volume is just equal to the surface charge density σ_d of the diffuse layer. Equating the electric field component in Eq. 2.30 to that in Eq. 2.36, we get

$$\left(\frac{\mathrm{d}\phi}{\mathrm{d}x}\right)_{x=0} = -\left(\frac{8RTc}{\varepsilon_0\varepsilon}\right)^{1/2} \sinh\left(\frac{f\phi_0}{2}\right) = \frac{\sigma_\mathrm{d}}{\varepsilon_0\varepsilon}. \tag{2.37}$$

The quantity ϕ_0 is just the *electric potential difference across the diffuse layer* adjacent to the membrane surface, measured with respect to the electric potential in the bulk solution. To grasp the meaning of the signs in Eq. 2.37, note that, when $(\mathrm{d}\phi/\mathrm{d}x)_{x=0}$ is positive, ϕ_0 is clearly negative with respect to its zero bulk value, just as $\sinh(f\phi_0/2)$; in this case, positive ions are attracted toward the membrane surface and σ_d is positive. If $|F\phi_0/(2RT)|$ is sufficiently smaller than unity to satisfy Eq. 2.32 at $x = 0$, this equation can be combined with Eq. 2.37 yielding

$$(\mathrm{d}\phi/\mathrm{d}x)_{x=0} = -\kappa\phi_0 = \sigma_\mathrm{d}/(\varepsilon_0\varepsilon). \tag{2.38}$$

This equation expresses a proportional relationship between ϕ_0 and the charge density σ_d in the adjacent diffuse layer. In the following, the approximate Eq. 2.38 is made use of, because it is much easier to handle than Eq. 2.37, in which σ_d is an implicit function of ϕ_0. However, it must be remembered that it holds strictly for $|F\phi_0/(2RT)| \ll 1$.

The last two members of Eq. 2.38 can be written as

$$\phi_0 = -\frac{l}{\varepsilon_0\varepsilon}\sigma_d \text{ with } l \equiv \kappa^{-1} = \left(\frac{\varepsilon_0\varepsilon RT}{2F^2c}\right)^{1/2}. \quad (2.39)$$

This equation has the same form as Eq. 2.18, which expresses the potential difference across a parallel-plate capacitor, with its plates placed at a distance $l = \kappa^{-1}$ from each other. In other words, the potential difference across the diffuse layer is equivalent to that across a capacitor whose charge density σ_d is hypothetically placed on an ideal plane at a distance $l = \kappa^{-1}$ from the membrane surface. Even though the diffuse layer charge is distributed in space, decreasing progressively as we depart from the membrane surface up to its extinction in the bulk phase, it is formally equivalent to a charge homogeneously distributed on an ideal planar surface, with a charge-free region interposed between this surface and the membrane surface. According to this approximation, the electric potential profile is modified, as shown in Fig. 2.8, since the electric potential across this charge-free region varies linearly with distance, instead of showing an exponential decay.

In this figure, the external and internal membrane surface planes are placed on the $x = 0$ and $x = d$ planes, respectively, and the quantities pertaining to the external and internal sides of the membrane are denoted by a prime and a double prime, as in Fig. 2.6. The charge densities, σ'_d and σ''_d, of the external and internal diffuse layers are distributed on the planes marked by two vertical dashed lines, at distances l' and l'' from the corresponding membrane surface planes, expressed by the following equations:

$$l' \equiv \left(\frac{\varepsilon_0\varepsilon_w RT}{2F^2c'}\right)^{1/2}, \quad l'' \equiv \left(\frac{\varepsilon_0\varepsilon_w RT}{2F^2c''}\right)^{1/2}, \quad (2.40)$$

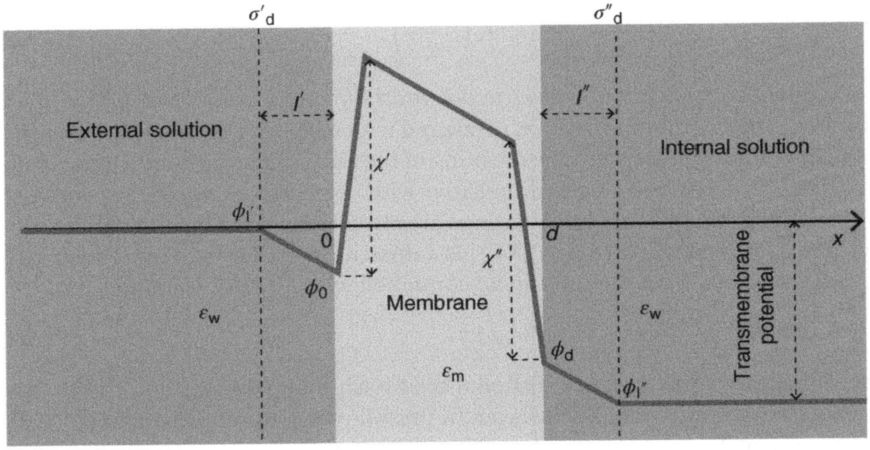

Figure 2.8 Electric potential profile across a neutral bilayer lipid membrane, for a negative transmembrane potential $\Delta\phi$, according to the *linearized* Gouy–Chapman theory.

where ε_w is the dielectric constant of the aqueous medium (approximately equal to 78), whereas c' and c'' are the concentrations of the 1,1-valent electrolyte in the external and internal bulk solutions, respectively. Equation 2.40 shows that the thickness of the diffuse layer decreases linearly with an increase in the square root of the electrolyte concentration in the corresponding bulk solution.

In this approximate representation of an electroneutral membrane, the ion charges are considered to be distributed on the two ideal planes $x = -l'$ and $x = d + l''$. The charge densities, σ'_d and σ''_d, on these two planes are equal in magnitude and opposite in sign, in view of the electroneutrality of the whole electrified interface (Eq. 2.20). However, the medium between these two oppositely charged planes cannot be treated as that between the plates of a parallel-plate capacitor, because the dielectric constant is not uniform throughout this medium. Thus, the dielectric constant within the two diffuse layers is approximately equal to that ($\varepsilon_w \cong 78$) of the aqueous phase. Incidentally, some molecular dynamics simulations point to a gradual decrease of ε_w from 78 to about 50 in passing from the bulk solution to the membrane surface (Jakobsson, 1997). It should also be noted that the ions adjacent to the membrane cannot bring their center of charge in direct contact with the membrane surface plane, due to their finite size. This determines the presence of an uncharged layer, occupied by partially oriented water molecules, which is in direct contact with the membrane surface. This layer, called *Stern layer*, has a thickness comparable with the solvated radius of the ions that are closest to the membrane surface (Delahay, 1965). The dielectric constant of the hydrocarbon tail region of the lipid bilayer is close to that ($\varepsilon_m \cong 2$) of liquid hydrocarbons, such as pentane or decane. The dielectric constant of the two polar head regions is roughly estimated between 10 and 20 (Flewelling and Hubbell, 1986), and hence is intermediate between that of water and that of the hydrocarbon tail region.

For simplicity, we ignore the two polar head regions, by assimilating them to the adjacent diffuse layer regions, as concerns their dielectric constants. The medium enclosed between the two planar charge distributions σ'_d and σ''_d can be considered as consisting of three distinct parallel-plate capacitors in series: the external diffuse layer, the hydrocarbon tail region, and the internal diffuse layer. In fact, an ideal infinitesimally sized observer located within any given slab of uniform dielectric constant sees on the left side of the slab an overall charge density equal in magnitude and opposite in sign to that on the right side. This is due to the fact that the electric field created by an infinitely extended planar charge distribution does not depend on the distance from the distribution. Hence, all planar charge distributions on the left side of each given slab exert the same effect inside the slab, as though they were lumped together on the left boundary of the slab. Analogous conclusions hold for the right side of the slab. Moreover, the sum of the planar charge densities located on the left side of each given slab must be necessarily equal in magnitude and opposite in sign to the sum of the planar charge densities on its right side, in view of the electroneutrality of the whole electrified interface.

On the basis of the previous considerations, we can apply the expression of Eq. 2.18 for the potential difference across a parallel-plate capacitor to the three capacitors in series that compose the interfacial region enclosed between the two bulk aqueous solutions bathing the membrane:

$$\phi_0 - \phi_{l'} = -\frac{l'}{\varepsilon_0 \varepsilon_w}\sigma'_d; \quad \phi_d - \phi_0 = -\frac{d}{\varepsilon_0 \varepsilon_m}\sigma'_d + \chi' - \chi'';$$
$$\phi_{l''} - \phi_d = -\frac{l''}{\varepsilon_0 \varepsilon_w}\sigma'_d. \tag{2.41}$$

In these equations, $\phi_{l'}$, ϕ_0, ϕ_d, and $\phi_{l''}$ are the electric potentials on the planes $x = -l'$, $x = 0$, $x = d$, and $x = (d + l'')$, respectively, as shown in Fig. 2.8. The potential difference across the hydrocarbon tail region, $\phi_d - \phi_0$, includes the surface dipole potentials χ' and χ''. The negative sign in front of χ'' accounts for the fact that it decreases along the positive direction of the x-axis.

From Eq. 2.41, it is apparent that the potential differences across the two diffuse layers are much smaller than that across the hydrocarbon tail region. In fact, at a physiological 0.1 M concentration of a 1,1-valent electrolyte, the diffuse layer thickness amounts to ~10 Å, and is, therefore, comparable with that (~30 Å) of the hydrocarbon tail region. Conversely, the dielectric constant of the diffuse layer is almost 40 times higher than that of the hydrocarbon tail region. In Figs. 2.6 and 2.8, the potential differences across the two diffuse layers are notably enlarged with respect to that across the hydrocarbon tail region, to make them visible on the same potential scale.

2.7 Measurement of the Charge Density of the Polar Heads of a Charged Lipid

Let us consider the more general case of a membrane consisting of lipids whose polar heads are charged, and let us denote by σ' and σ'' the charge densities on the external and internal membrane surfaces, respectively, as shown in Fig. 2.9. These charge densities are fixed and are considered as smeared out uniformly on the two membrane surfaces, $x = 0$ and $x = d$, as a first approximation.

The figure refers to the case in which both σ' and σ'' are negative, and σ' is more negative than σ''. The transmembrane potential is set equal to zero to point out that, when the polar heads are charged, two diffuse layers are necessarily formed to maintain the electroneutrality of the whole interface. Conversely, at zero transmembrane potential, no diffuse layers are present at a symmetrical membrane composed of electroneutral lipids.

In view of the electroneutrality of the whole electrified interface, the sum of all charge densities contained in it equals zero:

$$\sigma'_d + \sigma' + \sigma'' + \sigma''_d = 0. \tag{2.42}$$

The potential differences across each of the two diffuse layers and across the hydrocarbon tail region are then estimated as in Eq. 2.41. Thus, the sum of all

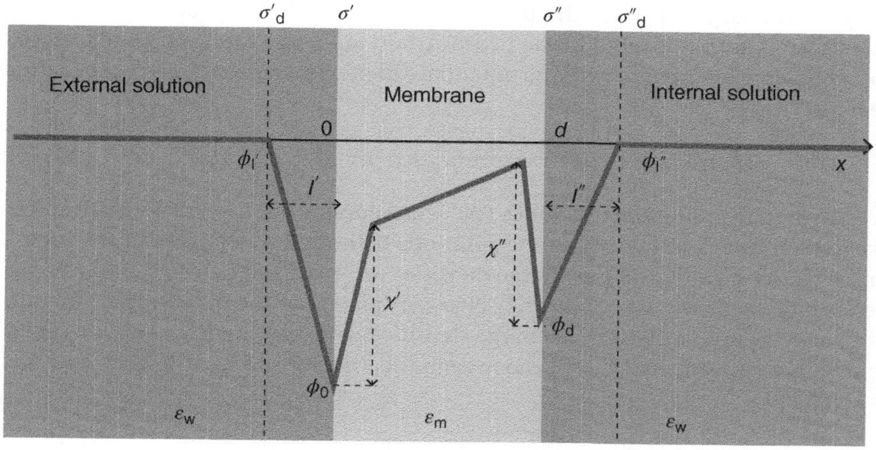

Figure 2.9 Electric potential profile across a charged bilayer lipid membrane, for a zero transmembrane potential $\Delta\phi$, according to the *linearized* Gouy–Chapman theory.

charge densities on the left of each given slab is considered as lumped together on its left boundary, while the sum of all charge densities on the right is considered as lumped together on its right boundary. The electric potential difference across each slab is then given by Eq. 2.18 for a parallel-plate capacitor:

$$\phi_0 - \phi_{l'} = -\frac{l'}{\varepsilon_0\varepsilon_w}\sigma'_d; \quad \phi_d - \phi_0 = -\frac{d}{\varepsilon_0\varepsilon_m}(\sigma'_d + \sigma') + \chi' - \chi'';$$
$$\phi_{l''} - \phi_d = -\frac{l''}{\varepsilon_0\varepsilon_w}(\sigma'_d + \sigma' + \sigma''). \tag{2.43}$$

The sum of these three electric potential differences yields the potential difference across the whole interface, $\phi_{l''} - \phi_{l'}$, namely the transmembrane potential referred to the electric potential in the external bulk solution:

$$\phi_{l''} - \phi_{l'} = -\frac{l'}{\varepsilon_0\varepsilon_w}\sigma'_d - \frac{d}{\varepsilon_0\varepsilon_m}(\sigma'_d + \sigma') + \chi' - \chi'' - \frac{l''}{\varepsilon_0\varepsilon_w}(\sigma'_d + \sigma' + \sigma''). \tag{2.44}$$

This potential difference can be measured experimentally by immersing two identical reference electrodes in the solutions that bath the two sides of the membrane.

If we know the fixed charge densities σ' and σ'' on the polar heads of the membrane, then it is also possible to determine the charge densities σ'_d and σ''_d of the corresponding diffuse layers, by combining Eq. 2.44 with the electroneutrality condition of Eq. 2.42 for the whole interface, and solving the resulting system of two algebraic equations with two variables. The overall charge density, $\sigma'_d + \sigma'$, on the external side of the membrane is equal in magnitude and opposite in sign to the overall charge density, $\sigma'' + \sigma''_d$, on its internal side, in view of Eq. 2.42. It is, therefore, evident that these two sides are correlated, in that a change in σ' also affects σ''_d and a change in σ'' also affects σ'_d.

By far the largest part of the potential difference, $\phi_{l''} - \phi_{l'}$, across the whole interfacial region is located in the hydrocarbon tail region and can be identified, to a good approximation, with the potential difference, $(\phi_d - \phi_0)$, across the latter region:

$$(\phi_{l''} - \phi_{l'}) \cong (\phi_d - \phi_0) = -\frac{d}{\varepsilon_0 \varepsilon_m}(\sigma'_d + \sigma') + \chi' - \chi''. \tag{2.45}$$

A change in the potential difference $(\phi_{l''} - \phi_{l'})$ applied between two identical reference electrodes immersed in the solutions bathing the two sides of the membrane causes a change in the charge densities $(\sigma'_d + \sigma')$ and $(\sigma'' + \sigma''_d)$, which can be considered as the two oppositely charged "plates" of the membrane. The rate of change of any of these two charge densities with varying $(\phi_{l''} - \phi_{l'})$ is called *differential capacitance*, C_m, of the membrane. In view of Eqs. 2.42 and 2.45, C_m is given by

$$C_m \equiv -\frac{d(\sigma' + \sigma'_d)}{d(\phi_{l''} - \phi_{l'})} = \frac{d(\sigma'' + \sigma''_d)}{d(\phi_{l''} - \phi_{l'})} = \frac{\varepsilon_0 \varepsilon_m}{d}. \tag{2.46}$$

In obtaining this equation, it was further assumed quite reasonably that the two surface dipole potentials, χ' and χ'', do not change with varying $(\phi_{l''} - \phi_{l'})$. The definition of C_m in Eq. 2.46 is perfectly analogous to that used for the capacitance of an electric capacitor. It must be stressed, however, that the differential capacitance C_m is the ratio of two infinitesimal quantities and, as such, depends to some extent on the potential difference applied across the membrane. Conversely, the capacitance of an electric capacitor does not depend on the voltage applied between its plates and is the ratio of two finite quantities. In particular, the thickness of a bilayer lipid membrane (BLM) decreases slightly with an increase in the absolute value of the charge densities, $(\sigma'_d + \sigma')$ and $(\sigma'' + \sigma''_d)$, on its two sides, due to electrostatic attraction between these equal but opposite charge densities.

In the following text, a bilayer lipid membrane spanning a small orifice in a hydrophobic septum that separates two aqueous compartments is briefly denoted by the widely adopted acronym BLM (*bilayer lipid membrane* or *black lipid membrane*) (cf. Section 6.3). If the BLM is formed by depositing into the orifice a droplet of a lipid solution in a volatile liquid hydrocarbon, it incorporates a non-negligible amount of hydrocarbon molecules. In this case, the effect of the electrostatic attraction between the opposite sides of the BLM is particularly evident. The compressibility of BLMs, termed *electrostriction*, causes the membrane thickness d to attain a maximum when the charge densities on its two sides are equal to zero, that is, for $(\sigma'_d + \sigma') = (\sigma'' + \sigma''_d) = 0$. In this case, the electroneutrality condition holds not only for the whole interface (Eq. 2.42), but also for each of the separate charge densities on the opposite sides of the membrane. Under these conditions, characterized by a maximum BLM thickness d, the experimentally measurable differential capacitance C_m attains a minimum, in view of Eq. 2.46. The compressibility of BLMs allows an estimate of the fixed charges σ' and σ'' of the polar heads, which do not vary with a change of $(\phi_{l''} - \phi_{l'})$.

Let us consider, for example, a symmetrical BLM bathed by two solutions of identical composition, and let us assume, for the sake of simplicity, that σ' and

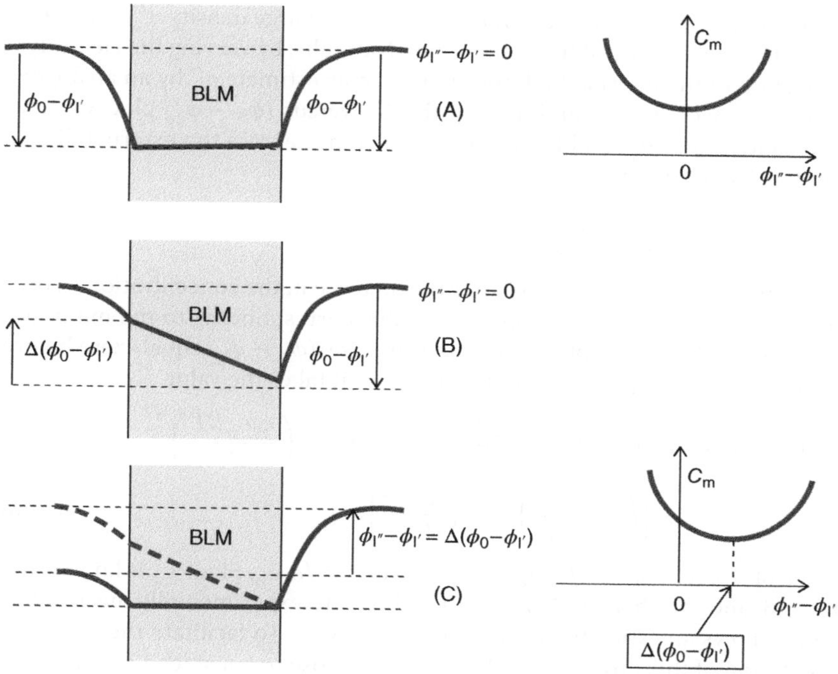

Figure 2.10 Electric potential profile (thick lines) across a symmetrical BLM at $\phi_{l''} - \phi_{l'} = 0$ before (A) and after an increase in the electrolyte concentration, c', in the external bulk solution (B), and subsequent adjustment of $\phi_{l''} - \phi_{l'}$ to attain a new minimum in the C_m versus $(\phi_{l''} - \phi_{l'})$ curve (C). Downward arrows denote negative quantities and upward arrows positive quantities.

σ'' are negative and equal (see Fig. 2.10, where the two surface dipole potentials χ' and χ'' are omitted, to simplify the potential profiles). When the applied potential $(\phi_{l''} - \phi_{l'})$ equals zero, one also has $(\sigma'_d + \sigma') = (\sigma'' + \sigma''_d) = 0$ and C_m assumes its minimum value in view of Eq. 2.46. More precisely, the plot of C_m against $(\phi_{l''} - \phi_{l'})$ has an approximately parabolic shape, with its minimum at $(\phi_{l''} - \phi_{l'}) = 0$ (Fig. 2.10A). Let us now increase the electrolyte concentration in the external solution from c' to $c' + \Delta c'$; this causes a decrease in the thickness, l', of the external diffuse layer and, consequently, in the absolute value of the potential difference, $(\phi_0 - \phi_{l'})$, across it. The change, $\Delta(\phi_0 - \phi_{l'})$, in $(\phi_0 - \phi_{l'})$ has a positive sign, as shown in Fig. 2.10B. This generates a negative potential difference, $(\phi_d - \phi_0)$, equal in magnitude but opposite in sign, across the hydrocarbon tail region, where the electric potential varies linearly with distance (Fig. 2.10B). Under these conditions, to annihilate the electric potential difference, $(\phi_d - \phi_0) = -\Delta(\phi_0 - \phi_{l'})$, across the hydrocarbon tail region and, hence, to attain the minimum of the C_m versus $(\phi_{l''} - \phi_{l'})$ plot, we must oppose it by an equal but opposite potential difference, $(\phi_{l''} - \phi_{l'}) = \Delta(\phi_0 - \phi_{l'})$, across the whole interfacial region, as appears from in Fig. 2.10C.

This procedure may be used to estimate the fixed charge density σ' of the polar heads on the external side of the membrane. (Naturally, by varying the electrolyte concentration in the internal bulk solution, we may estimate σ'' by an analogous procedure.) In fact, at the minimum of the C_m versus $(\phi_{l''} - \phi_{l'})$ plot, we have $\sigma' = -\sigma'_d$ and $\sigma'' = -\sigma''_d$, and the potential difference across the external diffuse layer in Eq. 2.43 becomes

$$\phi_0 - \phi_{l'} = -\frac{l'}{\varepsilon_0\varepsilon_w}\sigma'_d = \frac{l'}{\varepsilon_0\varepsilon_w}\sigma'. \tag{2.47}$$

This relation holds both before and after the increase in the concentration c', provided we constantly refer to the $(\phi_0 - \phi_{l'})$ value corresponding to the minimum of the C_m versus $(\phi_{l''} - \phi_{l'})$ plot. In the present case, $(\phi_{l''} - \phi_{l'})$ equals zero before the increase $\Delta c'$ in c', whereas after this increase it takes the value

$$(\phi_{l''} - \phi_{l'}) = \Delta(\phi_0 - \phi_{l'}) = \frac{\Delta l'}{\varepsilon_0\varepsilon_w}\sigma' \quad \text{with} \quad \Delta l' = \left(\frac{\varepsilon_0\varepsilon_w RT}{2F^2}\right)^{1/2} \times \left[\left(\frac{1}{c' + \Delta c'}\right)^{1/2} - \left(\frac{1}{c'}\right)^{1/2}\right]. \tag{2.48}$$

The slope of a plot of $(\phi_{l''} - \phi_{l'})$ against $\Delta l'$ equals $\sigma'/\varepsilon_0\varepsilon_w$. Since ε_w is known and equal to 78, and the changes $\Delta(\phi_0 - \phi_{l'})$ and $\Delta c'$ are experimentally measurable, this slope allows an estimate of the charge density σ'. To facilitate the graphical representation of this procedure in Fig. 2.10, we chose to consider a symmetrical BLM. However, it is evident that the procedure can also be applied to the case of an asymmetrical BLM, in which the transmembrane potential $(\phi_{l''} - \phi_{l'})$ corresponding to the minimum of the C_m versus $(\phi_{l''} - \phi_{l'})$ plot before the increase in c' is different from zero.

2.8 Electropermeabilization of Lipid Bilayers

If the transmembrane potential $\Delta\phi$ exceeds a certain value, the membrane undergoes rupture and its conductance increases drastically. This phenomenon is called *irreversible electrical breakdown*. The typical shape of the current flowing across a BLM subject to an electric potential pulse high enough to cause rupture is shown in Fig. 2.11. After an initial current decay between $t = t_0$ and $t = t_1$, due to the capacitive charging of the membrane, the current attains a low steady-state value, typical of the intact membrane (Chizmadzhev et al., 2004). Subsequently, the current increases showing fluctuations, and ultimately attains a high limiting value identical with that observed in the absence of the BLM. Sometimes, if the potential pulse is not too long, the BLM returns from the intermediate region of breakdown fluctuations to its initial intact state. This phenomenon is called *reversible electrical breakdown*.

Under irreversible breakdown conditions, the mean lifetime of the membrane, $\langle\tau\rangle$, defined as $t_2 - t_0$, undergoes a tenfold decrease per each 100 mV increase in $\Delta\phi$. The BLM stability decreases significantly with an increase in temperature, and

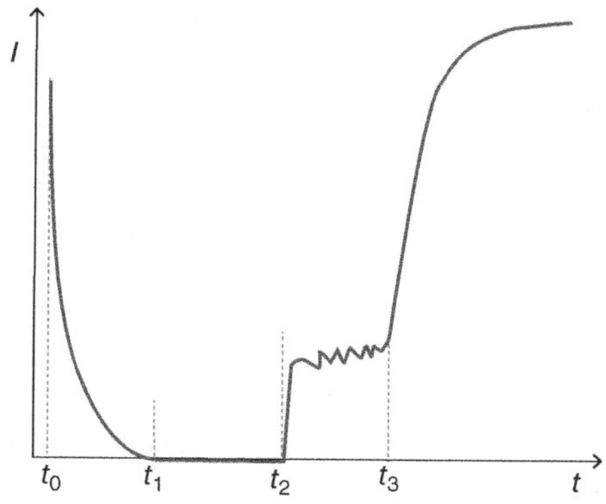

Figure 2.11 Typical stages of a current versus time curve during the irreversible electrical breakdown of a BLM: decaying capacitive current from t_0 to t_1; low steady-state current from t_1 to t_2; random fluctuations from t_2 to t_3; and tendency toward the maximum limiting current corresponding to the absence of the BLM after t_3.

is also dependent on the electrolyte composition and on the organic solvent used to form the BLM. The process leading to membrane breakdown is not accompanied by a noticeable increase in capacitance, which would denote membrane thinning. This strongly suggests that the permeabilization of a BLM in a strong electric field is not a large-scale process; rather, it is associated with the development of local defects. Different views on the instability and collapse of a lipid bilayer as a whole have also been proposed. Thus, the membrane has been considered as an elastic capacitor of constant permittivity subject to the pressure exerted by the electric field (Crowley, 1973) or as a flat layer of a nonconducting liquid interposed between two conducting liquids (Michael and O'Neill, 1970).

Irreversible breakdown is interpreted starting from the consideration of a structural defect of the membrane, such as a narrow pore. Imagine punching an infinitesimal circular hole in a membrane. The interfacial tension γ, namely the force exerted by the membrane per unit length of the circumference of the pore, will tend to enlarge it. The infinitesimal force acting on an infinitesimal arc dl of the circumference, of generic radius r, is given by $\gamma\, dl = \gamma r\, d\theta$, where $d\theta$ is the infinitesimal angle subtended by the arc dl. The infinitesimal work done by γ is given by the force $\gamma r\, d\theta$ times the displacement dr. The overall work performed by the membrane to form a pore of given radius r is, therefore, given by

$$W_\gamma = \int_{r=0}^{r}\int_{\theta=0}^{2\pi} \gamma r\, d\theta\, dr = \gamma \int_{r=0}^{r} r\, dr \int_{\theta=0}^{2\pi} d\theta = \pi r^2 \gamma. \qquad (2.49)$$

In the language of thermodynamics, the membrane represents our *closed system* (cf. Section 3.1), whereas the rest of the universe that interacts with the system is referred to as the *surroundings*. The work W_γ is positive because the force exerted by γ to contract the membrane surface around the pore has the same direction as the displacement. Consequently, the thermodynamic reversible work accomplished by the surroundings to exactly counterbalance this force, while the pore size increases, involves the application of a force on the pore rim that is directed toward the center of the pore, namely in the direction opposite to the displacement. The resulting work done by the surroundings on the system is, therefore, negative, and is given by $-\pi r^2 \gamma$.

In addition to this work, a further contribution to the overall mechanical work comes from the linear tension γ_l, namely the force that acts at a generic point of the pore rim, tangentially to it, which accomplishes a work equal to $-\int_{l=0}^{2\pi r} \gamma_l \, dl = -2\pi r \gamma_l$. The linear tension tends to contract the rim length and, therefore, does a negative work during the progressive expansion of the pore, since its direction is opposite to that of the displacement. The work done by the surroundings to counterbalance γ_l, so as to render the pore expansion thermodynamically reversible, is positive and equal to $2\pi r \gamma_l$. Therefore, the overall mechanical work W_{mech} is given by

$$W_{mech} = -\pi r^2 \gamma + 2\pi r \gamma_l. \tag{2.50}$$

In addition to the mechanical work, an electrical work W_{el} is also involved in the formation of a pore, because the lipid molecules of the bilayer are replaced by water molecules, which have a higher polarizability. Hence, such a replacement is favored by the application of a transmembrane potential $\Delta\phi$ and contributes to the enlargement of the pore radius. In other words, the progressive insertion of water molecules along the periphery of the pore moves the periphery outward, which amounts to a positive work done by the membrane system. Consequently, the surroundings must accomplish a negative work of equal magnitude to generate a pore under thermodynamically reversible conditions. The electrical work to discharge a capacitor of capacitance $C = q/\Delta\phi$ and initial charge density $|Q|$ on its plates is obtained by moving progressively infinitesimal charge elements dq from one plate to the other across the potential difference $\Delta\phi$ between the plates:

$$\int_{q=Q}^{0} \Delta\phi \, dq = \frac{1}{C}\int_{q=Q}^{0} q \, dq = \left|\frac{q^2}{2C}\right|_{Q}^{0} = -\frac{Q^2}{2C} = -C\frac{\Delta\phi^2}{2}. \tag{2.51}$$

The capacitance of a pore of radius r filled with lipid molecules is given by $C_m = \pi r^2 \varepsilon_0 \varepsilon_m / d$ (cf. Eq. 2.19), where ε_0 is the permittivity of free space, while ε_m and d are the dielectric constant and the thickness of the membrane. Similarly, the capacitance of the same pore filled with water molecules is given by $C_w = \pi r^2 \varepsilon_0 \varepsilon_w / d$, where ε_w is the dielectric constant of water. In view of Eq. 2.51, the work done by the system to discharge the capacitor of capacitance C_m and transmembrane potential $\Delta\phi$ and to charge the capacitor of capacitance C_w at the same $\Delta\phi$ value is given by

$$(C_w - C_m)\frac{\Delta\phi^2}{2} = \frac{\pi r^2 \varepsilon_0}{2d}\Delta\phi^2(\varepsilon_w - \varepsilon_m). \tag{2.52}$$

This work is equal in magnitude but opposite in sign to the electrical work, W_{el}, done by the surroundings.

The total reversible work spent by the surrounding to form a pore of radius r is, therefore, given by

$$W_{tot} = W_{mech} + W_{el} \equiv \Delta G = 2\pi r \gamma_l - \pi r^2(\gamma + \Delta C \Delta\phi^2/2)$$
$$\text{with } \Delta C = \varepsilon_0(\varepsilon_w - \varepsilon_m)/d, \tag{2.53}$$

where ΔC is the difference in capacitance per unit surface area. W_{tot} is just the total change in the *Gibbs energy*, $\Delta G(r, \Delta\phi)$, which accompanies the formation of a pore of radius r at a transmembrane potential $\Delta\phi$. The dependence of ΔG upon r is parabolic. The radius, $r^*(\Delta\phi)$, at which the parabola attains a maximum, is obtained by equating to zero the derivative of W_{tot} with respect to r:

$$r^*(\Delta\phi) = \gamma_l/(\gamma + \Delta C\Delta\phi^2/2). \tag{2.54}$$

Replacing r^* into Eq. 2.53 yields the maximum value attained by ΔG:

$$\Delta G_{max} = \pi\gamma_l^2/(\gamma + \Delta C\Delta\phi^2/2). \tag{2.55}$$

Curves of ΔG against r for different absolute values of $\Delta\phi$ are shown in Fig. 2.12.

At $\Delta\phi = 0$, the parabolic curve of W_{tot} against the pore radius r attains a maximum for $r^* = \gamma_l/\gamma$, as shown by highest curve in Fig. 2.12. If r is less than r^*, the pore will tend to close; conversely, if it is greater, it will tend to grow indefinitely, ultimately leading to membrane breakdown. The BLM is, therefore, a metastable system. If an increasing transmembrane potential $\Delta\phi$ is applied to the BLM, an increasing number of water molecules will tend to occupy the pore, displacing the lipid molecules, because of an increasing electrostatic interaction of the high dipole moment of the water molecules with the interfacial electric field. Hence, an increase in $|\Delta\phi|$ tends to decrease the critical pore radius r^* by making a negative contribution to the Gibbs energy of pore formation, ΔG (Eq. 2.53). This provides a qualitative explanation for the decrease in the membrane mean lifetime, $\langle\tau\rangle$, with an increase in the transmembrane potential.

Pores may be hydrophobic, if they are lined with the hydrocarbon tails of the lipids, in their usual vertical orientation with respect to the bilayer plane, or they may be hydrophilic, if they are lined with the lipid polar heads. In the latter case, the lipid molecules in contact with the pore must undergo an appropriate rotation (Fig. 2.13). In this regard, lipids can be classified into three categories (Chizmadzhev et al., 2004). Those with the cross-sectional area of the polar head very close to that of the two hydrocarbon tails, such as phosphatidylcholine, are represented as cylinders and tend to form hydrophobic pores. Lipids with the cross-sectional area of the polar head smaller than that of the hydrocarbon tails, such as phosphatidylethanolamine, are represented as cones with the polar

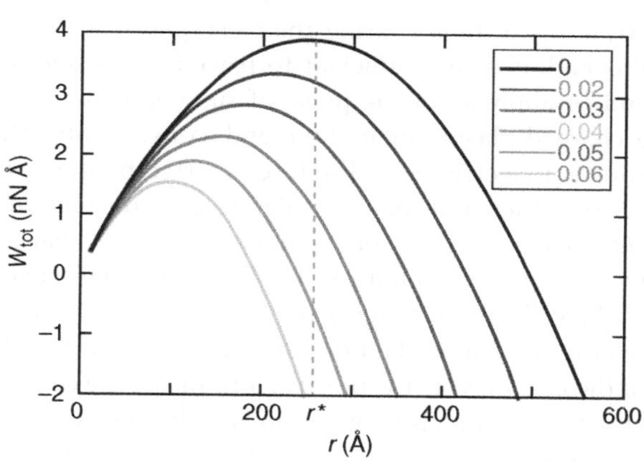

Figure 2.12 Dependence of the total work W_{tot} of pore formation upon pore radius for absolute values of $\Delta\phi$ increasing progressively from 0 to 0.06 V, as calculated from Eq. 2.53 for $\gamma_l = 5 \times 10^{-12}$ N, $\gamma = 2 \times 10^{-4}$ N/m, and $\Delta C = 0.17$ F/m^2.

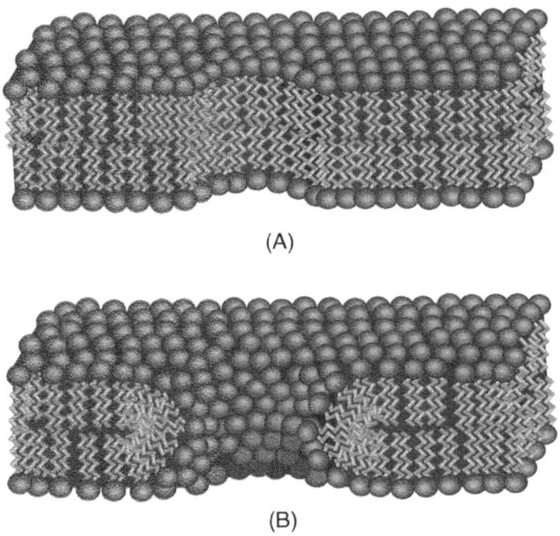

Figure 2.13 Schematic vertical cross section of (A) a hydrophobic pore and (B) a hydrophilic pore in a BLM. (Reprinted from Wikipedia Commons: Pore schematic.svg; author: MDougM.)

head located at the cone vertex. Finally, lipids with the cross-sectional area of the polar head larger than that of the tails are represented as cones with the polar head located at the cone base; a typical example is lysophosphatidylcholine, which has only one hydrocarbon tail. The latter lipids tend to form hydrophilic pores, but only provided the pore radius is large enough. In fact, hydrophilic pores with small radii are energetically disadvantageous and are not formed at all. It is hypothesized that, initially, only hydrophobic pores are formed spontaneously by lateral thermal fluctuations of the lipid molecules. Only when their radius exceeds some critical value, does a reorientation of the lipids become energetically favorable.

Electropermeabilization of membranes also takes place with cells and vesicles subjected to a sufficiently strong external electric field. Reversible electrical breakdown of cells and vesicles has important biochemical applications. During their reversible permeabilization, drugs, foreign proteins, and genes can penetrate them, thus altering their cytoplasm without affecting their viability. To this end, it is just sufficient to place a cell suspension between two electrodes and to apply a voltage pulse of suitable height and duration. Most cells have a charged plasma membrane and, consequently, they move under an external constant electric field. The force exerted by the electric field is balanced by the viscous drag, and the cells migrate at a constant velocity, a phenomenon called *electrophoresis*. The voltage pulse is so short that it does not cause the cells to migrate by electrophoresis, but the resulting electric field deforms them.

The cell can be roughly represented as a conducting sphere in a poorly conducting medium. The lines of force of the electric field converge to the surface of the cell, moving the positive ions inside the cell in the direction of the external electric field and the negative ions in the opposite direction. This creates a surface charge distribution that screens the external electric field, annihilating the electric field inside the cell, and generates an induced dipole moment with an orientation parallel to the undisturbed external lines of force. The magnitude of the induced dipole moment is proportional to the magnitude $|\boldsymbol{E}|$ of the external electric field according to the cell polarizability, $\alpha(t)$; this is time dependent and a complex function of the specific conductances of the cell interior and of the external solution, and also of the cell size and the pulse duration. The force acting on the cell is proportional to the product of the electric field by the induced dipole moment. In practice, for short pulses, the forces are parallel to the lines of force of the electric field and pull the cell at its poles in poorly conducting solutions, whereas they are perpendicular to the lines of force and squeeze the cell along the equator in highly conducting solutions. In both cases, the cell shape changes from spherical to ellipsoidal (Teissié, 2004).

The electric potential created by a uniform electric field $\boldsymbol{E}$ directed along the x-axis is given by $\phi = x(\mathrm{d}\phi/\mathrm{d}x) + \text{const} = -xE + \text{const}$, where E is the magnitude of the field (cf. Eq. 2.3). At any given point $\mathbf{P}$ immediately outside the surface of a spherical cell of radius r, x is given by $r\cos\theta$, where θ is the angle between the direction of the electric field and the normal to the cell surface (Fig. 2.14). Hence, the electric potential just outside the cell surface at point $\mathbf{P}$ is given by $-rE\cos\theta + \text{const}$. An approximate expression for the transmembrane potential at $\mathbf{P}$ induced by the external electric field $\boldsymbol{E}$ is given by

$$\Delta\phi_{\mathrm{E}}(\theta) = -frE\cos\theta, \tag{2.56}$$

where f is a parameter that depends on the specific conductances of the cell membrane, the cell interior, and the external solution. According to usage, $\Delta\phi_{\mathrm{E}}$ is the potential at the inside of the cell with respect to its outside. Noting that the electric field induced by $\boldsymbol{E}$ across the cell membrane, of thickness d, equals $-frE\cos\theta/d$, the maximum ratio of this electric field to the external one, E, is attained at $\theta = 0$ and amounts to fr/d. For typical values, $r = 5$ μm and $d = 5$ nm, of the cell radius and membrane thickness, the amplification factor fr/d equals $10^3 f$. If the specific conductance of the membrane equals zero, f equals 1.5. This justifies the occurrence of electropermeabilization. As a matter of fact, a membrane in not a perfect dielectric and several ions leak across it, causing a drop in $\Delta\phi_{\mathrm{E}}$.

In practice, the external electric field $\boldsymbol{E}$ tends to move cations toward the front hemisphere of a spherical cell and away from its back hemisphere with respect to its direction, while moving the anions in the opposite direction. In Fig. 2.14, the sphere is darker where cations prevail over anions. The electric field creates a transmembrane potential that is negative on the back hemisphere and positive on the front one. The field-induced transmembrane potential $\Delta\phi_{\mathrm{E}}$ is superimposed on the resting potential $\Delta\phi_{\mathrm{r}}$ of the plasma membrane, which is negative and clearly independent of θ. In Fig. 2.14, the arrows point from the negative to the positive potential value, and their length is a rough measure of the magnitude of the respective potential differences. At the cell equator (i.e., for $\theta = 90°$), $\Delta\phi = (\Delta\phi_{\mathrm{r}} + \Delta\phi_{\mathrm{E}})$ is not affected by the external field. When $|\Delta\phi|$ exceeds a threshold value $|\Delta\phi_{\mathrm{t}}|$, the portions of the cell surface enclosed in the solid angles

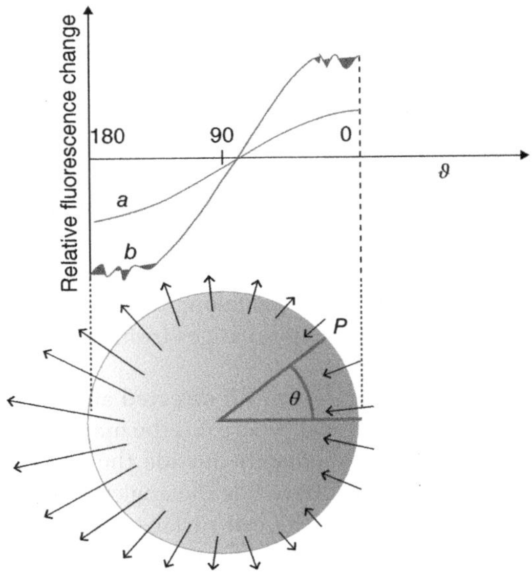

Figure 2.14 Representation of a spherical cell and the relative fluorescence change along its surface, for 0° < θ < 180°. The electric field ***E*** is directed from left to right. The arrows measure the transmembrane potential $\Delta\phi$, positive in the direction of the arrow, at different points of the cell surface. At the cell equator ($\theta = 90°$), only the resting transmembrane potential is present, negative toward the interior of the cell.

for which the $|\Delta\phi| > |\Delta\phi_t|$ condition is satisfied become electropermeabilized. Over these two portions, which straddle the θ values 0° and 180°, the curve of $\Delta\phi$ against θ ceases to be sinusoidal, due to a sharp increase in conductance. As a short electric field pulse is instantly applied, $\Delta\phi_E$ increases rapidly tending asymptotically to its maximum limiting value with a relaxation time that is a complex function of the membrane properties. With short pulses and large cells, $\Delta\phi_E$ may attain only a fraction of its maximum value at the end of the pulse.

Cell permeabilization can be followed by the inflow of exogenous molecules, such as dyes, upon detecting and quantifying the entrapped molecules. It can also be followed by the leakage of molecules or ions out of the cell, upon detecting their amount. Electropermeabilization can be monitored at the single cell level by submicrosecond imaging under a pulsed-laser microscope, using a fluorescence probe sensitive to $\Delta\phi$, as shown schematically in Fig. 2.14 (Kinosita et al., 1990). At low field strengths of the pulse, the fluorescence signal complies with the sinusoidal dependence of $\Delta\phi$ upon θ expressed by Eq. 2.56 (see curve *a* in Fig. 2.14). Conversely, at high field strengths, the signal flattens in the region of the poles, as shown by curve *b*. The $\Delta\phi_t$ threshold value depends on many parameters, such as cell size, duration and number of voltage pulses, and assay method. At any rate, it normally ranges from 200 to 500 mV. In general, a high-voltage pulse of short duration affects a large portion of the cell surface with a small permeability. Conversely, a voltage pulse slightly stronger than its threshold value, but of long duration (or a large number of consecutive pulses), affects a small surface area but determines a high local permeability. Electropermeabilization of cells favors their fusion, if they are sufficiently close during the pulse.

Electropermeabilization phenomena reported with cells are quite similar to those observed with unilamellar vesicles and with BLMs, thus indicating that the phospholipids are the target of the electric field. In both cell membranes and BLMs, a sudden interruption of a voltage pulse causes the conductance to decrease rapidly, within a few microseconds. However, the complete restoration of the intact behavior (*resealing*) may require hours in cell membranes, while it occurs within seconds in BLMs. This difference is ascribed to the presence of integral proteins in the cells.

References

Becucci, L., M. R. Moncelli, R. Herrero, R. Guidelli, *Langmuir* 16 (2000) 7694–7700.

Chapman, D. L., *Phil. Mag.* 25 (1913) 475–481.

Chizmadzhev, Y. A., J. Teissié, D. Walz in: *Bioelectrochemistry of Membranes*, D. Walz, J. Tiessié, G. Milazzo (Eds.); Birkhäuser Verlag, Basel, Switzerland, 2004.

Clarke, R. J., *Biochim. Biophys. Acta* 1327 (1997) 269–278.

Crowley, I. M., *Biophys. J.* 13 (1973) 711–717.

Delahay, P., *Double Layer and Electrode Kinetics*, Interscience Publishers, New York, 1965.

Flewelling, R. F., W. L. Hubbell, *Biophys. J.* 49 (1986) 541–552.

Gouy, L.G., *Compt. Rend.* 149 (1910) 654–657.

Hill, T. L., *An Introduction to Statistical Thermodynamics*, Addison Wesley, Reading, U.S.A., 1960.

Jakobsson, E., *Trends Biochem. Sci.* 22 (1997) 339–344.

Kinosita, K., M. Hibino, M. Shigemori, I. Ashikawa, H. Itoh, K. Nagayama, K. Ikegami in: *Science on Form*, S. Ishizaka (Ed.), KTK Publishers, Tokio, 1990.

Michael, D. H., M. E. O'Neill, *J. Fluid Mech.* 41 (1970) 571–580.

Mohilner, D. M. in: *Electroanalytical Chemistry*, A. J. Bard (Ed.), Vol.1, Marcel Dekker, New York, 1966.

Shadowitz, A., *The Electromagnetic Field*, Dover Publications, New York, 1975.

Teissié, J. in: *Bioelectrochemistry of Membranes*, D. Walz, J. Tiessié, G. Milazzo (Eds.), Birkhäuser Verlag, Basel, Switzerland, 2004.

3

Thermodynamics

In Chapter 2, the electrostatic properties and the electropermeabilization of bilayer lipid membranes were examined in the absence of biomolecules capable of imparting unique functions into them. Before analyzing these various functions in detail, it is of paramount importance to review some basic concepts of irreversible thermodynamics, especially in connection with active transport. To bring the pertinent terminology to mind, we first provide a brief overview of chemical thermodynamics.

3.1 Some Concepts of Chemical Thermodynamics

In thermodynamics, a *closed system* is a system that can exchange energy (as heat or work) with the surroundings, but not matter. In the conventional language of thermodynamics, a chemical species whose mole number can be varied independently of all other species of the system is called a *component*. Quite often, the total number of different species (which are more generally referred to as *constituents*) contained in a closed system is higher than the number of thermodynamic components. Consider, for example, the ions composing an electrolytic solution; it is evident that we cannot change the concentration of a single ionic species while leaving unaltered the concentrations of all the other constituents of the system. Consequently, ionic species cannot be considered as thermodynamic components of a thermodynamic system.

Let us consider an infinitesimal transformation of our closed system, involving an exchange of heat $\mathrm{d}Q$ and of work $\mathrm{d}W = (-P\,\mathrm{d}V + \mathrm{d}W')$ with the surroundings, where $-P\,\mathrm{d}V$ is an infinitesimal compression–expansion work and $\mathrm{d}W'$ is any infinitesimal work different from $-P\,\mathrm{d}V$. The first law of thermodynamics states that the corresponding infinitesimal change, $\mathrm{d}U$, of the *internal energy* U of our system is given by

$$\mathrm{d}U = \mathrm{d}Q - P\,\mathrm{d}V + \mathrm{d}W'. \tag{3.1}$$

Heat is conventionally positive when it is transferred from the surroundings to the system, because in this way it contributes to increasing the internal energy of the system. This explains the positive sign in Eq. 3.1. For the same reason, work

Bioelectrochemistry of Biomembranes and Biomimetic Membranes, First Edition. Rolando Guidelli.

is positive when it is done by the surroundings on the system. This explains why the compression–expansion work is preceded by the negative sign. Consider, for example, a gaseous system consisting of a cylinder equipped with a movable piston, on which the surroundings exert an external pressure higher than the internal one. This will cause an infinitesimal decrease $\mathrm{d}V$ in the system volume; hence, P (force per unit area of the piston) is positive, whereas $\mathrm{d}V$ is negative, and $-P\,\mathrm{d}V$ is positive and contributes to increasing U, as expected. According to the second law of thermodynamics, if a transformation is reversible, $\mathrm{d}Q$ is equal to $T\,\mathrm{d}S$, where S is the *entropy* of the system and T its *absolute temperature*. From Eq. 3.1 it follows that

$$\mathrm{d}U = T\,\mathrm{d}S - P\,\mathrm{d}V + \mathrm{d}W'. \tag{3.2}$$

Heat and work are two forms of energy in transit, in that it is possible to measure them only while they are flowing out of the system to the surroundings or vice versa; once the system has undergone a finite transformation that brings it from an initial to a final equilibrium state, there is no point in referring to work W or heat Q stored in the final state of the system. In fact, starting from the same initial state, the final state can be reached by following an infinite number of paths, characterized by different amounts of heat and work exchanged with the surroundings. Only the sum of the heat and of the various forms of work exchanged (i.e., the internal energy U) depends on the initial and final equilibrium states, such that it can be considered as a *thermodynamic state function* of the system (Zemansky, 1957, Kirkwood and Oppenheim, 1961). In practice, U is measured with respect to an initial equilibrium state set conventionally equal to zero, called the *standard state*. Same considerations apply to the entropy S, which is also a state function.

Since U is a state function, its infinitesimal change $\mathrm{d}U$ in a process is an *exact differential*. Therefore, if the infinitesimal process involves a change of entropy $\mathrm{d}S$, volume $\mathrm{d}V$, and $\mathrm{d}n_h$ in each of the mole numbers of the system components, we can write

$$\mathrm{d}U = \left(\frac{\partial U}{\partial V}\right)_{S,n_h}\mathrm{d}V + \left(\frac{\partial U}{\partial S}\right)_{V,n_h}\mathrm{d}S + \sum_h \left(\frac{\partial U}{\partial n_h}\right)_{V,S,n_j\neq n_h}\mathrm{d}V. \tag{3.3}$$

For the sake of brevity, the rate of change of the internal energy U with respect to the mole number of the hth component, while keeping constant the mole numbers $\{n_j\}$ of all other components as well as the volume V and the entropy S, is denoted by μ_h and is referred to as the *chemical potential* of the hth component:

$$(\partial U/\partial n_h)_{V,S,n_j\neq n_h} = \mu_h. \tag{3.4}$$

By comparing Eqs. 3.2 and 3.3, we obtain

$$(\partial U/\partial V)_{S,n_h} = -P, \tag{3.5}$$

$$(\partial U/\partial S)_{V,n_h} = T, \tag{3.6}$$

$$\sum_h \mu_h\,\mathrm{d}n_h = \mathrm{d}W'. \tag{3.7}$$

This allows us to write Eq. 3.3 in the more concise form:

$$dU = -P\,dV + T\,dS + \sum_h \mu_h\,dn_h. \tag{3.8}$$

Since maintaining the entropy S constant is much more difficult than maintaining the temperature T constant, the chemical potential is more conveniently defined in an alternative way, by introducing an auxiliary state function that combines the internal energy U with the entropy S, called the *Gibbs energy* or *free enthalpy*:

$$G = U + PV - TS. \tag{3.9}$$

Upon differentiating and taking Eq. 3.2 into account, we get

$$dG = dU + P\,dV + V\,dP - T\,dS - S\,dT = dW' + V\,dP - S\,dT. \tag{3.10}$$

Following a procedure analogous to that yielding Eq. 3.4, we obtain the alternative definition of μ_h:

$$d\mu_h = (\partial G/\partial n_h)_{T,P,n_j \neq n_h}. \tag{3.11}$$

Let us consider the particular case in which the system cannot exchange with the surroundings other forms of work besides the compression–expansion work. In this case, dW' equals zero and Eq. 3.7, which holds for a reversible process, becomes

$$\sum_h \mu_h\,dn_h = 0. \tag{3.12}$$

Since a reversible process can be considered as a sequence of infinitesimal equilibrium states, Eq. 3.12 applies to a system in equilibrium. Let us now examine two particularly significant transformations involving a change in the mole number of the system components, with no other work besides the compression–expansion work.

The first transformation refers to a system consisting of more than one phase (*polyphasic system*) and involves the passage of an infinitesimal quantity dn_h of moles of the hth component from a phase α to a phase β. In this case, Eq. 3.12 becomes

$$\mu_h^\alpha\,dn_h^\alpha + \mu_h^\beta\,dn_h^\beta = (\mu_h^\beta - \mu_h^\alpha)dn_h. \tag{3.13}$$

Here, account has been taken of the fact that the infinitesimal (positive) change, dn_h^β, in the mole number of the hth component in the β phase is necessarily equal in magnitude but opposite in sign to the infinitesimal change, dn_h^α, in the mole number of the same component in the α phase, since our system is closed. In Eq. 3.13, $-dn_h^\alpha$ and dn_h^β are denoted by dn_h. From Eq. 3.13, it follows that, at equilibrium, the chemical potential μ_h^α of a given hth component in the α phase is equal to the chemical potential, μ_h^β, of the same component in the β phase:

$$\mu_h^\beta = \mu_h^\alpha. \tag{3.14}$$

Obviously, Eq. 3.14 holds only if the *h*th component can exist in both the contacting phases α and β. It must be stressed that Eqs. 3.13 and 3.14 hold only if the two phases α and β are both included in the closed thermodynamic system under examination.

Let us now consider a transformation consisting of a generic chemical reaction:

$$a_1 A_1 + a_2 A_2 + \cdots = a_m A_m + a_{m+1} A_{m+1} + \cdots. \tag{3.15}$$

Here, a_h denotes the stoichiometric coefficient of the species A_h. The reaction may take place either in a single phase (*homogeneous reaction*) or at the boundary between two different phases of the same closed system (*heterogeneous reaction*). For convenience, let us move the terms in the first member of Eq. 3.15 to the second member, so as to express the reaction in the form of an equation:

$$\sum_h \nu_h A_h = 0. \tag{3.16}$$

The stoichiometric coefficients are now denoted by the symbol ν_h, and are positive for reaction products and negative for reactants. It is evident that the infinitesimal changes, $\{dn_h\}$, in the mole number of the A_h components participating in the chemical reaction are not independent of each other, but are proportional to the corresponding stoichiometric coefficients, according to an infinitesimal proportionality constant $d\lambda$:

$$dn_h = \nu_h \, d\lambda \quad \text{for all components.} \tag{3.17}$$

When λ becomes equal to unity, mole numbers of the reactants in Eq. 3.15 equal to the corresponding stoichiometric coefficients ($a_1 = -\nu_1, a_2 = -\nu_2, \ldots$) yield mole numbers of the products equal to the corresponding stoichiometric coefficients ($a_m = \nu_m, a_{m+1} = \nu_{m+1}, \ldots$). The parameter λ is called the *degree of advancement* of the reaction. Upon substituting the changes, dn_h, in the various infinitesimal mole numbers from Eq. 3.17 into Eq. 3.12, we obtain

$$\left(\sum_h \nu_h \mu_h\right) d\lambda = 0. \tag{3.18}$$

Under equilibrium conditions, this relationship holds for any arbitrary value of the infinitesimal quantity $d\lambda$; hence, the equilibrium condition can be written in the form

$$\sum_h \nu_h \mu_h = 0. \tag{3.19}$$

This result can be expressed by stating that, at equilibrium, the sum of the chemical potentials of the products (i.e., $\sum_{\nu_h>0} \nu_h \mu_h$) is equal to that of the reactants (i.e., $\sum_{\nu_h<0} |\nu_h| \mu_h$, where $|\nu_h|$ is the absolute value of ν_h).

Let us now consider a chemical system that does not exchange types of work W' different from the compression–expansion work with the surroundings, and let us further assume that it undergoes an *irreversible* transformation. In this case, Eq. 3.2 does not hold, and the first law of thermodynamics takes the form

$$dU = dQ_I - P_I\, dV, \tag{3.20}$$

in view of Eq. 3.1. In this equation, $P_I\, dV$ and dQ_I are the compression–expansion work and the heat exchanged irreversibly between the system and the surroundings. From the *entropy principle*, namely the principle of the increase of entropy, for an irreversible process the following inequality holds:

$$T\, dS = dQ_R > dQ_I. \tag{3.21}$$

Here, dQ_R is the heat that the system would exchange with the surroundings if the thermodynamic process were ideally carried out in a reversible way. This inequality can be better understood by considering, for example, a chemical reaction accompanied by release of heat to the surroundings (*exothermic reaction*); in this case, the heat exchanged has a negative sign. If this reaction is carried out irreversibly, the absolute value of the heat released by the reaction is greater than if it were released reversibly. However, the heat emitted irreversibly, taken with its negative sign, is less than that released reversibly. In other words, the irreversible exothermic reaction determines a dissipation of energy through the release of a *surplus* of heat, which is a degraded form of energy.

On the other hand, the infinitesimal change, dU, of the internal energy in passing from an initial equilibrium state to a final equilibrium state is exclusively a function of these two states, and hence is the same, independent of whether this passage occurs reversibly or irreversibly. Consequently, we can equate the expression of dU in Eq. 3.20, which refers to an irreversible process, to that in Eq. 3.8, which holds for a reversible process. This leads to the following equation:

$$T\, dS - dQ_I = -\sum_h \mu_h\, dn_h + P\, dV - P_I\, dV > 0. \tag{3.22}$$

In this equation, account is taken of the inequality of Eq. 3.21. P and P_I are the pressures exerted by the surroundings on the system under reversible and irreversible conditions, respectively, while dV is the infinitesimal volume change during the process. The first member of Eq. 3.22, when divided by T, is a measure of the *entropy production* due to the irreversible transformation. In this connection, it must be stressed that, for a process to take place spontaneously, it must proceed in the direction of an increase in entropy production, by satisfying the inequality of Eq. 3.22. In fact, a spontaneous process is necessarily irreversible and tends to a final equilibrium state; a reversible process is just a thermodynamic idealization, consisting of a thought sequence of equilibrium states.

Let us envisage, for simplicity, a transformation in which no chemical reaction or flow of matter between different phases of the system takes place, such that dn_h equals zero for all components. If the transformation is irreversible, the entropy principle states that the exchanged heat, dQ_I, is less than that,

$dQ_R = T\,dS$, which would be exchanged if the transformation were carried out reversibly. Thus, we have $P\,dV > P_I\,dV$, in view of Eq. 3.22. This inequality might be caused, for instance, by dissipative phenomena, such as friction. We may consider, for example, a system consisting of a gas contained in a cylinder subjected to an irreversible expansion, via the movement of a piston that encounters friction against the cylinder walls. If the piston movement were ideally frictionless, the infinitesimal expansion would take place reversibly, and the external pressure P required for the piston movement would be smaller than that inside the piston by only an infinitesimal amount; the reversible work would then be equal to $-P\,dV$. Conversely, in the presence of friction, the external pressure P_I required for the piston movement must be smaller than the internal pressure P by a finite amount. Considering that, in the ideal frictionless case, the internal pressure P is equal to the external one, we have $P_I\,dV < P\,dV$. From Eq. 3.22, it follows that the positive quantity $(P - P_I)dV$ contributes to the entropy production.

In irreversible chemical transformations, the contribution to the entropy production from the compression–expansion work is usually negligible with respect the other term, $-\sum_h \mu_h\,dn_h$, and will be disregarded in the following. Let us first apply the inequality of Eq. 3.22 to the flow of dn_h moles of the hth component from a phase α to a phase β within the same system. Since we have $dn_h^\beta = -dn_h^\alpha \equiv dn_h$, as in the case of the equilibrium condition of Eq. 3.13, Eq. 3.22 becomes

$$\sum_h \mu_h\,dn_h = \left(\mu_h^\beta - \mu_h^\alpha\right) dn_h < 0. \tag{3.23}$$

This implies that, for the hth component to pass spontaneously from the α to the β phase, that is, for dn_h^β to be > 0, μ_h^α must be $> \mu_h^\beta$. In other words, if there is a difference in the chemical potential of one component between two contacting phases, the component will tend to pass spontaneously from the phase where its chemical potential is higher to that where it is lower. This passage will cease when the chemical potentials of the given component become equal, satisfying the equilibrium condition of Eq. 3.14.

For a system in which the generic chemical reaction of Eq. 3.15 takes place, the inequality of Eq. 3.22 takes the form

$$\left(\sum_h \nu_h \mu_h\right) d\lambda < 0. \tag{3.24}$$

This implies that, for the reaction to proceed spontaneously from left to right, with an increase in the degree of advancement λ ($d\lambda > 0$), the "overall" chemical potential, $\sum_{\nu_h<0} |\nu_h| \mu_h$, of the chemical species in the left member of Eq. 3.15, that is, the species with negative stoichiometric coefficients ν_h, must be higher than that, $\sum_{\nu_h>0} \nu_h \mu_h$, of the species in the right member, that is, those with positive stoichiometric coefficients. The chemical reaction will cease when the overall chemical potential of the products becomes equal to that of the reactants, satisfying the equilibrium condition of Eq. 3.19.

In general, the chemical potential plays toward a chemical species a role analogous to that played by the electric potential toward a positive point charge or by the gravitational potential toward a body. Thus, a positive charge moves spontaneously in the direction of decreasing electric potential, while a body moves spontaneously in the direction of decreasing gravitational potential. In all these examples, the *potential* of a species is a measure of the degree of instability of the species and of its tendency to decrease its own potential, by moving toward the attainment of an equilibrium condition.

3.2 The Electrochemical Potential

All preceding thermodynamic relations apply strictly to thermodynamic components, namely to species whose concentration can be varied independently from that of all other species of the system. In this respect, thermodynamic components must be neutral, such as the atoms of a metal (not ion cores and electrons), solvent molecules, neutral solutes, and salts (not single ions, such as K^+ and Cl^-, but KCl). Thus, for example, in an aqueous solution of KCl, it is impossible to vary the concentration of potassium ions while leaving unaltered that of chloride ions. Nonetheless, it is often convenient to extend the definition of chemical potential to a charged species, thus abandoning the domain of thermodynamics. In analogy with the definition of the chemical potential of Eq. 3.11 for a neutral species, the chemical potential of a charged hth species, called *electrochemical potential* and denoted by $\widetilde{\mu}_h$, is defined as

$$\widetilde{\mu}_h = (\partial G/\partial n_h)_{\mathrm{T,P},n_j \neq n_h}. \tag{3.25}$$

The symbol of the electrochemical potential is usually supersigned with a tilde.

A charged species shares the nature of a neutral molecular species, by interacting with the surrounding molecules, with that of a point charge, by experiencing any change in the electric potential. From the properties of ideal solutions, the chemical potential of a neutral solute molecule that does not interact with other solute molecules, but only with the solvent molecules, is given by

$$\mu_h = \mu_h^{\emptyset} + RT \ln c_h. \tag{3.26}$$

Here, c_h is the molar concentration of the solute and $\mu_h^{\emptyset}$ is its *standard chemical potential*, which accounts for the interactions between the given solute and the solvent molecules that surround it. Quite often, solute molecules interact between themselves the more, the higher their concentration c_h is. This causes deviations from the behavior of a system of noninteracting solute molecules. These deviations are accounted for empirically by forcing the experimental chemical potential to satisfy an equation identical with Eq. 3.26, in which the concentration c_h is replaced by the *activity* a_h. The ratio of the empirical activity a_h to the experimental concentration c_h is called *activity coefficient* and is denoted by f_h. On the other hand, a pure point charge of charge number z_h, when passing from a phase α to a phase β of different electric potential, requires an electrical work $z_h e(\phi^\beta - \phi^\alpha)$

to be done *on* the system, where ϕ^{α} and ϕ^{β} are the electric potentials in the α and β phase, respectively. It is, therefore, logical to express the electrochemical potential $\tilde{\mu}_h$ as the sum of the chemical potential, μ_h, of a neutral species and of the electrical work:

$$\tilde{\mu}_h = \mu_h^{\emptyset} + RT \ln a_h + z_h F \phi. \tag{3.27}$$

The electrochemical potential is an extra-thermodynamic quantity and, as such, cannot be *measured*; it can only be *estimated* by having recourse to some model, just as all quantities not pertaining to the realm of thermodynamics. Its extra-thermodynamic nature is clearly apparent if we extend to electrochemical potentials the equilibrium condition of Eq. 3.14 for the passage of a component from a phase α to a phase β, which is thermodynamically significant for chemical potentials:

$$\tilde{\mu}_h^{\beta} = \tilde{\mu}_h^{\alpha}. \tag{3.28}$$

By replacing the expression of Eq. 3.27 for the electrochemical potential into Eq. 3.28 and rearranging terms, we obtain

$$\mu_h^{\emptyset,\beta} + RT \ln c_h^{\beta} - (\mu_h^{\emptyset,\alpha} + RT \ln c_h^{\alpha}) = z_h F(\phi^{\alpha} - \phi^{\beta}). \tag{3.29}$$

The superscripts α and β refer to quantities in the α and β phases, respectively. In Eq. 3.29, as well as in the following, we use concentrations in place of activities, for the sake of simplicity. If the solvent in the α phase is different from that in the β phase, $\mu_h^{\emptyset,\alpha}$ and $\mu_h^{\emptyset,\beta}$ are also different. In this case, when a charged species passes from the α to the β phase, it experiences different short-range interactions with the new solvent as well as a different electric potential, and it is impossible to separate the chemical from the electrical contribution to the electrochemical potential. Therefore, it is also impossible to measure the potential difference $(\phi^{\alpha} - \phi^{\beta})$ between the two phases from Eq. 3.29.

If the interactions of an ionic species with the solvent in the β phase are much more attractive than those in the α phase, then $\mu_h^{\emptyset,\beta} \ll \mu_h^{\emptyset,\alpha}$. In this case, at constant $(\phi^{\alpha} - \phi^{\beta})$, the species accumulates in the β phase, and $c_h^{\beta} \gg c_h^{\alpha}$ in view of Eq. 3.29. This situation is encountered at the boundary between the lipid bilayer of a membrane and the aqueous solution that baths it. The standard potential of an inorganic ion in the aqueous solution is much lower than that in the lipid bilayer, and the ion partitions preferentially into the aqueous phase. On the other hand, the phases that bath the two sides of the plasma membrane of a living cell, namely the cytosol and the extracellular medium, are both characterized by a large predominance of water molecules, and can be definitely considered as two different aqueous phases. Consequently, the standard chemical potential $\mu_h^{\emptyset}$ of an *h*th ionic species is practically the same in these two phases. If the ion has the possibility of translocating from the extracellular to the cytosolic side of the membrane or vice versa, due to the presence of an ion channel or ion carrier, the equilibrium condition of Eq. 3.29 simplifies yielding

$$\phi^{\beta} - \phi^{\alpha} = (RT/z_h F) \ln(c_h^{\alpha}/c_h^{\beta}). \tag{3.30}$$

This equation, called *Nernst equation*, brings us again into the realm of thermodynamics, because the interactions of the ion with the solvent are now almost the same in the two different phases. For example, this equation is used to determine the electric potential difference across a plasma membrane, if the concentration of potassium ions is known both inside and outside the cell and valinomycin, an ion carrier highly selective for potassium ions, is incorporated in the membrane.

All properties of chemical potentials can be extended to electrochemical potentials, in spite of their extra-thermodynamic nature. This approach is often convenient whenever one ultimately returns into the realm of thermodynamics by imposing the electroneutrality condition to the whole system, for example, by setting the sum of positive and negative charges within the system equal to zero:

$$\sum_h \left(z_h^+ n_h^+ + z_h^- n_h^-\right) = 0. \tag{3.31}$$

Here, n_h^+ and n_h^- are the mole numbers of the positive and negative species within the system, and z_h^+ and z_h^- are the corresponding charge numbers. This *electroneutrality condition* holds *locally* within any infinitesimal volume element of a bulk conducting phase, such as an electrolytic solution, where the charge of anions matches that of cations, or a bulk metal, where the negative charge of conduction electrons matches that of the corresponding metal ion cores. Most importantly, it holds for any system interposed between two conducting phases, such as a membrane interposed between two bulk aqueous solutions or a lipid bilayer interposed between a bulk metal electrode and a bulk aqueous solution, albeit *no longer locally*. These systems, referred to as *electrified interfaces*, present a charge separation within them, namely they are locally charged, but their overall charge equals zero, in accordance with Eq. 3.31. If, at the end of an extra-thermodynamic process dealing with electrochemical potentials and occurring within an electrified interface, the electroneutrality condition of Eq. 3.31 is ultimately applied, one is brought back to the domain of thermodynamics, where only neutral species such as salts or metal atoms have right of citizenship.

3.3 Thermodynamics of Irreversible Processes

Classical thermodynamics draws its indisputable authority from the fact that it renounces any considerations of the structure of matter and confines its attention to systems in equilibrium or undergoing idealized, reversible changes (namely, infinite sequences of equilibrium states). Under these conditions, the properties of state are well defined. All *real* processes are described by inequalities (see Eqs. 3.23 and 3.24) that simply indicate the direction of change, but no indication is provided for the rate of change and, in fact, classical thermodynamics ignores the variable *time*. In this respect, classical thermodynamics is not suited for the description of living systems, which are always displaced from equilibrium and are characterized by flows of matter and energy both within the system and between the system and its surroundings. For living systems, equilibrium conditions imply death.

Irreversible thermodynamics strives to extend the principles of classical thermodynamics to systems in nonequilibrium that are characterized by irreversible flows of matter and energy (De Groot, 1966). It replaces the inequalities of classical thermodynamics with equalities and deals with rates and flows, by explicitly introducing the variable time. However, a penalty must be paid while doing so, inasmuch as the principles of irreversible thermodynamics are not valid under all conditions, and their range of validity must be determined empirically. In this respect, this discipline does not retain the indisputable authority of classical thermodynamics.

The entire framework of irreversible thermodynamics starts from the expression of Eq. 3.22 for the entropy production:

$$T\,\mathrm{d}S - \mathrm{d}Q_{\mathrm{I}} = -\sum_h \widetilde{\mu}_h\,\mathrm{d}n_h = -\sum_j \left(\widetilde{\mu}_j^{\beta} - \widetilde{\mu}_j^{\alpha}\right)\mathrm{d}n_j^{\beta} - \left(\sum_k \nu_k \widetilde{\mu}_k\right)\mathrm{d}\lambda. \tag{3.32}$$

Here, the summation is more generally extended to electrochemical potentials of charged species, by temporarily relaxing the constraint of the electroneutrality condition. In the third member of Eq. 3.32, it is pointed out that the summation in the second member refers, in general, to a given number of flows of neutral or charged species from one phase α to a different phase β, and to a given number of chemical reactions. If these different processes are not coupled between themselves (the meaning of *coupling* is discussed later), each of them contributes to an increase in the entropy production. Thus, in the case of an infinitesimal passage of moles of the jth species from the α to the β phase ($\mathrm{d}n_j^{\beta} > 0$), the electrochemical potential of this species in the α phase, $\widetilde{\mu}_j^{\alpha}$, must be necessarily higher than that, $\widetilde{\mu}_j^{\beta}$, in the β phase. Analogously, the sum of the electrochemical potentials of the reactants of any given reaction, $\sum_{\nu_k<0}|\nu_k|\widetilde{\mu}_k$, each multiplied by the absolute value of its negative stoichiometric coefficient, must be higher than the sum of the electrochemical potentials of the corresponding reaction products, $\sum_{\nu_k>0}\nu_k\widetilde{\mu}_k$, each multiplied by its positive stoichiometric coefficient. Many other irreversible processes may contribute to entropy production, but are not be considered here. One irreversible expansion process was briefly examined in connection with Eq. 3.22.

In irreversible thermodynamics, the time variable is introduced by differentiating Eq. 3.32 with respect to time t at constant temperature T:

$$\begin{aligned}\frac{\mathrm{d}}{\mathrm{d}t}(T\,\mathrm{d}S - \mathrm{d}Q_{\mathrm{I}}) &= T\frac{\mathrm{d}S_{\mathrm{in}}}{\mathrm{d}t} = -\sum_h \widetilde{\mu}_h \frac{\mathrm{d}n_h}{\mathrm{d}t}\\ &= -\sum_j \left(\widetilde{\mu}_j^{\beta} - \widetilde{\mu}_j^{\alpha}\right)\frac{\mathrm{d}n_j^{\beta}}{\mathrm{d}t} - \left(\sum_k \nu_k\widetilde{\mu}_k\right)\frac{\mathrm{d}\lambda}{\mathrm{d}t}.\end{aligned} \tag{3.33}$$

In this equation, the infinitesimal entropy production is denoted by $\mathrm{d}S_{\mathrm{in}} = \mathrm{d}S - \mathrm{d}Q_{\mathrm{I}}/T$, and the positive electrochemical potential differences $-(\widetilde{\mu}_j^{\beta} - \widetilde{\mu}_j^{\alpha})$

and $-\sum_k \nu_k \widetilde{\mu}_k$, responsible for the flow of matter and for the degree of advancement λ of chemical reactions, respectively, are considered as time independent. These electrochemical potential differences are called *driving forces*, while the respective time derivatives, dn_j^{β}/dt and $d\lambda/dt$, are referred to as *conjugate flows*. Equation 3.33 can be written in a more concise form:

$$\Phi \equiv T\, dS_{in}/dt = \sum_h X_h J_h, \tag{3.34}$$

which includes any type of driving force, X_h, and any type of conjugate flow, J_h. The quantity $T\, dS_{in}/dt$ is called *dissipation function* and is often symbolized by Φ. In general, the dissipation function can be considered as a measure of the rate of dissipation of the Gibbs energy of the system, or of the rate at which the ability of the system to perform useful work decreases over time due to its conversion into heat.

Equation 3.34 is not valid on a general basis and its range of validity must be established empirically, wherever possible. A well-known dissipative process is the *Joule effect*, according to which the current I that flows along a resistor of resistance R under a potential difference $V = IR$ determines a heat loss equal to I^2R to the surroundings. With the terminology of Eq. 3.34, $V = IR$ is the driving force X_h, the current I is the conjugate flow J_h, and $(IR)I = I^2R$ is the contribution to the dissipation function due to the Joule effect. This example demonstrates that, if a system is characterized by only one driving force and its conjugated flow, as in the case of an electric heater, the whole Gibbs energy is dissipated as heat without producing useful work. Useful work can only be performed via a suitable interaction (*coupling*) between different flows and forces, in such a way that at least one of the coupled processes makes a negative contribution to the dissipation function. Naturally, in view of the entropy principle, the whole dissipation function must always remain positive.

For many years it has been empirically verified that, for a system involving a single force and a single flow, the flow is proportional to the force over a relatively broad range of forces and flows. Hence, we can write

$$J_h = L_h X_h. \tag{3.35}$$

J_h is the flow of an extensive quantity, X_h is the conjugate driving force, that is, a difference in the corresponding intensive quantity, and L_h is a proportionality constant having units of conductance. It was also known that a system with different forces and conjugate flows may give rise to a coupling between nonconjugate flows and forces. Moreover, if flows are sufficiently slow and the system is not displaced too far from equilibrium, the dependence of flows upon nonconjugate forces is also linear. These empirical observations may be described by expressing each single flow J_h as the sum of the products of all driving forces $\{X_j\}$ by the corresponding proportionality coefficients L_{hj}:

$$J_h = L_h X_h + \sum_{j \neq h} L_{hj} X_j. \tag{3.36}$$

In this equation, the coefficient L_h is called *straight coefficient*, since it relates the flow to its conjugate driving force. The coefficients L_{hj} are termed *cross coefficients* or *coupling coefficients*, since they relate a flow to nonconjugate driving forces. These coefficients are functions of the system state, but are independent of the driving forces.

In two fundamental works published in 1931, Lars Onsager related the phenomenological linear Eq. 3.36, which stems from empirical observations, to the dissipation function Φ in Eq. 3.34, which relies on a questionable assumption, in a groundbreaking synthesis that has strengthened the validity of irreversible thermodynamics. Onsager (1931) demonstrated that provided the flows and the conjugate forces are chosen appropriately (a nonobvious choice at those times) by ascribing to each flow its actual conjugate force so as to satisfy Eq. 3.34, then the following relationship holds:

$$L_{hj} = L_{jh} \quad \text{for all } h \text{ and } j. \tag{3.37}$$

These *reciprocal relations* allow a significant reduction in the number of cross coefficients that should be experimentally determined. Thus, if we have n flows and the corresponding n conjugate forces, the number of coefficients present in the relations of Eq. 3.36 is equal to n^2, such that the evaluation of these coefficients becomes complicated, even for relatively simple systems. Upon applying Onsager's reciprocal relations, the number of coefficients is reduced to $n(n+1)/2$. Another important consequence is that the experimental verification of the validity of the reciprocal relations confirms that the system is not too far from equilibrium and can be described by a combination of linear relations between flows and forces (Schultz, 1980, Walz, 1979).

Let us consider, for simplicity, a system characterized by two flows and two conjugate forces; it is described by the following equations:

$$J_1 = L_1X_1 + L_{12}X_2; \quad J_2 = L_{12}X_1 + L_2X_2. \tag{3.38}$$

If L_{12} equals 0, the two flows are uncoupled and we have

$$\Phi = J_1X_1 + J_2X_2 = L_1X_1^2 + L_2X_2^2 > 0. \tag{3.39}$$

In this system, all the energy stored in the two forces X_1 and X_2 is dissipated into heat. As a rule, each driving force X_h is assigned a positive sign. In the absence of coupling, its conjugate flow J_h is also assigned a positive sign, so that the straight coefficient L_h is positive. It follows that, in the absence of coupling, the product, J_hX_h, of a driving force by its conjugate flow is always positive and makes a positive contribution to the dissipation function Φ.

On the other hand, if the two flows J_1 and J_2 are coupled, we have

$$\Phi' = J_1X_1 + J_2X_2 = L_1X_1^2 + L_2X_2^2 + 2L_{12}X_1X_2. \tag{3.40}$$

Since X_1 and X_2 are positive, when $L_{12} < 0$ the coupling between the two flows causes the dissipation function Φ' to be less than that, Φ, in the absence of coupling, and the difference between the two is converted into useful work, instead of being dissipated. In particular, if the absolute value of $L_{12}X_2$, $|L_{12}X_2|$, is $> L_1X_1$, from Eq. 3.38 it follows that $J_1 < 0$. By analogous arguments, if $|L_{12}X_1| > L_2X_2$, $J_2 < 0$. Since the straight coefficient L_1 is necessarily positive, the fact that the flow J_1 is negative, and hence of opposite sign with respect to its conjugate force X_1, implies that the nonconjugate force X_2 moves the flow J_1 in the direction opposite to that in which it would be moved by its conjugate force, thus performing useful work. In this example, the energy of the flow J_2, which proceeds in the direction imposed by its conjugate force X_2, is converted into useful work. The *efficiency of energy conversion* is defined by

$$\eta = \frac{\text{power output}}{\text{power input}} = \frac{-J_1X_1}{J_2X_2} = 1 - \frac{\Phi}{J_2X_2}. \tag{3.41}$$

For an ideal reversible process, Φ equals zero and the efficiency η is unitary.

To better visualize the extent of coupling between two given flows J_1 and J_2, it is expedient to write Eq. 3.38 in the form

$$J_1 = L_1\left(X_1 + \frac{qX_2}{Z}\right); \quad J_2 = L_2(X_2 + qZX_1), \tag{3.42}$$

with

$$q \equiv \frac{L_{12}}{(L_1L_2)^{1/2}}; \quad Z \equiv \left(\frac{L_1}{L_2}\right)^{1/2}. \tag{3.43}$$

The parameter q may be positive or negative depending on the sign of L_{12}. We consider only the most significant case in which L_{12} (and, hence, q) is ≤ 0; q is referred to as *degree of coupling*, inasmuch as it measures the extent of coupling between the two flows. When $q = 0$, from Eq. 3.42 it follows that $J_1 = L_1X_1$ and $J_2 = L_2X_2$, and the two flows are completely uncoupled, with each flow depending exclusively upon its own conjugate driving force.

The other limiting case is attained when $q = -1$, which corresponds to maximum coupling. In this case, Eq. 3.42 becomes

$$J_1 = L_1\left(X_1 - \frac{X_2}{Z}\right); \quad J_2 = L_2(X_2 - ZX_1), \tag{3.44}$$

and yields the following ratio of the two flows:

$$\frac{J_1}{J_2} = \frac{L_1(ZX_1 - X_2)}{L_2Z(X_2 - ZX_1)} = -\left(\frac{L_1}{L_2}\right)^{1/2} = -Z. \tag{3.45}$$

This implies that the ratio of the two flows is independent of the ratio of the two forces. Hence, if one of the two flows is fixed, the other is uniquely determined and is independent of its conjugate force. In other words, the two flows are completely coupled. This ideal extreme situation is expressed by stating that the system operates as a *constant current source*. For $q = -1$, the dissipation function becomes

$$\Phi = J_1X_1 + J_2X_2 = J_1X_1 - J_1X_2/Z = J_1(X_1 - X_2/Z). \tag{3.46}$$

This function becomes vanishingly small for $X_1 - X_2/Z = 0$, resulting in no Gibbs energy dissipation. This might lead to the conclusion that, under these conditions, the system performs exclusively useful work. As a matter of fact, this is not the case, because Eq. 3.42 shows that, under these conditions, the two flows are both equal to zero. Nonetheless, it is important to note that the two flows are not equal to zero in the trivial case in which each of the two corresponding conjugate forces vanishes; rather, each of the two forces exactly opposes the other, thus annihilating its conjugate flow. In other words, the system may be regarded as in equilibrium, if we consider that $\Phi = 0$, but it may also be regarded as in nonequilibrium, if we consider that the two driving forces are different from zero. The two processes may also be viewed as *quasistatic*, that is, so slow as not to induce any change in a reasonable lapse of time; alternatively, their coupling may be viewed as a constraint inside the system, which prevents the two processes from attaining the equilibrium that they would reach if the constraint were removed, namely if the two processes were uncoupled.

An example of this situation is provided by the Na^+, K^+-ATPase ion pump. Under physiological conditions, Na^+, K^+-ATPase pumps three sodium ions from the cytoplasmic side of the plasma membrane to the extracellular side, where their electrochemical potential is higher than on the cytoplasmic side, while it pumps two potassium ions from the extracellular to the cytoplasmic side, where their electrochemical potential is higher than on the extracellular side. Therefore, both ionic flows proceed against their electrochemical potential gradients (namely, *uphill*), driven by the coupled reaction of hydrolysis of adenosine triphosphate (ATP) to adenosine diphosphate (ADP) and inorganic phosphate (P_i); this reaction is spontaneous, since it proceeds along its own driving force, namely *downhill*. If the electrochemical potential gradients of Na^+ and K^+ ions were artificially reversed by making the sodium ion concentration much higher on the cytoplasmic side of the pump and the potassium ion concentration much higher on its extracellular side, and if the ATP concentration were decreased in favor of the ADP concentration, then the Na^+, K^+-ATPase would operate in the reverse direction, by synthesizing ATP from ADP and P_i. If the Na^+ and K^+ flows were perfectly coupled to ATP hydrolysis ($q = -1$), an intermediate situation might also be envisaged, with the driving force of ATP hydrolysis exactly counterbalanced by the electrochemical potential gradients of Na^+ and K^+ ions between the two sides of the membrane. In practice, however, coupling is never complete and, in order to annihilate the Na^+ and K^+ downhill flows, ATP hydrolysis must proceed, albeit slowly, to satisfy the condition $\Phi' > 0$.

To better visualize the course of two coupled processes, it is convenient to plot the $J_1/(ZJ_2)$ ratio against ZX_1/X_2 for different values of the degree of coupling q, as shown in Fig. 3.1. In fact, from Eq. 3.42, it follows that

$$\frac{J_1}{J_2} = \frac{L_1}{L_2}\frac{(X_1 + qX_2/Z)}{(X_2 + qZX_1)} = Z\frac{ZX_1 + qX_2}{qZX_1 + X_2} \quad \rightarrow \quad \frac{J_1}{ZJ_2} = \frac{ZX_1/X_2 + q}{qZX_1/X_2 + 1}. \tag{3.47}$$

From this equation, it is apparent that the plot of $J_1/(ZJ_2)$ against ZX_1/X_2 depends exclusively upon q. Figure 3.1 shows plots relative to different negative values of q. For $q = 0$, the two processes are completely uncoupled and J_1 varies linearly with X_1. This is evident inasmuch as the linear segment corresponding to $q = 0$

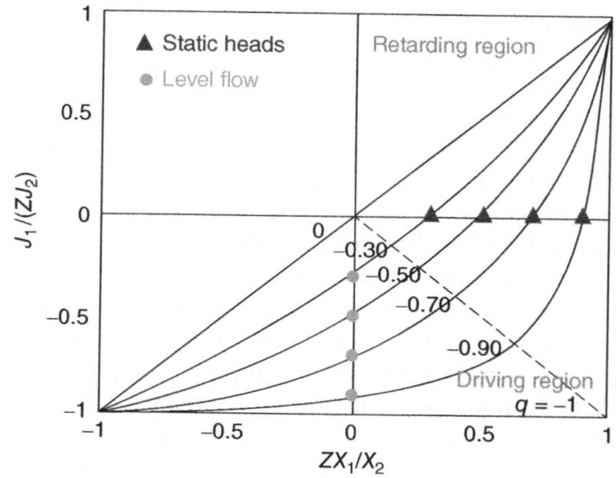

Figure 3.1 Plot of $J_1/(ZJ_2)$ against ZX_1/X_2 for different values of the degree of coupling q.

is the geometry locus of the points characterized by equal values of abscissa and ordinate. In other words, we have

$$\frac{J_1}{ZJ_2} = \frac{ZX_1}{X_2} \rightarrow \frac{J_1}{J_2} = Z^2\frac{X_1}{X_2} = \frac{L_1X_1}{L_2X_2} \rightarrow \begin{Bmatrix} J_1 = L_1X_1 \\ J_2 = L_2X_2 \end{Bmatrix}. \tag{3.48}$$

The lower right quadrant is the region in which the J_1 flow is negative, while its conjugate force X_1 is positive. In this case, process 1 proceeds uphill, driven by process 2, which proceeds downhill, that is, with both J_2 and X_2 positive. For this reason, the lower right quadrant is referred to as the *driving region*. The upper right quadrant, called the *retarding region*, is the region where J_1 and X_1 are both positive, so that process 1 proceeds downhill. From the figure, it is evident that, for a constant positive X_1/X_2 ratio, a gradual decrease of q from 0 to −1 causes the $J_1/(ZJ_2)$ ratio to pass from a positive value equal to ZX_1/X_2 to progressively less positive values (as long as we remain in the retarding region), until it vanishes and then takes negative values in the driving region. In other words, the progressive increase in the absolute value of the degree of coupling causes the J_1 flow first to decrease with respect to the value that it would have in the absence of coupling, while still remaining positive and proceeding downhill, and then to invert its sign, thus proceeding uphill.

Consider the case in which X_1 is initially zero, but process 1 is coupled to process 2, and hence q is negative. In this case, the system is located on the negative vertical semiaxis of Fig. 3.1, which corresponds to the so-called *level flow*, where J_1 is negative, notwithstanding the fact that its conjugate force equals zero. Moreover, the closer q is to its limiting −1 value, the more negative J_1 is. If the system is left free to evolve without external constraints, it will move along the curve corresponding to its negative q value, across the driving region. Along this curve, the X_1/X_2 ratio increases progressively. In other words, X_1 increases at the expense of X_2 until the positive horizontal semiaxis is reached, where J_1 vanishes. A point on this semiaxis is referred to as a *static head* and, in view of Eq. 3.42, satisfies the following condition:

$$J_1 = 0; \quad X_1 = -qX_2/Z; \quad J_2 = L_2(1 - q^2)X_2. \tag{3.49}$$

An example of the aforementioned system evolution may again be provided by Na^+, K^+-ATPase, upon envisaging the following nonphysiological situation of level flow. In this case, Na^+ ions flow from the cytoplasmic to the extracellular side of the membrane, while K^+ ions flow in the opposite direction, thanks to the coupled ATP hydrolysis, even though the electrochemical potential gradients of both K^+ and Na^+ ions across the membrane are initially equal to zero. The driving force provided by ATP hydrolysis continues moving sodium ions toward the extracellular medium and potassium ions toward the cytoplasm, increasing progressively their electrochemical potential gradients across the membrane. Ultimately, these electrochemical potential gradients match exactly the driving force of ATP hydrolysis and the Na^+ and K^+ flows cease, in correspondence to the static head. From Eq. 3.49, it is apparent that, when the Na^+ and K^+ flows cease (i.e., when J_1 equals 0), ATP hydrolysis continues, albeit more slowly, since the coupling between the two processes is not total ($J_2 \neq 0$ for $q > -1$). Na^+, K^+-ATPase tends to this static-head condition until some passive transport of Na^+ and/or K^+ ions across the membrane increases the pump activity, to restore the optimum electrochemical potential gradients of these ions.

The efficiency of energy conversion η, defined in Eq. 3.41, may be written as follows (Kedem and Kaplan, 1965):

$$\eta \equiv \frac{\text{power output}}{\text{power input}} = \frac{-J_1X_1}{J_2X_2} = -\frac{J_1}{ZJ_2}\frac{ZX_1}{X_2}. \tag{3.50}$$

From this equation, it is evident that, for a given degree of coupling q, the efficiency attains its maximum unitary value when $J_1/(ZJ_2)$ and ZX_1/X_2 have the same absolute value but opposite sign. This condition is fulfilled along the dashed segment that bisects the lower right quadrant in Fig. 3.1. Consequently, when a system moves from a state of level flow to one of static head across the driving region following a curve of constant q, it attains its maximum efficiency exactly halfway. Conversely, its efficiency is zero at the beginning ($X_1 = 0$) and at the end ($J_1 = 0$) of this path. It should be noted that the concept of efficiency of energy conversion is particularly significant for an energy converter, such as an electrical battery, which aims at attaining the maximum flow J_1 of electrons (the current) with a minimum of the driving force determined by the chemical reaction occurring within the battery. Conversely, many biological systems, such as ion pumps, aim at maintaining a state of static head, where $\eta = 0$.

3.4 Coupling of Primary and Secondary Active Transport in Biomembranes

Each type of membrane is the site of coupling between different active transports. This complex situation can be appreciably simplified having recourse to the concept of *master pump*. Each membrane has a master pump that uses a primary energy source to translocate one or more inorganic ions, thus generating, or more frequently maintaining, their electrochemical potential gradients across the membrane. Each master pump must fulfill three requirements, that is, it must have high *degree of coupling, low dissipation,* and *high capacity.*

The degree of coupling q is a measure of the fraction of the Gibbs energy of the primary energy source that is actually converted into the electrochemical potential gradient of the translocated ion(s), that is, the total amount of energy available for conversion into useful work. The remaining fraction is dissipated into heat.

By dissipation we mean the nonspecific flow of molecules across the membrane down their electrochemical potential gradient, namely a flow not determined by any membrane protein. A low dissipation is efficiently realized by pumps translocating inorganic ions, because the latter can hardy permeate membranes. The only exception is represented by the active transport of hydrogen ions, which are relatively permeant in view of their small size. However, in the case of hydrogen ions, the dissipation rate, measured by the so-called *leakage current,* is low; in fact, it is proportional to the hydrogen ion concentration, which is extremely low at physiological pH. Thanks to this low concentration of hydrogen ions, their electrochemical potential gradient exhibits a low dissipation. In practice, the physiological pH is maintained by the buffering activity of different weak acids and bases present in the solutions bathing the membrane.

The high capacity is the requirement according to which the ionic electrochemical potential gradient generated or maintained by the master pump must be much higher than that of the ions or molecules whose synport or antiport is coupled to that of the pumped ion. Nonetheless, the capacity of a pump should be estimated not only from the electrochemical potential gradient of the pumped ions in itself, but also from their concentration values on both sides of the membrane. Let us consider, for example, the Ca^{2+}-ATPase of the sarcoplasmic reticulum, which pumps calcium ions from the cytoplasm of muscle cells, where the Ca^{2+} concentration is about 10^{-7} M, to the lumen of the sarcoplasmic reticulum, where it is about 10^{-3} M. The transmembrane potential of the sarcoplasmic reticulum is quite low, since this membrane has an appreciable nonspecific permeability for ions. Hence, the electrochemical potential gradient of Ca^{2+} ions is almost completely determined by their concentration gradient and amounts to $RT\ln(10^{-3}\ \text{M}/10^{-7}\ \text{M})$, which is notably high. Nonetheless, this high Ca^{2+} concentration gradient cannot be exploited to translocate, by secondary active transport, molecules whose concentration in the lumen of the sarcoplasmic reticulum or in the cytoplasm is of the order of 0.1 M, due the very low Ca^{2+} concentration. In addition, Ca^{2+} ions play an important regulatory role in muscle cell contraction and relaxation; consequently, the hypothetical coupling of their flow with that of another molecular species would severely compromise this role.

The concept of master pump is useful, because it may suggest the probable mechanism of all transport systems in a given membrane, once its master pump has been identified. One may wonder why all transport across a membrane is associated with a single master pump. The explanation resides in the fact that the secondary active transport resulting from the coupling of two flows is more effective than that carried out exclusively by ion pumps operating independently from each other. Ionic electrochemical potential gradients store energy “packets” smaller than the chemical energy stored in the ATP molecules that fuel ion pumps; hence, they are more readily exploitable in case of need and with less dissipation of energy. By choosing the most convenient stoichiometry in the coupling of the two flows, the energy required for the transport can be reduced to a minimum, with an increase in the absolute value of the degree of coupling. Cotransporters and countertransporters are, therefore, more efficient than independent ion pumps, and they also exert a function of control.

From the *coupling stoichiometry* of the transporter, that is, the ratio of the number of driving ions to that of the driven molecular species per transport cycle, and from their charges one can determine whether the secondary active transport is electrogenic or electroneutral. For example, with a coupling stoichiometry of two sodium ions to one glucose molecule, Na^+/glucose cotransport represents an electrogenic process. Such charge movements across the membrane generate small electrical currents and, therefore, electrophysiological methods may be used to measure the activity of electrogenic transporters. Let us now examine some examples of association of secondary active transport with a master pump in different types of membranes, subdivided into four classes, based on differences in their energetics. These differences came to maturity in the course of evolution, since the different organisms have adopted different strategies to tackle their specific environment.

3.4.1 Plasma Membranes of Animal Cells

In the majority of these membranes, the role of master pump is played by the Na^+,K^+-ATPase. This ion pump transports three sodium ions from the cytosol to the extracellular fluid and two potassium ions in the opposite direction (Fig. 3.2). The translocation of both ionic species takes place against their electrochemical potential gradients, at the expense of the Gibbs energy of ATP hydrolysis. In neurons, the transmission of nerve impulses involves the concerted opening and closing of voltage-gated sodium and potassium channels, which move the corresponding ions down their electrochemical potential gradients. The Na^+,K^+-ATPase prevents the electrochemical potential gradients of these two ions from decreasing, by pumping them against their gradients. The electrochemical potential gradient of sodium ions is also exploited in other ways. Thus, in animal plasma membranes, a *sodium/glucose cotransporter* couples the endergonic flow of one glucose molecule into the cytosol against its concentration gradient to the exergonic flow of two sodium ions in the same direction, down their electrochemical potential gradient. This serves to enrich the inside of the cell with the glucose nutrient. A further example is offered by the *sodium/calcium exchanger*, which maintains a low level of calcium in the cytosol of the cells of cardiac and skeletal muscles, where the Ca^{2+} concentration is about 10^{-7} M, by coupling the exergonic flow of three sodium ions into the cell to the endergonic

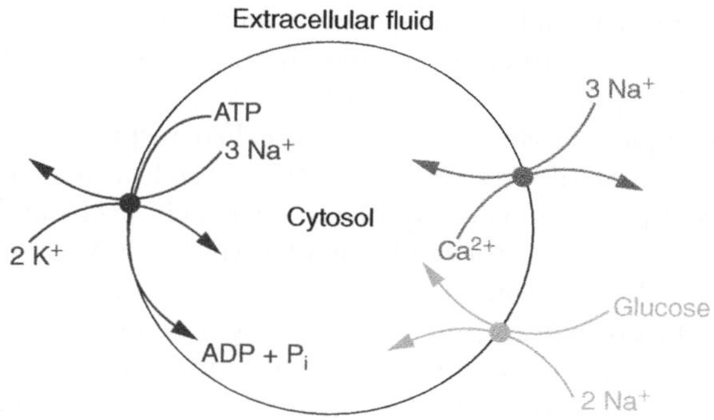

Figure 3.2 Scheme of the coupling of the Na^+, K^+-ATPase to secondary active transport across the plasma membrane of animal cells.

flow of one Ca^{2+} ion into the extracellular fluid, where their concentration is about 5 mM.

The concentration of K^+ ions is about 0.2 M in the cytosol and about two orders of magnitude lower in the extracellular fluid, whereas that of Na^+ ions is about 0.2 M in the extracellular fluid and about two orders of magnitude lower in the cytosol. Hence, the concentration gradients of Na^+ and K^+ ions are roughly equal across the plasma membrane, but in opposite directions. Despite this similarity, there is no clear evidence of coupling of the exergonic flow of K^+ ions toward the extracellular fluid to the endergonic flow of another species. This is due to the fact that the whole driving force that moves ionic species across membranes is their electrochemical potential gradient, which includes not only the chemical potential gradient, but also the gradient of the electrostatic potential energy, $z_h F \Delta\phi$, where $z_h F$ is the charge of one mole of the ion and $\Delta\phi$ is the transmembrane potential. The $\Delta\phi$ value of plasma membranes at rest usually ranges from −60 to −80 mV, as measured from inside the cell, and, hence, decreases the absolute value of the electrochemical potential gradient of K^+ ions while increasing that of Na^+ ions, under otherwise identical conditions. The low Gibbs energy of the exergonic flow of K^+ ions toward the extracellular fluid prevents its exploitation for secondary active transport.

Not all secondary active transporters of the plasma membrane utilize Na^+ as the driving ion. Some use an existing proton gradient. For example, the H^+/oligopeptide cotransporter (PepT), found in the small intestine, couples the exergonic movement of protons into the cell to the endergonic transport of di- and tripeptides in the same direction.

3.4.2 Inner Mitochondrial Membrane, Thylakoid Membrane, and Bacterial Plasma Membrane

These membranes, which share a common evolutionary origin, do not contain cholesterol. A difference in the electrochemical potential of the proton across these membranes drives their functions. This proton gradient is generated by a chain of redox proteins embedded in the membrane, referred to as the *electron-transport chain*. In case of need, these membranes use the proton electrochemical potential gradient to generate ATP via the integral protein *ATP synthase.* In the inner mitochondrial membrane, a system of three multiprotein complexes, called *proton-pumping oxidoreductases,* acts as the master pump. These three complexes, which form the so-called respiratory electron-transport chain, are in charge of the oxidation of the NADH produced in the mitochondrial matrix by the reactions of glycolysis, citric acid cycle, and oxidation of fatty acids (cf. Section 1.2.2). The oxidation of NADH to NAD^+ by molecular oxygen is coupled to a flow of protons from the matrix to the intermembrane space (Rawn, 1989).

Electron transport occurs along a sequence of redox couples incorporated in integral or peripheral proteins and characterized by progressively increasing reduction potentials. Each redox couple accepts electrons from the redox couple with lower reduction potential that precedes it in the series and donates them to the redox couple with higher reduction potential that follows. The first complex of the respiratory chain, the *NADH–ubiquinone reductase,* catalyzes the reduction of the small, liposoluble molecule ubiquinone (UQ) to ubiquinol (UQH_2) by NADH, which is oxidized to NAD^+; this exergonic process is accompanied by the translocation of three protons (Fig. 3.3). The second complex, the *ubiquinone–cytochrome c reductase,* catalyzes the oxidation of UQH_2 to UQ by *ferricytochrome c* (Cyt c_{ox}), a small peripheral protein located in the intermembrane space, which is reduced to *ferrocytochrome c* (Cyt c_{red}); this process is accompanied by the translocation of four protons, while the UQ molecules are again made available for NADH–ubiquinone reductase. The third complex, the *cytochrome c oxidase,* catalyzes the oxidation of Cyt c_{red} to Cyt c_{ox} by molecular oxygen, which is reduced to water; this process is accompanied by the translocation of four protons, while Cyt c_{ox} is again made available for ubiquinone–cytochrome *c* reductase.

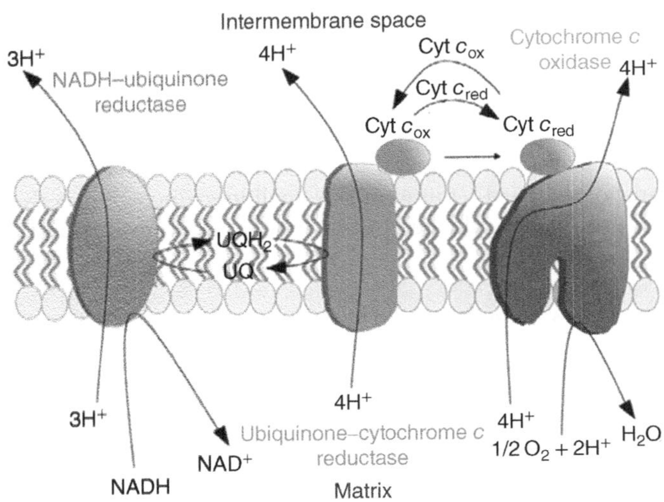

Figure 3.3 Scheme of the respiratory electron-transport chain.

The overall process driven by the three proton-pumping oxidoreductases is, therefore, the exergonic oxidation of NADH to NAD^+ by oxygen that is reduced to water, while 11 protons are transported from the matrix to the intermembrane space against their electrochemical potential gradient. This fuels an electrochemical potential gradient of protons, the *proton-motive force*, which is exploited in various ways. Thus, *ATP synthase*, a protein embedded in the inner mitochondrial membrane, couples the exergonic proton flow from the intermembrane space to the matrix along its channel with the highly endergonic synthesis of ATP from ADP and inorganic phosphate. This synthesis takes place on the matrix side of the membrane. One may wonder why proton-pumping oxidoreductases have such a high capacity with respect to protons, in spite of the very low physiological concentration of the latter. The reason is that biological solutions are highly buffered, such that the capacity of the system is not due to the hydrogen ion concentration, but rather to that of buffer species, which release or take up protons according to need.

Proton pumping by the respiratory chain leaves a proton deficiency in the matrix, which determines a negative transmembrane potential on the matrix side of the membrane. This transmembrane potential activates the *ATP/ADP translocase*, which couples the flow of ATP^{4-} from the matrix to the intermembrane space with the flow of ADP^{3-} in the opposite direction (Fig. 3.4). Since ATP bears four negative charges and ADP only three, the ion exchange carried out by this countertransporter amounts to translocating one negative charge down its electrochemical potential gradient, and consequently occurs spontaneously. This serves to keep the negative transmembrane potential of the inner mitochondrial membrane under control, through a balance between the pumping activity of the respiratory electron-transport chain and the ion exchange by the ATP/ADP translocase. Another transport system embedded in the inner mitochondrial membrane is the *phosphate/hydroxyl ion countertransporter*, which couples the phosphate flow from the intermembrane space to the matrix with that of hydroxyl ions in the opposite direction. In view of the ionic product for water, this countertransporter can also be considered as a phosphate/proton cotransporter. Since the phosphate ion at physiological pH is in its diprotonated form, $H_2PO_4^-$, the phosphate/hydroxyl ion countertransporter is electrically silent. The flow of hydroxyl ions to the intermembrane space proceeds downhill and is, therefore, exergonic. This is not necessarily the case for the phosphate flow to the matrix. If on the one hand the phosphate concentration is lower in the matrix, due to its consumption for the ATP synthesis by ATP synthase, on the other hand, its negative charge is opposed by the negative transmembrane potential. The ATP/ADP translocase acts synergically with the phosphate/hydroxyl ion countertransporter to provide the matrix with the reactants required by the ATP synthase to synthesize ATP.

A network of secondary active transport systems driven by an electron-transport chain is also present in the plasma membrane of bacteria. Individual bacteria use multiple electron-transport chains, often simultaneously. For example, *Escherichia coli* (when growing aerobically by using glucose as an

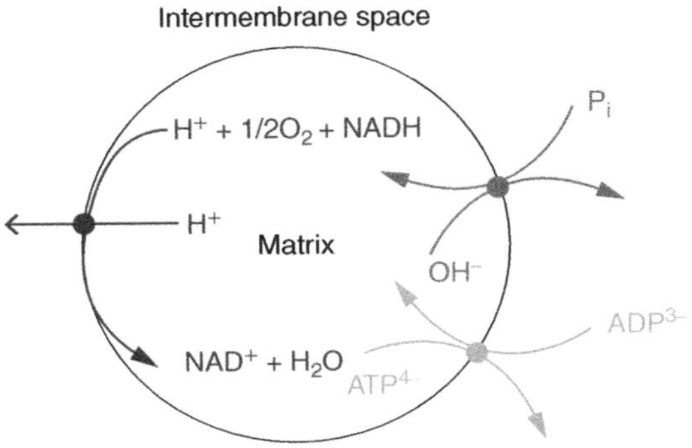

Figure 3.4 Scheme of the coupling of the proton-pumping oxidoreductases to secondary active transport across the inner mitochondrial membrane.

energy source) uses two different NADH–quinone reductases and two different quinol oxidases, for a total of four different electron-transport chains operating simultaneously. Gram-negative bacteria, such as *E. coli*, are enveloped by two membranes, an outer and an inner membrane, with a *periplasmic space* between them, which is analogous to the intermembrane space of mitochondria. The electron-transport chain of these bacteria pumps protons from the cytosol, enveloped by the inner membrane, into the periplasmic space, maintaining a high proton electrochemical potential there. In the inner membrane of *E. coli*, a cotransporter, called *lactose permease*, couples the endergonic flow of lactose into the cytosol to the exergonic flow of protons in the same direction; the stoichiometric ratio between lactose molecules and protons is 1:1.

3.4.3 Membranes of Plant and Fungal Cells

These membranes differ from the plasma membranes of animal cells because they lack Na^+, K^+-ATPase. Instead, they have a P-type ATPase translocating protons that acts as a master pump. The proton gradient created by this ATPase drives the transport of molecules and exerts other functions. This fundamental difference between the plasma membranes of plant and animal cells reflects a fundamental difference between their styles of life. As distinct from animal cells, plant cells do not have a liquid circulatory system at their disposal to stock sodium ions from outside; this makes it necessary to replace the sodium pump with a proton pump. Moreover, the plant cells have a rigid cell wall to support the plasma membrane during periods of osmotic imbalance. On the contrary, the animal cells maintain the osmotic balance by regulating the inner concentration of electrolytes, such as sodium or potassium, so as to balance the concentration of external electrolytes.

Master proton pumps powered by light are embedded in the plasma membrane of photosynthetic bacteria, while in plants they are held inside organelles called chloroplasts, which are most abundant in leaf cells. In the thylakoid membrane of chloroplasts, light drives the conversion of water to oxygen and of nicotinamide adenine dinucleotide phosphate ($NADP^+$) to its reduced form, NADPH, with transfer of protons across the membrane. In this case, the overall redox reaction is an endergonic reduction, while in mitochondria it is an exergonic oxidation. Nonetheless, even in this case, the electron-transport chain consists of a sequence of redox couples incorporated in the integral or peripheral proteins of the thylakoid membrane and characterized by progressively increasing reduction potentials (Rawn, 1989). This is due to the action of two complex multiprotein systems that operate in series, designated *photosystem II* (PS II) and *photosystem I* (PS I). Each photosystem consists of an *antenna complex,* containing a high number of chlorophyll (Chl) molecules, and a *reaction center,* which contains two particular Chl molecules, called the *special pair*. The Chl molecules of the antenna complex absorb light energy and transfer it to the special pair, whose electron is photoexcited to a high energy level. This electron is then released to a first electron-transport chain similar to that in the inner mitochondrial membrane. The redox reactions of this transport chain transport protons from the stroma to the lumen of the thylakoid membrane (cf. Fig. 1.8).

The photosynthetic process starts from PS II, which is surrounded by a membrane-spanning Chl–protein complex called *chlorophyll a/b-binding light-harvesting complex* (LHC). This protein contributes to funneling photochemical energy into the PS II special pair, together with the PS II antenna complex (Fig. 3.5). As soon as the PS II special pair releases one excited electron, it receives a further electron to photoexcite from a peripheral water-splitting enzyme, associated with PS II and located on the lumenal side of the thylakoid membrane, which oxidizes water molecules to molecular oxygen and releases protons to the lumen. The photoexcited electron is released to *pheophytin a,* a molecule that differs from Chl by the substitution of Mg^{2+} in the porphyrin ring with two hydrogen atoms. From reduced pheophytin *a,* the electron passes to a plastoquinone molecule (PQ_A) bound to PS II, and from this to PQ_B, a quinone reversibly associated with PS II. After accepting a further electron from PQ_A and two protons from the stromal side of the thylakoid membrane, PQ_B is fully reduced to plastoquinol, PQH_2. This is released into a larger pool of liposoluble plastoquinones, whose reduced species move within the membrane

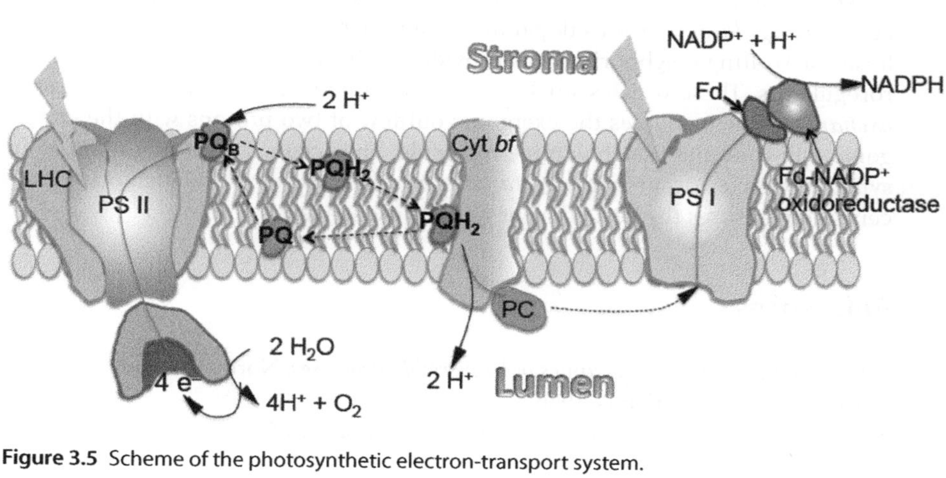

Figure 3.5 Scheme of the photosynthetic electron-transport system.

from PS II to the membrane-spanning protein *cytochrome bf complex* (Cyt *bf*). Here, they are reoxidized to PQ molecules and made again available for PS II. The electrons released by PQH_2 to Cyt *bf* are transferred along a series of redox couples contained in the latter protein to a copper-containing peripheral protein, plastocyanine (PC), located on the lumenal side. PC plays toward PS I a role analogous to that played by the water-splitting enzyme toward PS II, by filling with one electron the electron hole left in the special pair of PS I by a photoexcited electron. The latter electron is transferred along a series of three Fe–S proteins contained in PS I to *ferredoxin* (Fd), a fourth Fe–S electron acceptor soluble in the stroma. The reduced form of Fd binds to and reduces *ferredoxin*– $NADP^+$ *oxidoreductase*, a flavoenzyme associated with the stromal side of the membrane. Finally, the reduced form of the latter oxidoreductase reduces $NADP^+$ to NADPH. The overall *light reaction* taking place in the thylakoid membrane is, therefore, the reduction of $NADP^+$ to NADPH by H_2O, with O_2 formation.

In bacteria, yeast, and plants, the exergonic inflow of one proton is coupled to the outflow of one sodium ion by a Na^+/proton *exchanger*, which provides increased salt tolerance by removing sodium in exchange for extracellular protons, while it is coupled to the inflow of one sucrose molecule by a *sucrose/proton cotransporter*. A *nitrate/proton cotransporter* requires the inflow of two protons to move one nitrate ion into the cell.

3.4.4 Membranes of the Vesicular System

These membranes envelope lysosomes and the secretory vesicles deriving from the Golgi apparatus. They incorporate a V-type ATPase translocating protons, which acts as a master pump by pumping protons into the vesicle. The proton electrochemical potential gradient created by this ATPase drives processes within the membrane and makes the interior of the vesicle acidic, a property often important for vesicle functions. The vesicles released from axon terminals of neurons liberate *neuroregulators*, such as dopamine or another catecholamine, which diffuse locally activating neighboring cells provided with specific receptors for the neuroregulators. These vesicles stock the catecholamine via a *catecholamine/proton exchanger*, which couples the exergonic outflow of two protons with the endergonic inflow of one catecholamine molecule. The accumulation of glutamate in synaptic vesicles occurs via a *glutamate/proton exchanger*; a low chloride concentration ensures the best efficiency of this exchanger.

References

De Groot, S. R., *Thermodynamics of Irreversible Processes*, North-Holland, Amsterdam, Netherlands, 1966.

Kedem, O., S. R. Kaplan, *Trans. Faraday Soc.* 61 (1965) 1897–1911.

Kirkwood, J. G., I. Oppenheim, *Chemical Thermodynamics*, McGraw-Hill, New York, 1961.

Onsager, L., *Phys. Rev.* 37 (1931) 405–426; 38 (1931) 2265–2279.

Rawn, J. D., *Biochemistry*, N. Patterson Publishers, Burlington, U.S.A, 1989.

Schultz, S. G., *Basic Principles of Membrane Transport*, Cambridge University Press, London, 1980.

Walz, D., *Biochim. Biophys. Acta* 505 (1979) 279–353.

Zemansky, M. W., *Heat and Thermodynamics*, McGraw-Hill, New York, 1957.

4

Passive Transport

Biological membranes are not permeable to inorganic ions; nonetheless, the movement of ions in and out of a cell across its plasma membrane is critical for its normal activity. Ions do not pass through the plasma membrane by simple diffusion; rather, their transport is mediated by hydrophilic pores formed by clustered protein molecules, called *ion channels*. Ion transport through these channels is an example of passive transport because it proceeds with a decrease in Gibbs energy, and the movement of ions occurs down their electrochemical potential gradient. Passive transport is also called *facilitated diffusion*, because the proteins make the solute movement across a membrane possible.

4.1 How do Ion Channels Look Like?

An ion channel spans the whole membrane, providing a passage through which ions can traverse the membrane. However, an ion channel is not simply a "hole" in the membrane. An ion channel is extremely selective for the ion allowed through, in that distinct channels are available for each inorganic ion. An ion channel consists of multiple subunits of integral membrane proteins that form a pore through the membrane. In its simplest form, the ion channel is always open and the flow of ions is bidirectional. The specificity for a particular ion stems from the interactions between the ion and specific amino acid residues that line the lumen of the pore and act as a selective filter. Ions pass through the pore in *single file*, namely one by one, and at a rate of several thousand ions per millisecond.

Many ion channels are gated, that is, regulated to be in either an open or a closed conformation. In the so-called *excitable cells* such as neurons and muscle cells, some channels open or close in response to changes in the potential difference across the membrane (the transmembrane potential) and are said to be *voltage-gated* channels. Many other ion channels open or close in response to the binding of a small signaling molecule termed ligand, and are referred to as *ligand-gated* channels. Some of these channels are gated by extracellular ligands, while some others by intracellular ones. In both cases, the ligand is not the molecule that is transported when the channel opens. In view of their peculiarity, voltage-gated ion channels are particularly suited for functional investigations by electrochemical techniques. In a resting neuron, voltage-gated ion channels are in the closed conformation and ions cannot pass through, whereas they open in response to neuron stimulation.

The best-investigated channels are the voltage-gated potassium channels, although their mechanism of action is likely to be similar to that of other voltage-gated ion channels. Other important voltage-gated ion channels are sodium and calcium channels. Sodium and calcium channels consist of four *domains* of high internal homology, connected by conformationally flexible loops. Each domain consists of six sections (labeled S1 through S6), which appear to be membrane-spanning α-helixes (Fig. 4.1). The four domains form a tetrameric cluster circularly arranged around a transmembrane pore. Potassium channels have a similar structure, in which the four domains, each consisting of six membrane-spanning α-helixes, are not bound together and are referred to as *subunits*. The four subunits, arranged around a central K^+-conducting pore, may be identical, forming a *homotetramer*. There are also potassium channels formed by related but not identical protein subunits, giving rise to *heterotetrameric* complexes. In view of the great variety of potassium channels, they are grouped into subfamilies. Heterotetramers are only formed from subunits of the same *subfamily* and exhibit hybrid properties intermediate between those of the corresponding homotetramers. Conversely, subunits from different subfamilies cannot combine to form heterotetramers.

The narrowest portion of the transmembrane pores of ion channels is termed the *selectivity filter*. Potassium channels remove four water molecules from the first hydration shell of a K^+ ion to allow its passage through the selectivity filter, leaving at most one water molecule at the front and one at the back of the ion. The selectivity filter is formed by a sequence of five amino acid residues, termed

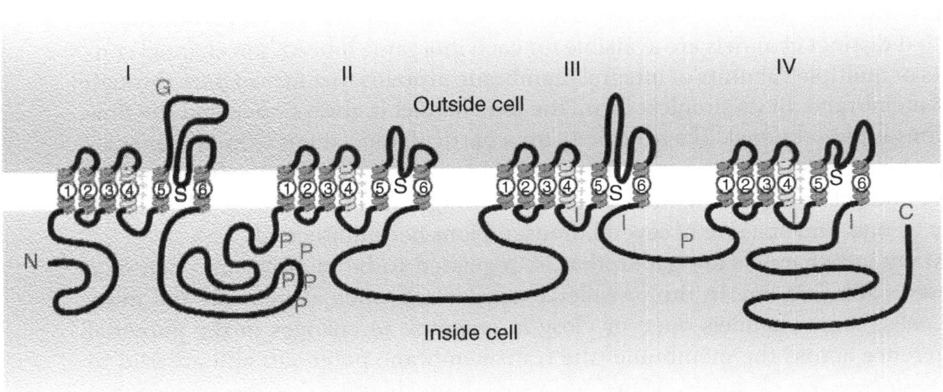

Figure 4.1 The four domains of a sodium channel. G, P, S, and I denote sites of glycosylation, phosphorylation, ion selectivity, and inactivation, respectively. Positive (+) charges in S4 are important for transmembrane voltage sensing. Source: Cthuljew, https://commons.wikimedia.org/wiki/File:Sodium-channel.svg. Used under CC BY-SA 3.0 https://creativecommons.org/licenses/by-sa/3.0/deed.en.

Figure 4.2 Selectivity filter of the KcsA potassium channel. Only two of the four subunits of the tetramer are displayed for the sake of clarity. Potassium ions occupy the S2 and S4 sites, whereas the oxygen atoms of water molecules occupy the S1 and S3 sites. The tips of the P-loops that interact with these species are carbonyl groups. (Zhou et al., 2001). Source: https://commons.wikimedia.org/wiki/File:2RH1.png.

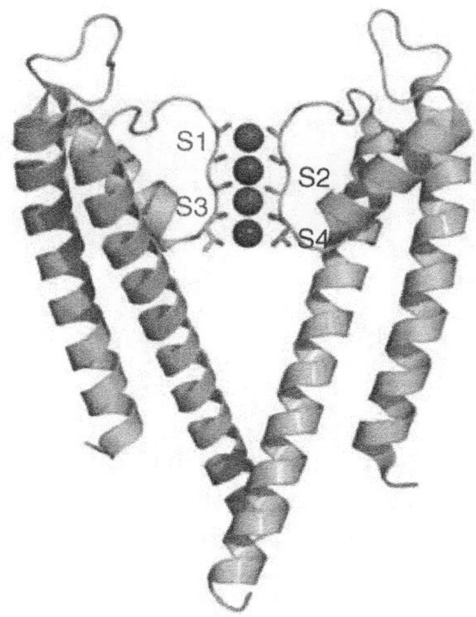

the *signature sequence*, provided by each of the loops connecting the segments of each of the four subunits, called *P-loops*. These loops turn the oxygens of their main-chain carbonyl groups toward the center of the selectivity filter, where they replace the water oxygens removed from the hydration shell (Fig. 4.2).

Sodium ion cannot pass through the selectivity filter of a potassium channel because its size is too small to be effectively coordinated by the carbonyl oxygens of the filter, which are too far apart; moreover, the energy required for its partial dehydration is higher than that of potassium ion, because of its closer distance from the water molecules of its hydration shell. In sodium channels, sodium ion is much more permeant than potassium ion. This is due to the fact that the selectivity filter of the sodium channel is slightly larger than that of the potassium channel, and may accommodate a sodium ion with three water molecules, one of them at its side. This is not possible for potassium ions, because of their larger size. In calcium channels, sodium and potassium ions are about three orders of magnitude less permeant than calcium ions. This ion selectivity cannot involve dehydration, since Ca^{2+} is more heavily hydrated than Na^{+}, and the unhydrated diameters of Ca^{2+} and Na^{+} are almost identical. As distinct from the P-loops of the selectivity filter of the sodium channel, those of the calcium channel always contain a glutamate residue at the same position, giving rise to a ring of four negative charges in the selectivity filter. Due to its higher positive charge, calcium ion prevails over monovalent cations in competing for this negatively charged ring. This explanation is supported by the observation that monovalent cations can pass readily through calcium channels in the absence of calcium ions.

In sodium and potassium channels, the passage through the pore is governed by two different types of gates: the *activation gate* and the *inactivation gate*. At the resting potential (ranging from −60 to −80 mV), the inactivation gate is open while the activation gate is closed, preventing ion flux through the ion channel, which is said to be in the *closed state*. As a consequence of an external stimulus,

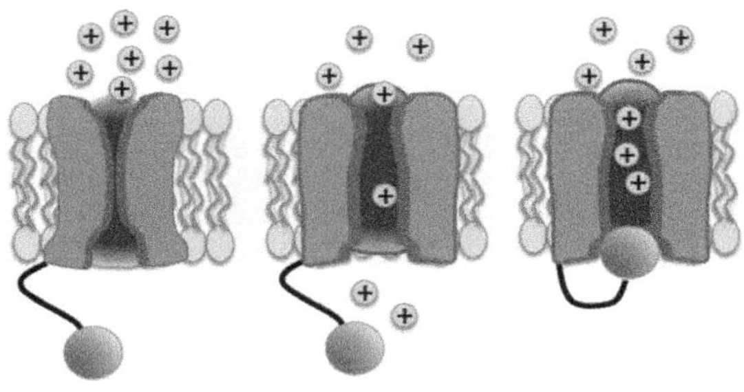

Figure 4.3 Scheme of a closed, open, and inactive ion channel, with the ball-and-chain model of the inactivation gate.

the activation gate may open, with the inactivation gate still open, allowing an ion flux through the ion channel, which is said to be in the *open state*. After a short and well-defined time period, which is shorter for sodium channels than for potassium channels, the inactivation gate closes, while the activation gate is still open, and the ions are prevented from flowing through the channel, which is said to be in the *inactive state* (Fig. 4.3). As distinct from the closed and open states, the inactive state is resistant (*refractory*) to changes in the transmembrane potential. Then, the ion channel passes to the closed state, in which the inactivation gate is again open and the activation gate closed, through an intermediate state in which both gates are closed. The inactive state prevents a direct passage from the open to the closed state. The importance of the inactive state will be apparent when the action potential is examined.

A creditworthy molecular model for inactivation is the *ball-and-chain model*. According to this model, inactivation is produced by a mobile portion of a domain or subunit of the ion channel, which swings into the inner mouth of the open channel pore, so as to block it. The ball, tethered to the rest of the protein through a loop (the chain), contains a concentration of positively charged residues. The hydrophobic residues are in the center of the ball, while the charged residues are on its exterior. For a particular conformation, the ball may be transiently held in contact with the negatively charged pore of a cation-selective channel, preventing ion movement through it. Since there are four domains or subunits in an ion channel, only one ball is expected to block the pore.

Voltage-gated ion channels must also contain a region of protein designated *voltage sensor*, which detects any change of transmembrane potential capable of inducing their passage from the closed to the open state through a movement of charge (the *gating charge*), strictly connected to the opening of the inactivation gate while the activation gate is closed. Nowadays, it is generally believed that this charge movement resides in one of the six segments that compose each domain of sodium or calcium channels, or each subunit of potassium channels. This segment, S4, is highly conserved in voltage-gated cation channels and exhibits an unusual array of positively charged arginine and lysine residues. Each positive charge in S4 is separated from its nearest neighbors by about 5 Å, measured along the axis of the helix, and is believed to interact attractively with the corresponding negative charges of the surrounding S1, S2, or S3 segments. According to

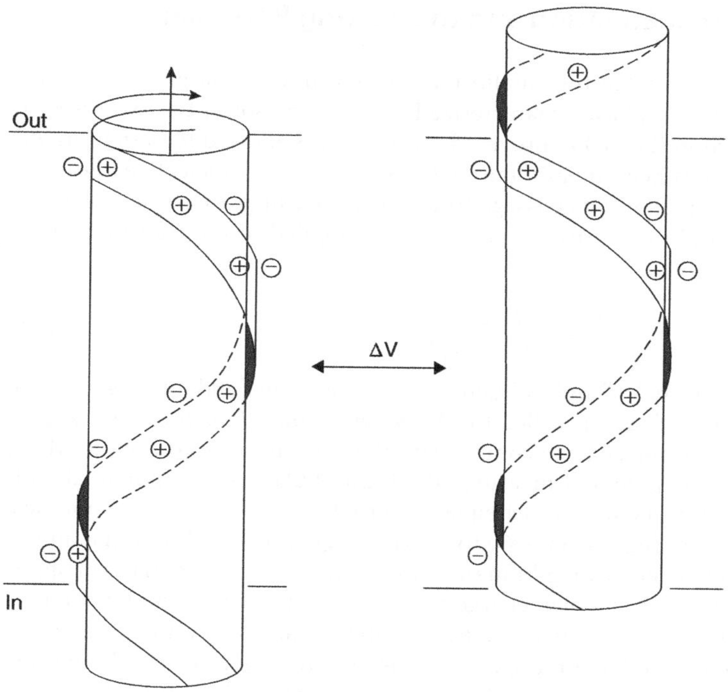

Figure 4.4 Scheme of the sliding helix model of voltage sensor. The arginine residues of the S4 segment determine a helix of positive charges, which form ion pairs with an array of negative charges on other segments. Source: Catteral (1986). Reproduced with permission of Elsevier.

the *sliding helix* model of gating, the S4 segments respond to a positive shift of the transmembrane potential by rotating by about 60° and moving toward the extracellular side of the membrane by 4.5 Å, so as to bring each positive charge into alignment with the next negative charge (Fig. 4.4). The helix is assumed to move rigidly outward, with no changes in the distances between the positively charged residues within the helix. This rotation would have the effect of exposing a negative charge at the intracellular side of the membrane and a positive charge at its extracellular side, which amounts to moving one charge across the whole membrane (Catteral, 1986). Since each channel has four S4 segments, this would involve the outward movement of four positive charges per channel.

It is not clear how this movement of the S4 segment is connected to the opening of the inactivation gate. While the opening of the inactivation gate, triggered by the gating charge movement, is induced by an external positive shift in the transmembrane potential, the subsequent passages from the open to the inactive state and then to the closed state do not require any external input and usually occur in a few milliseconds. In other words, inactivation is generated by the channel control machinery.

4.2 The Nernst Equation and the Resting Potential

In Section 3.2, we saw that the equilibrium condition for a charged species that exists in two different phases characterized by the same solvent is expressed by the Nernst equation (Eq. 3.30), provided this species is allowed to pass from one phase to the other. When a membrane contains an ion channel selective toward an hth ionic species, it allows its passage from the intracellular to extracellular side and vice versa, and the Nernst equation can be applied to the transmembrane potential $\Delta\phi$:

$$\Delta\phi_h \equiv \phi^{\text{in}} - \phi^{\text{ex}} = \frac{RT}{z_h F} \ln \frac{c_h^{\text{ex}}}{c_h^{\text{in}}} \equiv \frac{kT}{z_h e} \ln \frac{c_h^{\text{ex}}}{c_h^{\text{in}}}. \quad (4.1)$$

Here, the superscripts "ex" and "in" denote the extracellular and intracellular side of the membrane and $\Delta\phi_h$ is called the *Nernst potential of the hth ionic species.* This equation is written so as to comply with the convention according to which the transmembrane potential measures the intracellular electric potential with respect to the extracellular one, set equal to zero. For many plasma membranes, $\Delta\phi$ is negative and ranges from −60 to −80 mV. Equation 4.1 shows that an ion tends to accumulate on the membrane side where the electric potential is of opposite sign with respect to its own charge $z_h e$. From a different viewpoint, we may note that, if no external electric potential is applied across the membrane and $\Delta\phi$ is initially zero, the hth ionic species tends to move from the membrane side where its concentration is higher to that where it is lower, with a purely *diffusional motion.* After attainment of equilibrium, the concentration gradient responsible for diffusion is exactly matched by the electric potential gradient gradually built up, which is responsible for a purely *migrational motion* in the opposite direction.

If the membrane contains exclusively a single type of ion-selective channel, at equilibrium it determines a Nernst transmembrane potential $\Delta\phi_h$ expressed by Eq. 4.1, which is independent of the number density of the ion channel and of its specific features. The transmembrane potential can be measured by immersing two identical Ag/AgCl reference electrodes in the two solutions bathing the membrane and by applying to them a potential difference equal and opposite to $\Delta\phi$ with a battery, using a Poggendorff bridge. To better understand how this system works, let us consider a system consisting of two aqueous solutions of KCl of different concentrations, separated by a membrane containing a K^+-selective ion channel; two Ag/AgCl reference electrodes, directly connected by a metal wire without the interposition of a battery, are immersed in the two aqueous solutions. This system behaves similar to a concentration cell. The electric potential in the solution with higher KCl concentration (the *in* solution) is initially more negative than that in the solution with lower KCl concentration (the *ex* solution), in view of Eq. 4.1. Hence, electrons tend to move from the Ag/AgCl electrode in the *in* solution to that in the *ex* solution along the metal wire. This takes place through Ag oxidation to AgCl at the reference electrode immersed in the *in* solution. This oxidation reaction subtracts Cl^- ions from the *in* solution, because these anions cannot permeate the membrane. To maintain the electroneutrality of the bulk *in* solution, K^+ ions will then pass to the *ex* solution through the K^+-selective ion channel. This is possible since, simultaneously, the electrons coming from the reference electrode in the *in* solution reach the reference electrode in the *ex* solution, determining the electroreduction of AgCl therein, with formation of Ag metal and Cl^- ions. These anions are released into the *ex* solution, where they counterbalance the positive charge of the incoming K^+ ions. This process will continue up to annihilation of the initial Nernst potential difference between the two Ag/AgCl electrodes. To block this spontaneous process from the very beginning, it is necessary to oppose to the Nernst potential difference an equal potential difference, for example, by inserting a variable voltage battery between the two reference electrodes.

Let us now imagine applying to the system previously considered a transmembrane potential different from the Nernst equilibrium value for the hth ion species. In this case, the ionic charge that flows along a single ion-selective channel per unit time, that is, the current i, is proportional to the difference, $\Delta\phi - \Delta\phi_h$, between the applied potential and the equilibrium value. In other words, the channel current satisfies Ohm's first law. The ratio, γ, of the current i to $(\Delta\phi - \Delta\phi_h)$

$$i/(\Delta\phi - \Delta\phi_h) = \gamma \tag{4.2}$$

is termed *single-channel conductance* (De Felice, 1997). It depends on the geometry and internal structure of the ion channel as well as on the physicochemical forces that act on the ion as it moves along the lumen of the channel. The proportional relationship of Eq. 4.2 holds strictly only if the ion concentration is the same on both sides of the membrane. If this condition is not fulfilled, it holds approximately over a narrow $\Delta\phi$ range straddling $\Delta\phi_h$. In the latter case, at transmembrane potentials $\Delta\phi$ sufficiently far from $\Delta\phi_h$, i becomes directly proportional to $\Delta\phi$ (cf. Section 4.5); moreover, the resulting conductance γ turns out to be proportional to the ion concentration on the membrane side from which the applied potential moves it toward the opposite side; hence, γ varies from one side to the other.

In general, a biological membrane contains different types of ion channels: some of them are selective toward inorganic cations such as Na^+, K^+, and Ca^{2+}, while some others are selective toward anions such as Cl^-. This situation is more complicated, and the transmembrane potential $\Delta\phi$ that is established in the absence of an external applied potential is not a true equilibrium potential; it is referred to as a *resting potential*. Before deriving its expression, we must consider that ion channels are not always open, but open and close randomly, regardless of being voltage or ligand gated. It is this randomness that explains most of the interesting properties of excitable membranes, such as action potential and propagation of nerve impulses. We can think of random opening and closing as a gate on the channel pore: the pore may be either open or closed. It is important to estimate the *probability p* of the pore being in the open state. To this end, let us figure a purely thought experiment in which we take a very high number of snapshots of the pore: in some of these snapshots, the pore is open, while in some others it is closed. By realizing that there is no regular sequence with which open and closed states follow one another, we must conclude that the process is truly random. The ratio of the number of snapshots with the pore open to their total number provides the probability p. Therefore, the *effective conductance* of a single channel is not given by γ, which expresses its conductance if it were always open, but rather by the $p\gamma$ product. If we wish to determine the *specific conductance* G of a given ion channel, namely the conductance per unit area of the membrane due to all the ion channels of the same type, we must multiply the effective conductance $p\gamma$ of a single channel by its number density N:

$$G = Np\gamma. \tag{4.3}$$

From a statistical point of view, this product can be conceptually viewed either as the product of N by $(p\gamma)$ or as the product of (Np) by γ. The first way of viewing the specific conductance G consists in imagining taking a very high number of snapshots of a single ion channel and calculating the effective conductance of this channel, as previously described; this is then multiplied by the number (N) of channels of the same type present on the unit surface of the membrane. The second way of viewing G consists in imagining taking a single snapshot of a membrane surface, large enough to contain a very high number of ion channels, counting the number of open channels and dividing this number by the total number of channels in the snapshot; this provides Np, which is then multiplied by the conductance γ that a single channel would have if it were always open. A further snapshot of the same membrane surface will show that some of the ion channels that were open in the previous snapshot are now closed, while some that were closed are now open. However, if the number of channels on the membrane surface is high enough to be statistically significant, the ratio of the number of open channels to their total number will remain the same. These two different ways of viewing the probability p, namely by following the behavior of a single channel for a sufficiently long time, or else by observing the state of a statistically significant number of channels only once, are perfectly equivalent.

Let us now consider a membrane containing two types of ion channels, one selective toward a monovalent cation (+), and the other selective toward a monovalent anion (−). Let us further assume that the solutions that bath the two sides of the membrane contain different concentrations of the salt (+−) of the two ions. In this case, the transmembrane potential $\Delta\phi$ is modulated by both ion channels to an extent proportional to their relative specific conductances, G_+ and G_-. More precisely, $\Delta\phi$ can be considered as a weighted average of the *Nernst potentials*, $\Delta\phi_+$ and $\Delta\phi_-$, of the ions (+) and (−), as expressed by Eq. 4.1, where the weights are the specific conductances, G_+ and G_-, of the corresponding ion channels:

$$\Delta\phi = \frac{G_+\Delta\phi_+ + G_-\Delta\phi_-}{G_+ + G_-} = \frac{N_+p_+\gamma_+\Delta\phi_+ + N_-p_-\gamma_-\Delta\phi_-}{N_+p_+\gamma_+ + N_-p_-\gamma_-}. \tag{4.4}$$

This amounts to assuming that, in the immediate vicinity of the mouth of each ion-selective channel, this tends to establish the equilibrium situation with respect to the corresponding ion to an extent proportional to its effective conductance.

The bulk electrolytic solutions that bath the membrane satisfy the condition of *local electroneutrality*, according to which an infinitesimal volume element of the solution, large enough to contain a statistically significant number of ions, is such that the total charge of the anions is equal in magnitude, but opposite in sign, to that of the cations. To fulfill this condition, the cation *flux* (i.e., the number of cations that flow across the unit surface of the membrane separating the two solutions per unit time) must move a charge equal in magnitude, but opposite in sign, to that moved by the anion flux in the same unit time. Clearly, this stationary condition is not an equilibrium condition. Equilibrium will be attained only when, ultimately, the salt concentration will become equal on both sides of the membrane, as a consequence of the continuous flux of cations and anions from the solution of higher to that of lower concentration.

In general, the specific conductances of two different ion channels are different. It is this difference that determines the resting potential. For example, let us consider a membrane containing potassium and chloride channels interposed between two KCl solutions of different concentrations. Let us further assume that the transmembrane potential $\Delta\phi$ is initially zero. Since the specific conductance of the potassium channel is higher than that of the chloride channel, potassium ions will tend to move from the solution of higher to that of lower KCl concentration more rapidly than chloride ions. By so doing, they will leave behind them, on the membrane side bathed by the solution of higher concentration, a very small deficiency of cations, which will make this membrane side negative with respect to the opposite side. In a very short time, a stationary situation will, therefore, be attained, in which the transmembrane potential thus built up slows down the potassium flux and speeds up the chloride flux, so as to equate them. The transmembrane potential that is formed under these stationary conditions is the *resting potential* and is expressed by Eq. 4.4. To better grasp its significance, let us replace the expressions of the Nernst potentials $\Delta\phi_+$ and $\Delta\phi_-$ from Eq. 4.1 into Eq. 4.4, yielding

$$\Delta\phi \equiv \phi^{\text{in}} - \phi^{\text{ex}} = \frac{kT}{e}\frac{G_+ - G_-}{G_+ + G_-}\ln\frac{c^{\text{ex}}}{c^{\text{in}}}. \tag{4.5}$$

Here, $c^{\text{ex}} \equiv c_+^{\text{ex}} = c_-^{\text{ex}}$ and $c^{\text{in}} \equiv c_+^{\text{in}} = c_-^{\text{in}}$ are the common values of the cation and anion concentrations in the extracellular and intracellular bulk solutions, respectively. It is evident that, if $G_+ > G_-$ and $c^{\text{in}} > c^{\text{ex}}$, the transmembrane potential $\Delta\phi$ is negative. In fact, initially, potassium ions move faster than chloride ions from the intracellular to the extracellular side, leaving behind a defect of positive ions, until the stationary state is attained.

The sum of the two specific conductances G_+ and G_- can be considered as the *total conductance* of the membrane:

$$G = G_+ + G_-. \tag{4.6}$$

This conductance should not be confused with the conductance of a homogeneous electronic or ionic conductor, which is defined as the ratio of the current I that flows along the conductor to the potential difference V across it. In point of fact, G does not give rise to any current. However, we may observe that the product of the specific conductance, for instance, of the potassium channel, G_+, by the Nernst potential $\Delta\phi_+$ of potassium ions expresses the current density, j_+, due to the flux of potassium ions along the potassium channel. Analogously, the product $G_-\Delta\phi_-$ of the specific conductance of the chloride channel by the Nernst potential of chloride ions expresses the current density j_- due to the flux of chloride ions along the chloride channel. The resting potential $\Delta\phi$ of Eq. 4.4 can then be written as

$$\Delta\phi = \frac{G_+\Delta\phi_+ + G_-\Delta\phi_-}{G_+ + G_-} = \frac{j_+ + j_-}{G_+ + G_-} = \frac{\text{total current density}}{\text{total conductance}}. \tag{4.7}$$

From this point of view, this expression of $\Delta\phi$ formally satisfies Ohm's law. There is, however, a fundamental difference with respect to Ohm's law as applied to an electrolytic solution. In the case of Eq. 4.7, the cationic current density j_+ proceeds in the same direction as the anionic current density j_-, and hence no electric current is generated. Conversely, in the case of an electrolytic solution, the cationic and anionic currents driven by the electric field proceed in opposite directions, thus summing their effects and generating a net electric current.

Let us now envisage a physiologically more relevant situation in which the membrane contains three different types of ion channel, selective toward Na^+, K^+, and Cl^- ions. It is well known that the K^+ ion concentration in the cytoplasm is much higher than in the extracellular medium, while the opposite is true for the Na^+ concentration. If only the three inorganic ions Na^+, K^+, and Cl^- were present, the Cl^- concentration would be the same inside and outside of the cell and equal to the sum of the Na^+ and K^+ concentrations, to ensure the local electroneutrality of the intra- and extracellular solutions. However, the cytoplasm also contains bulky negatively charged molecules, such as ATP and ADP, which cannot permeate the plasma membrane and partially compensate the charge of potassium and sodium ions. This explains why the chloride concentration is higher in the extracellular medium. Even in this case, the transmembrane potential is the weighted average of the Nernst potentials of the three ion channels:

$$\Delta\phi = \frac{G_K\Delta\phi_K + G_{Na}\Delta\phi_{Na} + G_{Cl}\Delta\phi_{Cl}}{G_K + G_{Na} + G_{Cl}}. \tag{4.8}$$

For realistic values of the intra- and extracellular concentrations of these three ions, the corresponding Nernst potentials take the following values:

$$\Delta\phi_K = \frac{RT}{F}\ln\frac{2.5\,\text{mM}}{200\,\text{mM}} = -110\,\text{mV};\quad \Delta\phi_{Na} = \frac{RT}{F}\ln\frac{200\,\text{mM}}{10\,\text{mM}} = +75\,\text{mV};$$
$$\Delta\phi_{Cl} = -\frac{RT}{F}\ln\frac{200\,\text{mM}}{55\,\text{mM}} = -33\,\text{mV}. \tag{4.9}$$

In these equations, the numerator contains the extracellular concentration and the denominator the intracellular one. In practice, K^+ ions tend to pass from the intra- to the extracellular side of the membrane, leaving behind a defect of positive charges and causing the intracellular side to become negative ($\Delta\phi_K < 0$). Conversely, Na^+ ions tend to pass from the extra- to the intracellular side of the membrane, creating an excess of positive charges on the intracellular side and causing this side to become positive ($\Delta\phi_{Na} > 0$). Cl^- ions also tend to pass from the extra- to the intracellular side of the membrane; however, being negatively charged, they contribute to creating a negative transmembrane potential, just as K^+ ions.

It should be noted that the ionic charge Q_{in} on the intracellular side of the membrane required to create the aforementioned Nernst potentials is extremely small. This is given by the product of the capacitance C of the membrane by the corresponding Nernst potential. Naturally, a charge equal in magnitude and opposite in sign must be present on the extracellular side of the membrane, to ensure the electroneutrality of the whole electrified interface. For a typical cell of small size, C amounts to about $10\,\text{pF} \equiv 10^{-11}$ F. At the Nernst potentials in Eq. 4.9, the charge Q_{in} on the intracellular side is then given by

$$Q_K = -1.1 \times 10^{-12}\ \text{C}; \quad Q_{Na} = +7.5 \times 10^{-13}\ \text{C}; \quad Q_{Cl} = -3.3 \times 10^{-13}\ \text{C}. \tag{4.10}$$

These very small charges are localized in the immediate vicinity of the inner surface of the membrane, so as to ensure the local electroneutrality of the bulk cytosol. Naturally, these are only hypothetical charges, in that the actual charge depends on the specific conductances, $G_h = N_h p_h \gamma_h$ (cf. Eq. 4.3), of the three different ion channels.

As a rule, during a rapid process, the number density N_h of an ion channel and its single-channel conductance γ_h remain constant over time, while its probability p_h of being open is time dependent. Thus, a time variation of the probabilities relative to potassium, sodium, and chloride channels may allow the transmembrane potential to take any value between the extreme values of the Nernst potentials, which in Eq. 4.9 range from −110 to +75 mV. Since these ion channels are voltage gated, the time dependences of their different probabilities p_h may be interlinked, giving rise to a well-defined transmembrane potential versus time curve that, cyclically reiterated over time, generates a "transmembrane potential wave" termed *action potential*.

4.3 A First Approach to the Action Potential

An action potential is a short-lasting event in which the transmembrane potential of a cell rapidly rises and falls, yielding a well-defined sequence of transmembrane potential changes over time, which constitutes a real "electric potential wave." Action potentials occur in several types of animal cells, called *excitable cells*, which include neurons, muscle cells, and endocrine cells. In neurons, action potentials play a central role in cell-to-cell communication, and are also known as *nerve impulses* or *spikes*. Their temporal sequence is called *spike train*, and a neuron that emits an action potential is said to *fire*. In other types of cells, the main function of action potentials is to activate intracellular processes. In muscle cells, for example, the action potential is the first step in a chain of events that leads to muscle contraction. In β cells of pancreas, action potentials provoke the release of insulin.

In neurons, the action potential propagates very rapidly along the central elongated portion of this cell, called *axon*. The change of the transmembrane potential over time, recorded at any section of the axon membrane, has a peaked shape. To justify this shape, we may consider a notably simplified and idealized axon membrane, containing only one sodium and one potassium channel, as well as a *leak channel*. Leak channels are formally introduced to account for the small, nonspecific flow of inorganic ions across membranes, which are not totally impermeable to ions even in the absence of ion channels. As expected, the small leak-specific conductance γ_L is not ion selective and is practically time independent. The corresponding leak current, I_L, is given by $\gamma_L(\Delta\phi - \Delta\phi_L)$, where $\Delta\phi_L$ is a transmembrane potential, usually close to zero, at which the leak current vanishes. The change of $\Delta\phi$ over time implies a change in the charge Q located on one side of the membrane and in the equal but opposite charge, $-Q$, on the other side, similar to what happens when one changes the voltage across the plates of an electric capacitor. Thus, in addition to the current I_K due to the potassium channel, the current I_{Na} due to the sodium channel and the leak current I_L, we must also consider the purely capacitive current I_c, due to the time dependence of the charge Q. Denoting by C the differential capacitance of the membrane, I_c is given by

$$C \equiv \frac{dQ}{d\Delta\phi}; \quad I_c \equiv \frac{dQ}{dt} = \frac{dQ}{d\Delta\phi}\frac{d\Delta\phi}{dt} = C\frac{d\Delta\phi}{dt}. \tag{4.11}$$

The total current I is given by

$$\begin{aligned} I &= I_c + I_K + I_{Na} + I_L \\ &= C\frac{d\Delta\phi}{dt} + \gamma_K p_K(\Delta\phi - \Delta\phi_K) + \gamma_{Na} p_{Na}(\Delta\phi - \Delta\phi_{Na}) \\ &\quad + \gamma_L(\Delta\phi - \Delta\phi_L) = 0. \end{aligned} \tag{4.12}$$

In this expression, the driving force of the current induced by a given ion channel is just the difference between the transmembrane potential $\Delta\phi$ and the corresponding Nernst potential (Aidley, 1998; De Felice, 1997; Hille, 2001). The probability p_L of the leak channel being open is set equal to unity, since it is reasonable to assume that this fictitious channel is always open. Moreover, the total current I is set equal to zero, by assuming that a stationary condition is established almost instantaneously, in which the different components of the current compensate each other. The solution of the differential equation 4.12 is given by

$$\begin{aligned} \Delta\phi &= \frac{\gamma_L \Delta\phi_L + \gamma_K p_K \Delta\phi_K + \gamma_{Na} p_{Na} \Delta\phi_{Na}}{\gamma_L + \gamma_K p_K + \gamma_{Na} p_{Na}} \\ &\quad \times \left[1 - \exp\left(-\frac{\gamma_L + \gamma_K p_K + \gamma_{Na} p_{Na}}{C} t\right)\right] + \Delta\phi(t = 0). \end{aligned} \tag{4.13}$$

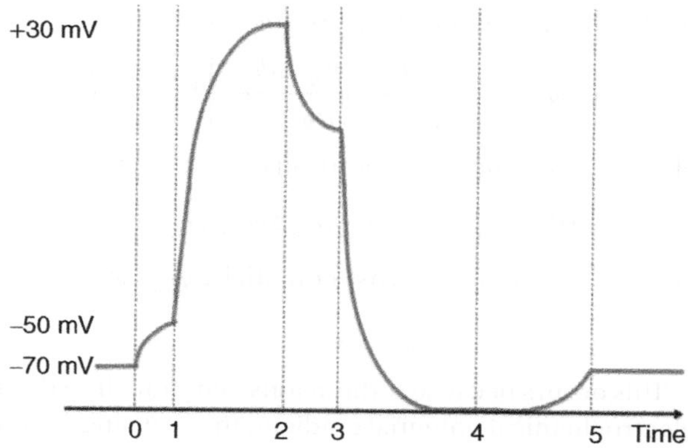

Figure 4.5 Plot of the transmembrane potential against time during a hypothetical action potential determined by a single sodium channel and a single potassium channel.

To schematize the sequence of events that determines the action potential, we assume that the probabilities p_K and p_{Na} vary discontinuously at times t_0, t_1, t_2, t_3, and t_4 in Fig. 4.5. Imagine monitoring the transmembrane potential at a well-defined section of an axon membrane. Before an external stimulus capable of generating the action potential reaches this position, the transmembrane potential is at its resting value. Both probabilities p_K and p_{Na} are very close to zero, but p_K is greater than p_{Na}. In other words, sodium and potassium channels are both almost closed, but sodium channel is "more closed" than potassium channel. Under these conditions, $\Delta\phi$ is practically time independent; setting $d\Delta\phi/dt$ and $\gamma_{Na}=0$ in Eq. 4.12, the resting potential is given by

$$\Delta\phi(t = t_0) = \frac{\gamma_K p_K \Delta\phi_K + \gamma_L \Delta\phi_L}{\gamma_K p_K + \gamma_L}. \tag{4.14}$$

The contribution to $\Delta\phi$ from the potassium channel is greater than that from the leak channel. Hence, the resting potential is relatively close to $\Delta\phi_K$, assuming a value of about −70 mV.

The arrival of the stimulus at time $t=t_0$ (Fig.4.5) initially induces a more marked closure of the potassium channel. The change, $\Delta\Delta\phi$, of the transmembrane potential $\Delta\phi$ over time is then given by Eq. 4.13, in which p_K and p_{Na} are now both equal to zero:

$$\Delta\Delta\phi = \Delta\phi_L[1 - \exp(-\gamma_L(t - t_0)/C)]. \tag{4.15}$$

Consequently, $\Delta\Delta\phi$ tends asymptotically to the $\Delta\phi_L$ value with the time constant:

$$\tau = C/\gamma_L. \tag{4.16}$$

Before the transmembrane potential may reach the almost zero value of $\Delta\phi_L$, at time $t=t_1$ it assumes a value of about −50 mV, at which the voltage-gated sodium channel opens, bringing its probability p_{Na} close to unity. The difference between this potential and the resting potential (in the present case, +20 mV) is termed *threshold* potential. Under these conditions, setting $p_K=0$ in Eq. 4.13, we obtain the further potential change of $\Delta\phi$ over time:

$$\Delta\Delta\phi = \frac{\gamma_L \Delta\phi_L + \gamma_{Na} p_{Na} \Delta\phi_{Na}}{\gamma_L + \gamma_{Na} p_{Na}} \left[1 - \exp\left(-\frac{\gamma_L + \gamma_{Na} p_{Na}}{C}(t - t_1)\right)\right]. \quad (4.17)$$

Hence, $\Delta\Delta\phi$ tends asymptotically to the value

$$\Delta\Delta\phi = (\gamma_L \Delta\phi_L + \gamma_{Na} p_{Na} \Delta\phi_{Na})/(\gamma_L + \gamma_{Na} p_{Na}) \quad (4.18)$$

close to the sodium Nernst potential $\Delta\phi_{Na}$, with the time constant

$$\tau = C/(\gamma_L + \gamma_{Na} p_{Na}). \quad (4.19)$$

This occurs because sodium ions penetrate into the axon, by flowing down their electrochemical potential gradient, thus causing the intracellular side of the membrane to become temporarily more positive than the extracellular one, up to the attainment of a $\Delta\phi$ value close to +30 mV. At time t_2, the positive shift of $\Delta\phi$, termed *depolarization* (in the present case, +100 mV), induces an increase in the probability p_K. In other words, the potassium channel opens, causing the transmembrane potential $\Delta\phi$ to become controlled by all three channels and to be expressed by the whole Eq. 4.13. Since the effective conductance, $p_K\gamma_K$, of the potassium channel prevails over that, $p_{Na}\gamma_{Na}$, of the sodium channel and $\Delta\phi_K$ is negative, in practice $\Delta\phi$ shifts toward more negative values with the time constant

$$\tau = C/(\gamma_L + \gamma_K p_K + \gamma_{Na} p_{Na}). \quad (4.20)$$

At time $t = t_3$, the sodium channel becomes *inactive* and the transmembrane potential undergoes a further shift toward more negative values. This phase of the action potential, characterized by the attainment of a maximum of $p_K\gamma_K$ followed by its gradual decrease (Fig. 4.6), is called membrane *repolarization*. When, during its decrease, the transmembrane potential attains its pristine resting value, the potassium channel has not yet recovered the low probability p_K typical of the latter potential. In fact, p_K is slightly higher, causing $\Delta\phi$ to become temporarily more negative than the resting potential. This phenomenon is called membrane *hyperpolarization*. The period, of about 2–3 ms, during which $\Delta\phi$ is more negative than its resting value, is termed *after hyperpolarization* period. It is also called *refractory period*, because it is the time interval during which the sodium channel is in the inactive state, such that a second action potential cannot be initiated, no matter how large a stimulus may be. Finally, at time $t = t_4$, p_K decreases further, allowing $\Delta\phi$ to return to its resting value, at time $t = t_5$.

In this connection, it must be recalled that the sodium channel has two gates, the activation gate (*a* gate) and the inactivation gate (*i* gate), both located on the intracellular side of the membrane (cf. Fig. 4.3). At the resting potential, the *a* gate is closed while the *i* gate is open, and the sodium channel is *closed*. During depolarization, both gates are open and the channel is *open*. After depolarization, the *i* gate closes in about 1 ms, while the *a* gate is still open and the channel is *inactive*. On *afterhyperpolarization*, the sodium channel passes from the inactive to the closed state, through an intermediate state in which both gates are closed, and is ready to open at the next action potential. This allows the membrane

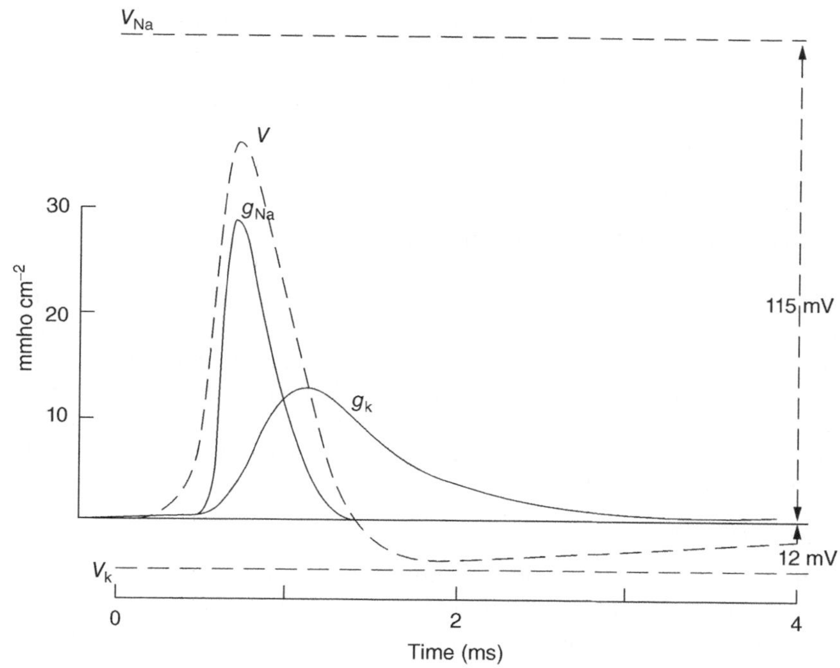

Figure 4.6 Plot of the transmembrane potential (V) against time, as a neuron fires an action potential (right-hand scale). The figure also shows the time dependence of the specific conductances, g_{Na} and g_K, of Na^+ and K^+ channels during the action potential (left-hand scale). Source: Reprinted from Hodgkin (1958), The Croonian Lecture: Ionic movements and electrical activity in giant nerve fibers, by permission of the Royal Society.

depolarization to move along the axon exclusively downhill with respect to the stimulus that has generated the action potential. In fact, the sodium ions that have penetrated the axon at the monitored section of the axon membrane during depolarization diffuse along the neuron cell in both directions, namely both downhill and uphill. However, their uphill diffusion has no consequence because the uphill sodium channels cannot open, since they are still in the inactive state. Conversely, the downhill sodium channels are in the closed state and are, therefore, ready to open following the downhill depolarization movement. This allows the action potential to move downhill, causing a $\Delta\phi$ versus t curve identical with that in Fig. 4.6 to be monitorable at a downhill section of the axon membrane.

If we were able to monitor $\Delta\phi$ along the axon membrane at a given time t, we would obtain a curve of $\Delta\phi$ against distance entirely analogous to the $\Delta\phi$ versus t curve in Fig. 4.6. This is a typical feature of waves. Not only sodium channels, but also potassium channels pass from the open to a voltage-insensitive inactive state, during repolarization. Some of them undergo a fast inactivation, which occurs from 1 – 3 ms after activation. This control mechanism allows neurons to regulate their firing frequency. A typical example of fast inactivation is provided by the *Shaker* potassium channel.

The continuous flow of sodium ions into the cell and of potassium ions out of the cell, as a consequence of the action potential, would ultimately annihilate the electrochemical potential gradients of both ions, if the Na^+,K^+-ATPase in the cell membrane did not maintain these gradients constant by pumping these ions in opposite directions, at the expense of ATP hydrolysis.

4.4 Single-channel Open Probability

The probability p of a single channel being open at a given instant can only assume any of two values, that is, the value 1 if the channel is open and the value 0 if it is closed. On the other hand, if we record a single-channel current for a time period τ long enough to allow it to open and close randomly a very high number of times, the total charge q flown during τ divided by τ is a measure of the *mean current* $\langle i \rangle$. This, in turn, is the product of the current i that would flow during τ, if the channel were always open, by its open probability p:

$$\langle i \rangle = q/\tau = pi. \tag{4.21}$$

Since, in practice, the time period τ cannot be made infinitely long, the p value obtained from Eq. 4.21 fluctuates around an average value with varying τ, the more the shorter τ is. The mean probability p may assume any value between 0 and 1.

Analogous fluctuations are exhibited by the total current generated by a high number of channels of the same type spanning a membrane, such as that recorded across a plasma membrane by the patch-clamp technique in the whole-cell configuration (cf. Section 6.3). If, for example, 90 channels out of a total number of 100 channels are open at a given instant, then the probability p equals 0.9. Since, however, the channels open and close randomly, the percentage of open channels at a given instant of the recording is not exactly 90%, but rather fluctuates around this value, which is the *mean value of the probability p*. If the total number N of channels is 10 instead of 100, the mechanism that underlies the fluctuations determined by the random opening and closing of the channels remains the same. However, it is intuitive that the difference between the maximum and the minimum value of the recorded current for a total of 10 channels is statistically less than that for a total of 100 channels. In other words, the *absolute fluctuation*, that is, the *noise*, decreases with decreasing N. However, if we divide the difference between the maximum and the minimum value of the current by its average value, it is equally intuitive that the resulting *average fluctuation* decreases with increasing N.

If we have only two channels and the probability p of each of them being open is 0.9, the probability of both being open is clearly given by the product of the two single probabilities, namely (0.9)(0.9) = 0.81 or 81%. The probability of one of the two channels, chosen at will, being open and the other closed is (0.9)(0.1) = 0.09 or 9%; in fact, the probability of a single channel being closed is the complement to unity of its probability of being open, since a single channel can only be either open or closed. However, we must also consider that the two channels are distinguishable. If we denote each of a set of channels by a progressive integer number, starting from 1, and its open or closed state by the subscript O or C, respectively, then in the case of two channels we have two different possibilities, expressed by 1_O2_C and 1_C2_O. It follows that the probability of one of the two channels being open and the other closed, irrespective of which is closed and which is open, is given by $0.09 \times 2 = 0.18$ or 18%. Finally, the probability of both channels being closed is $(0.1)(0.1) = 0.01$ or 1%. Only upon considering the two possibilities, 1_O2_C and 1_C2_O, does the sum of the probabilities relative to all possible ways in which we can find the two channels amount to unity, as expected for the total probability.

Let us now consider the general case of N channels per unit area of the membrane, and let us estimate the probability of N_0 selected channels, out of the N channels, being open, with the remaining $(N - N_0)$ closed (De Felice, 1997). Let us again denote by p the probability of each single channel being open. The overall probability of all the N_0 selected channels being open is the product of the single probabilities p, and is therefore given by p^{N_0}. Analogously, the probability of the remaining $(N - N_0)$ channels being closed is given by the product of the probabilities of each channel being closed, that is, $(1 - p)^{N-N_0}$. The probability of the N_0 selected channels being open and the remaining $(N - N_0)$ channels being closed is, then, given by the product of the two groups of probabilities:

$$p^{N_0}(1 - p)^{N-N_0}. \tag{4.22}$$

This, however, is not the total probability of N_0 channels being open, irrespective of their positions, and of the remaining $(N - N_0)$ channels being closed, again irrespective of their positions. In fact, there are many different ways of gathering N_0 channels into one group, by picking them up from a total number N of channels, considered as structurally identical but distinguishable in view of their distinct positions in the membrane. These different ways are called *configurations*.

To calculate the number of configurations, let us imagine ordering the N channels into a linear array of N boxes in all possible ways. To fill the first of the N boxes, we have N choices, as many as the N channels. To fill the second box, we are left with $(N - 1)$ choices, because the channel already placed in the first box is no longer available. In the same way, we have $(N - 2)$ choices of filling the third box, and so on, until all boxes are filled. The total number of choices, namely the number of *permutations*, is therefore given by

$$N(N - 1)(N - 2) \cdots 1 \equiv N!. \tag{4.23}$$

This product, which also expresses the total number of differently filled linear arrays, is called *N factorial* and is denoted by an exclamation mark.

Let us now imagine picking up the channels that occupy the first N_0 boxes of the $N!$ filled linear arrays and gathering them into a first group, considered to be in the open state. The channels that occupy the last $(N - N_0)$ boxes of the $N!$ filled linear arrays form the second group, considered to be in the closed state. At this point, one might be tempted to consider $N!$ as the number of configurations.

As a matter of fact, this conclusion is erroneous, because $N!$ overestimates the number of configurations. Thus, if we swap the positions of any two channels (e.g., channels j and k) that occupy the first N_0 boxes and participate in the first group, this swapping does not change their configuration. In fact, even though the two channels are distinguishable, the simple fact of ascribing first channel j and then channel k to the first group, or first channel k and then channel j, does not change the configuration: it is only the order in which we ascribe the two open channels to the first group that changes. To eliminate the unwanted distinguishability of the order in which the open channels occupying the first N_0 boxes of the $N!$ filled linear arrays are ascribed to the first group, we must divide $N!$ by the number of permutations of the first N_0 boxes, namely $N_0!$. By analogous arguments, swapping the position of any two closed channels occupying the last $(N-N_0)$ boxes of the $N!$ filled arrays does not cause a change in configuration. To eliminate the distinguishability of the order in which the closed channels occupying the last $(N-N_0)$ boxes of the $N!$ filled arrays are ascribed to the second group, we must further divide $N!$ by the number of permutations of these boxes, namely $(N-N_0)!$. The total number of ways of picking up N_0 open channels out of a total number N, that is, the number of configurations, is, therefore, given by

$$\frac{N!}{N_0!(N-N_0)!}. \tag{4.24}$$

The probability of N_0 channels being open and $(N-N_0)$ channels closed, out of N channels, irrespective of the choice of the N_0 open channels, is given by the product of the probability of N_0 *preselected* channels being open and the remaining channels closed, as expressed by Eq. 4.22, by the total number of possible ways of carrying out this selection, as expressed by Eq. 4.24:

$$\psi(N,N_0,p) \equiv \frac{N!}{N_0!(N-N_0)!}p^{N_0}(1-p)^{N-N_0}. \tag{4.25}$$

This expression is the probability of finding a *well-defined number* N_0 of open channels, out of a total number N of channels.

If we carry out the summation of ψ over all possible N_0 values ranging from 0 to N, we obtain the sum of the probabilities of finding our system of N channels in all its possible configurations:

$$\sum_{N_0=0}^{N} \frac{N!}{N_0!(N-N_0)!}p^{N_0}(1-p)^{N-N_0}. \tag{4.26}$$

This sum of all possible probabilities is expected to be equal to unity, as already shown for the simple case of two channels. To verify this, it is sufficient to use the general expression of the expansion of the Nth power of a binomial $(x+y)$, according to the binomial theorem of elementary algebra:

$$(x+y)^N = \sum_{N_0=0}^{N} \frac{N!}{N_0!(N-N_0)!}x^{N_0}y^{N-N_0}. \tag{4.27}$$

Two well-known particular cases of this general expression are the second and third power of a binomial:

$$(x+y)^2 = x^2 + 2xy + y^2; \quad (x+y)^3 = x^3 + 3x^2y + 3y^2x + y^3. \tag{4.28}$$

It is evident that Eq. 4.26 is equal to the power expansion of Eq. 4.27 for the particular case of $x = p$ and $y = 1 - p$. Carrying out this substitution, we obtain

$$[p + (1-p)]^N = 1^N = 1, \tag{4.29}$$

as expected.

If a reasonable guess can be made about the number density N of the channels in a membrane, Eq. 4.25 can be used to estimate the probability p from the fluctuations of the current I recorded as a function of time for a sufficiently long period τ. Each different current level I' attained by the fluctuating current is equal to $N_0'i$, where N_0' is the number of open channels generating that level, and i is the current flowing across a single open channel. The ratio, ρ, of the time covered by the given current level I' to the whole recording period τ is proportional to the probability of finding the number N_0' of open channels out of the total number N of channels, as given by the ψ expression in Eq. 4.25. By keeping N constant, ψ can be plotted against N_0 for different p values. The resulting curves show a maximum for an N_0 value, N_0^*, corresponding to the mean current $\langle I \rangle = N\langle i \rangle = Npi = N_0^* I$, where $\langle i \rangle$ is the mean value of the single-channel current, as expressed by Eq. 4.21. The maximum lies at $N_0 = N/2$ for $p = 0.5$ and shifts toward progressively higher N_0 values, the higher p is (Fig. 4.7). The probability p can be estimated from the ψ versus N_0 plot, calculated at constant p and N, which parallels more closely the experimental ρ versus I' plot, provided the channel number density N is known with sufficient accuracy.

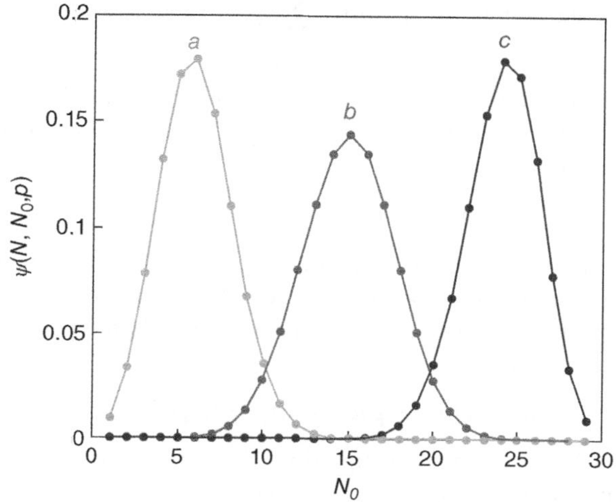

Figure 4.7 Plot of $\psi(N, N_0, p)$ against N_0 for $N = 30$ and $p = 0.2$ (a), 0.5 (b), and 0.8 (c).

4.4.1 The Variance

A procedure for estimating the probability p from current fluctuations, without the necessity of knowing N *a priori*, is based on the use of the *variance*. If we imagine repeating the measurement of a given quantity n times, we obtain n different values, $x_1, x_2, x_3, \ldots, x_n$, which fluctuate around their mean value $\overline{x}$:

$$\overline{x} = \frac{1}{n}\sum_{k=1}^{n} x_k. \tag{4.30}$$

The variance σ_x^2 of this series of measurements is defined as the average of the square deviations from the mean value $\overline{x}$:

$$\sigma_x^2 \equiv \frac{1}{n}\sum_{k=1}^{n} (x_k - \overline{x})^2. \tag{4.31}$$

By expanding the square under the summation sign and taking Eq. 4.30 into account, the variance can be expressed in the following form:

$$\begin{aligned}\sigma_x^2 &\equiv \frac{1}{n}\sum_{k=1}^{n} (x_k - \overline{x})^2 = \frac{1}{n}\sum_{k=1}^{n} x_k^2 - \frac{2\overline{x}}{n}\sum_{k=1}^{n} x_k + \overline{x}^2 \\ &= \frac{1}{n}\sum_{k=1}^{n} x_k^2 - 2\overline{x}^2 + \overline{x}^2 = \frac{1}{n}\sum_{k=1}^{n} x_k^2 - \overline{x}^2.\end{aligned} \tag{4.32}$$

Let us now consider a single ion channel that allows the flow of a current i, when it is open. If we apply the concept of variance to the measurement of a single-channel current, the following analogies hold:

$$\overline{x} \rightarrow \langle i \rangle = pi; \quad \frac{1}{n}\sum_{k=1}^{n} x_k^2 \rightarrow pi^2. \tag{4.33}$$

In particular, the quantity $\sum_k x_k^2/n$, which expresses the mean of the square of the single measurements, is given by the product of the square of the current i, which flows when the channel is open, by its probability p of being open. Substituting from Eq. 4.33 into Eq. 4.32 yields the current variance due to its random opening and closing:

$$\sigma_i^2 = pi^2 - p^2 i^2 = p(1-p)i^2. \tag{4.34}$$

If we now consider the mean current $\langle I \rangle$, resulting from a set of N channels and given by the equation $\langle I \rangle = Npi$, the corresponding variance, σ_I^2, is obtained by multiplying the variance for the single-channel current by the total number N of channels:

$$\sigma_I^2 = Np(1-p)i^2 = Npi^2 - N^2p^2i^2/N = i\langle I \rangle - \langle I \rangle^2/N. \tag{4.35}$$

According to this equation, the variance is proportional to the product of the probability, p, of the single channels being open by the probability, $(1-p)$, of the single channels being closed. It is evident that the $p(1-p)$ product is a maximum when p equals 0.5. This is an intuitive result, in that the current is noisy to the maximum extent possible when half the channels are open, while the other half are closed. Conversely, the variance vanishes when the channels are either all open or all closed.

The last member of Eq. 4.35 allows the values of both the total number N of channels and their open probability p to be extracted from an experimental plot of the variance σ_I^2 against $\langle I \rangle$. Such a plot can be obtained, for example, by measuring the current following the perturbation of the channels embedded in a membrane as a function of time, and by repeating this measurement a very high number of times, through the reiterated application of the same perturbation. In the case of a voltage-gated channel, such as the sodium channel, the perturbation can be induced by a potential jump causing the membrane depolarization. The sodium channel responds to this perturbation by opening very rapidly and by then closing within a few milliseconds. Figure 4.8(A) shows a series of successive records of the sodium current, following potential steps from the resting potential to -5 mV;

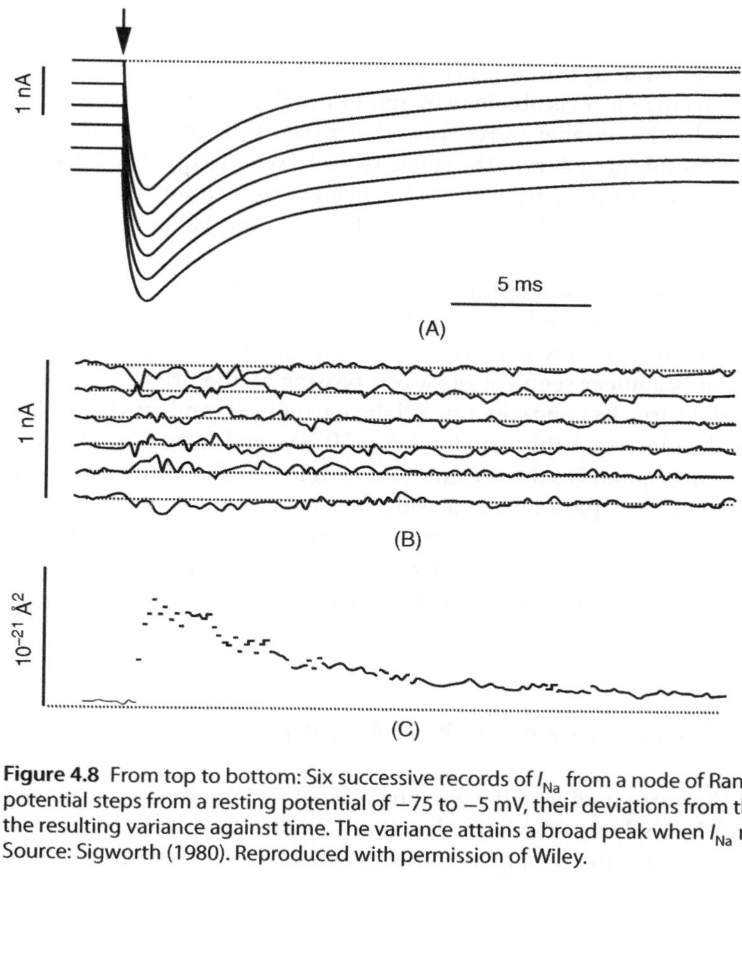

Figure 4.8 From top to bottom: Six successive records of I_{Na} from a node of Ranvier during potential steps from a resting potential of -75 to -5 mV, their deviations from the mean, and the resulting variance against time. The variance attains a broad peak when I_{Na} reaches a peak. Source: Sigworth (1980). Reproduced with permission of Wiley.

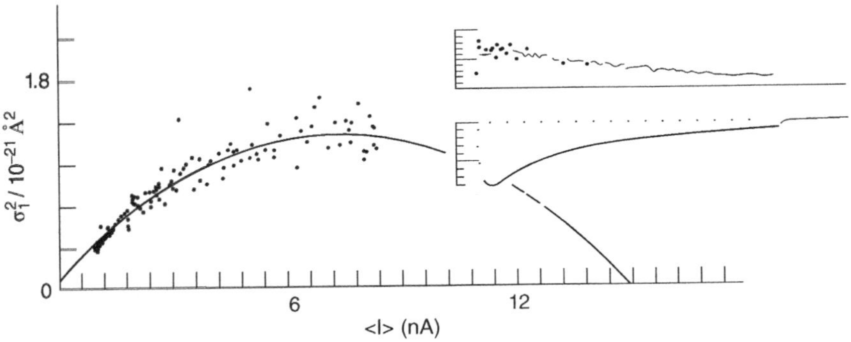

Figure 4.9 Plot of the variance, σ_I^2, against the Na$^+$ current, $\langle I \rangle$, from the same voltage clamp records as in Fig. 4.8. The solid curve is the best fit of the variance experimental data by Eq. 4.35. If all Na channels in this node opened simultaneously, they would determine a current $\langle I \rangle$ of 15 nA. Insets show the time course of variance (upper trace) and mean current (lower trace), with scale factors of 2×10^{-22} Å^2 and 1 nA per small division, respectively. Source: Sigworth (1980). Reproduced with permission of Wiley.

these curves are shifted along the current axis to avoid their almost complete overlapping. The mean current $\langle I \rangle$ is obtained by summing all the current records and dividing the sum by the number of records. The curve of the mean current $\langle I \rangle$ against time is then subtracted from the single records, so as to obtain a series of curves of the deviation of the current from its mean value as a function of time (Fig. 4.8(B)). The variance σ_I^2 is then determined for each time value, as shown in Fig. 4.8(C).

Finally, the variance is plotted against $\langle I \rangle$, as shown in Fig. 4.9. According to Eq. 4.35, this plot is expected to have a parabolic shape. For $\langle I \rangle = 0$, the variance equals zero. For small $\langle I \rangle$ values, the term $\langle I \rangle^2/N$ is negligible with respect to $\langle I \rangle$, such that the plot is a linear segment of slope i passing by the origin. At higher $\langle I \rangle$ values, the plot turns its concavity toward the current axis and, in view of Eq. 4.35, is expected to intersect this axis for $i\langle I \rangle = \langle I \rangle^2/N$, and hence for $\langle I \rangle = Ni$. The latter value corresponds to the extreme situation in which all N channels are simultaneously open. In experimental measurements of current fluctuations, this situation is never attained. However, upon fitting the experimental curve by Eq. 4.35, we can obtain the N value that provides the best fit. Using the expression $\langle I \rangle = Npi$, from the i and N values so derived, the average probability p is finally obtained.

4.5 The Goldman–Hodgkin–Katz Equation

In Section 4.2, the current i due to the opening of single channels was introduced for the first time by applying Ohm's first law (Eq. 4.2). This approach was used to derive an expression for the resting potential, by making use of the concepts of single-channel conductance γ and of the probability p of the ion channel being in the open state. Before the groundbreaking discovery of the patch-clamp technique, the potential dependence of the current I flowing across biomembranes and their experimental models was interpreted on the basis of the concepts of diffusion, driven by chemical potential gradients, and of migration, driven by electrostatic potential energy gradients. The most popular equation based on this approach, which disregards the fact that I is induced by the presence of ion channels or pores, was originally derived by Goldman (1943) and later rederived by Hodgkin and Katz (1949), who also introduced a number of simplifying assumptions. The resulting equation, which is still in use to determine ion permeability across membranes, is referred to as the *Goldman–Hodgkin–Katz (GHK) equation.* The expression of the resting potential stemming from this equation is commonly called *diffusion potential* or *reversal potential.*

We have seen that, according to irreversible thermodynamics, the number of moles of an hth species that pass from a phase α to a different phase β per unit time, dn_h^{β}/dt, that is, their flow, is proportional to the conjugate driving force, expressed by the difference, $-(\widetilde{\mu}_h^{\beta} - \widetilde{\mu}_h^{\alpha})$, between the electrochemical potentials of the given species in the two phases. By analogous arguments, the driving force acting on a particle of the hth species that moves unidirectionally in a homogeneous medium from a point x_1 to a point $x_1 + dx$ along the x-axis, can be considered as proportional to $-[\widetilde{\mu}_h(x_1 + dx) - \widetilde{\mu}_h(x_1)]/dx$, namely, to the opposite of the gradient, $d\widetilde{\mu}_h/dx$, of the electrochemical potential of the species. The negative sign is due to the fact that the particle moves from points where its electrochemical potential is higher to points where it is lower. This driving force is exactly counterbalanced by the viscous drag of the medium, such that the particle moves at a constant *drift velocity*, v_h, proportional to $-d\widetilde{\mu}_h/dx$:

$$v_h = -u_h d\widetilde{\mu}_h/dx. \tag{4.36}$$

The proportionality constant u_h is termed *mobility.*

The *flux* J_h of the hth species is defined as the number of moles of the given species that cross the unit area of an ideal plane normal to the direction x of motion per unit time. Since v_h is the distance covered by the species per unit time, the number of moles that cross the unit area per unit time are those contained in an ideal parallelepiped of unit base and height v_h. Denoting by c_h the molar concentration of the hth species, the flux J_h is, therefore, given by

$$J_h = (v_h \times \text{unit area})c_h = v_h c_h = -u_h c_h (d\widetilde{\mu}_h/dx). \tag{4.37}$$

We saw in Section 3.2 that the electrochemical potential of an hth species is expressed by the following equation:

$$\widetilde{\mu}_h = \mu_h^{\o} + RT\ln c_h + z_h F\phi, \tag{4.38}$$

where $\mu_h^{\o}$ accounts for short-range interactions of the given species with the solvent. In a homogeneous medium, $\mu_h^{\o}$ does not vary with distance. Hence, substituting $\widetilde{\mu}_h$ from Eq. 4.38 into Eq. 4.37 yields

$$J_h = -u_h c_h \frac{\mathrm{d}\widetilde{\mu}_h}{\mathrm{d}x} = -u_h c_h \left(RT \frac{\mathrm{d}\ln c_h}{\mathrm{d}x} + z_h F \frac{\mathrm{d}\phi}{\mathrm{d}x} \right) = -u_h RT \frac{\mathrm{d}c_h}{\mathrm{d}x} - u_h c_h z_h F \frac{\mathrm{d}\phi}{\mathrm{d}x}. \tag{4.39}$$

This equation, referred to as the *Nernst–Planck equation*, shows that the flux of a charged species in a homogeneous medium is composed of a contribution proportional to the concentration gradient (*diffusional flux*) and of one proportional to the electric potential gradient (*migrational flux*).

For a neutral species (i.e., for $z_h = 0$), the flux is exclusively diffusional. The proportional relationship between the flux of a neutral species and its concentration gradient is called *Fick's first law of diffusion*, after the scientist who proposed it in 1885 (Fick, 1855):

$$J_h = -D_h(\mathrm{d}c_h/\mathrm{d}x). \tag{4.40}$$

Historically, the proportionality constant is called *diffusion coefficient* and denoted by D_h. It is interesting to note that the proportional relationship between the flux of a charged species and the electrostatic potential energy gradient, under conditions in which the concentration gradient is vanishingly small, was derived introducing the mobility u as the proportionality constant. At that time, the concept of electrochemical potential had not been developed yet, just as the resulting relationship between the mobility u of a charged species and its diffusion coefficient D:

$$D_h = u_h RT. \tag{4.41}$$

This relation was disclosed by Albert Einstein in 1905 in his paper on Brownian motion and is referred to as the *Einstein relation* (Einstein, 1905).

The GHK relation starts from the Nernst–Planck Eq. 4.39. This is applied to a species permeating a membrane under the assumption that (i) the membrane is considered as a homogeneous medium; (ii) each permeating species moves within the membrane without interacting with other permeating species; and (iii) the electric potential gradient $\mathrm{d}\phi/\mathrm{d}x$ (and, hence, the electric field $E_x = -(\mathrm{d}\phi/\mathrm{d}x)$) within the membrane is constant. Incidentally, a more general solution of the Nernst–Planck relation, which holds for arbitrary mixtures of electrolytes without assumptions about the electric potential and concentration profiles, was derived by Schlögl (1954); however, the final expressions are quite difficult to apply in practice. If we express u_h as a function of D_h in Eq. 4.39 and we then multiply both members of the resulting equation by $\exp(z_h F\overline{\phi}/RT)$, after rearranging the terms, we obtain

$$\begin{aligned} J_h \mathrm{e}^{z_h f \overline{\phi}(x)} &= -D_h \left(\frac{\mathrm{d}\overline{c}_h}{\mathrm{d}x} + \overline{c}_h z_h f \frac{\mathrm{d}\overline{\phi}}{\mathrm{d}x} \right) \mathrm{e}^{z_h f \overline{\phi}(x)} \\ &= -D_h \frac{\mathrm{d}}{\mathrm{d}x} (\overline{c}_h \ \mathrm{e}^{z_h f \overline{\phi}(x)}) \quad \text{with } f \equiv \frac{F}{RT}. \end{aligned} \tag{4.42}$$

In this equation, concentration and electric potential within the membrane are overlined. In the following, they are to be related to the corresponding quantities immediately outside the membrane, which are denoted by non-overlined symbols.

Let us now integrate Eq. 4.42 across the whole membrane, upon denoting its extracellular surface by $x=0$ and its intracellular surface by $x=d$, where d is the membrane thickness. Under steady-state conditions, the permeating species cannot accumulate at any point within the membrane. Consequently, the flux J_h, albeit varying over time, is spatially uniform, that is, it does not vary over distance across the membrane. Hence, when integrating Eq. 4.42 over x, it will be brought outside the integral sign:

$$\begin{aligned} J_h \int_{x=0}^{x=d} e^{z_h f \overline{\phi}}\, dx &= -D_h \left[\overline{c}_h(d) e^{z_h f \overline{\phi}(d)} - \overline{c}_h(0) e^{z_h f \overline{\phi}(0)}\right] \\ &= -D_h \left[\overline{c}_h(d) e^{z_h f \Delta\overline{\phi}} - \overline{c}_h(0)\right] \quad \text{with } \Delta\overline{\phi} \equiv \overline{\phi}(d) - \overline{\phi}(0). \end{aligned} \tag{4.43}$$

In writing the third member of this equation, the potential $\overline{\phi}(0)$ was set equal to zero, thus referring the electric potential within the membrane to that at its external surface, according to usage in electrophysiology. Upon assuming that the electric potential varies linearly within the membrane, its value for $0<x<d$ is given by $\overline{\phi}(x) - \overline{\phi}(0) = \overline{\phi}(x) = \Delta\overline{\phi}\, x/d$. By replacing this expression in Eq. 4.43 and changing the variable of integration, we obtain

$$\begin{aligned} J_h \int_{x=0}^{x=d} e^{z_h f \Delta\overline{\phi}\, x/d}\, dx &= J_h \frac{d}{z_h f \Delta\overline{\phi}} \int_{x=0}^{x=d} e^{z_h f \Delta\overline{\phi}\, x/d}\, d\left(\frac{z_h f \Delta\overline{\phi}}{d} x\right) \\ &= J_h \frac{d}{z_h f \Delta\overline{\phi}} \left[e^{z_h f \Delta\overline{\phi}} - 1\right] \\ &= -D_h \left[\overline{c}_h(d) e^{z_h f \Delta\overline{\phi}} - \overline{c}_h(0)\right]. \end{aligned} \tag{4.44}$$

It follows that

$$J_h = -\frac{D_h z_h f \Delta\overline{\phi}}{d} \frac{\overline{c}_h(d) e^{z_h f \Delta\overline{\phi}} - \overline{c}_h(0)}{e^{z_h f \Delta\overline{\phi}} - 1}. \tag{4.45}$$

To relate the concentration $c_h(0)$ adjacent to the external surface, $x=0$, of the membrane on the aqueous side, with that, $\overline{c}_h(0)$, on the membrane side, we assume that the passage of the hth species across $x=0$ is so fast with respect to its passage across the whole membrane length as to consider it as in quasi-equilibrium. We may then apply the equilibrium condition, by equating the electrochemical potentials of the species immediately outside and inside the external boundary of the membrane:

$$\tilde{\mu}_h(0) = \mu_h^{\circ} + RT \ln c_h(0) + z_h F \phi(0) = \overline{\tilde{\mu}}_h(0) = \overline{\mu}_h^{\circ} + RT \ln \overline{c}_h(0) + z_h F \overline{\phi}(0). \tag{4.46}$$

Rearranging terms, we have

$$RT\ln\frac{\bar{c}_h(0)}{c_h(0)} = \mu_h^{\circ} - \bar{\mu}_h^{\circ} + z_hF[\phi(0) - \bar{\phi}(0)]. \tag{4.47}$$

The standard chemical potential, μ_h°, of the hth species in solution accounts for short-range interactions with water, while that, $\bar{\mu}_h^{\circ}$, inside the membrane, in close proximity to the external boundary, accounts for interactions with the polar heads of the lipid bilayer. The $\bar{c}_h(0)/c_h(0)$ ratio is called *partition coefficient* and is denoted by the symbol β_h. It is assumed that β_h is independent of the concentration of the species in solution, $c_h(0)$.

The passage of the species across the internal boundary of the membrane is also assumed to be in quasi-equilibrium, leading to a relation analogous to Eq. 4.47:

$$RT\ln\frac{\bar{c}_h(d)}{c_h(d)} = \mu_h^{\circ} - \bar{\mu}_h^{\circ} + z_hF[\phi(d) - \bar{\phi}(d)]. \tag{4.48}$$

In writing this equation, we have implicitly assumed that the difference, $\mu_h^{\circ} - \bar{\mu}_h^{\circ}$, between the standard chemical potentials across the internal boundary of the membrane, is the same as that across the external one. This assumption is reasonable, but not entirely obvious, since biological membranes are asymmetrical. Consequently, the polar heads exposed to the external solution are generally different from those exposed to the internal one.

The treatment is further simplified by assuming that the partition coefficient of the given species at the internal boundary of the membrane, $\bar{c}_h(d)/c_h(d)$, is identical with that at the external boundary, $\beta_h \equiv \bar{c}_h(0)/c_h(0)$:

$$\bar{c}_h(d)/c_h(d) = \bar{c}_h(0)/c_h(0) = \beta_h. \tag{4.49}$$

From Eqs. 4.47–4.49, it follows that this assumption also implies that the potential difference, $\phi(0) - \bar{\phi}(0)$, across the external boundary of the membrane is identical with that, $\phi(d) - \bar{\phi}(d)$, across the internal boundary. With this assumption, the potential difference, $\Delta\phi \equiv \phi(d) - \phi(0)$, between two points in the aqueous solution immediately adjacent to the two sides of the membrane is considered as identical with that, $\Delta\bar{\phi} \equiv \bar{\phi}(d) - \bar{\phi}(0)$, between two points within the membrane and adjacent to the two opposite sides of it. In view of Eq. 4.49, the flux J_h in Eq. 4.45 may be expressed as a function of the concentrations $c_h(0)$ and $c_h(d)$ and of the transmembrane potential $\phi(d) - \phi(0)$:

$$J_h = -P_h z_h f\Delta\phi\,\frac{c_h(d)e^{z_hf\Delta\phi} - c_h(0)}{e^{z_hf\Delta\phi} - 1} \quad \text{with } P_h \equiv \frac{D_h\beta_h}{d};$$
$$\Delta\phi \equiv \phi(d) - \phi(0) = \Delta\bar{\phi}. \tag{4.50}$$

P_h is termed *permeability*. Note that, in this expression, the potential differences across the diffuse layers adjacent to the two membrane sides are ignored.

The current density j_h, that is, the charge that flows across the unit area of the membrane per unit time, is given by the flux, that is, the number of moles of the permeating ion that flow across the unit area of the membrane per unit time, times the charge, $z_h F$, associated to 1 mol of the ion:

$$j_h = -z_h F J_h = P_h z_h^2 \frac{F^2}{RT} \Delta\phi \frac{c_h(d) e^{z_h f \Delta\phi} - c_h(0)}{e^{z_h f \Delta\phi} - 1}. \tag{4.51}$$

The change of sign in passing from J_h to j_h is due to the fact that the current, considered as a flow of positive charges, is taken as positive if it flows from inside to outside the membrane, by convention. Now, if a positive ion ($z_h > 0$) flows from inside to outside the membrane, its flux is negative, because it moves in the direction opposite to that of the x-axis, while the corresponding current is positive.

Equation (4.51) is referred to as the GHK *equation* for the current. This equation ignores the fact that ions permeate a membrane by moving along ion channels. Even the assumption of a constant electric field within the membrane is rough. The lumen of ion-selective channels presents sites where the ion is transiently bound and then released. Every favorable region for residence is represented by a valley, from which the ion must escape to advance further. Conversely, every hindrance, electrostatic or steric, to advancement within the channel can be viewed as a hill, which represents the minimum amount of energy that the ion must possess to overcome this obstacle. This sequence of potential-energy valleys and hills is also present in the absence of an electric field. If the potential energy barriers distributed along the channel are numerous and approximately of the same height, the assumption of a constant electric field within the membrane postulated in the GHK equation is rather satisfactory. In addition, the assumption of a concentration-independent diffusion coefficient D_h implies that the movement of the permeating ion, at any point within the membrane, is influenced exclusively by its own electrochemical potential gradient at that point, and is independent of the presence of other ions of the same or different species. In spite of this, Eq. 4.51 holds satisfactorily for membranes with a low density of fixed charges. The GHK equation allows a straightforward estimate of the permeability P_h of the permeating ion, since the quantities involved ($\Delta\phi$, $c_h(0)$ and $c_h(\mathrm{d})$) can be determined experimentally under controlled conditions.

When $c_h(d)$ equals $c_h(0) \equiv c_h$, Eq. 4.51 becomes

$$j_h = \frac{P_h z_h^2 F^2 c_h}{RT} \Delta\phi. \tag{4.52}$$

This equation satisfies Ohm's law, in that the current is proportional to the potential difference across the membrane throughout the whole $\Delta\phi$ range (Fig. 4.10, curve *a*). This is no longer true if $c_h(0)$ is different from $c_h(d)$ (Fig. 4.10, curves *b* to *e*). From Eq. 4.51 it is apparent that the current density j_h due to the permeating ion vanishes when the following condition is satisfied:

$$\frac{c_h(d)}{c_h(0)} = e^{-z_h f \Delta\phi} \rightarrow \Delta\phi = \frac{RT}{z_h F} \ln \frac{c_h(0)}{c_h(d)} \equiv \Delta\phi_h, \tag{4.53}$$

that is, when the transmembrane potential equals the Nernst potential, $\Delta\phi_h$, of the permeating hth ion. If the ion is positive and $\Delta\phi$ is sufficiently positive to satisfy

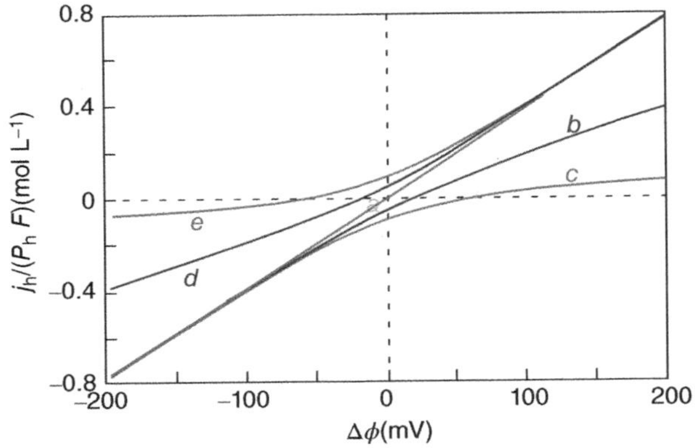

Figure 4.10 Curves of $j_h/(P_hF)$ against $\Delta\phi$ calculated from Eq. 4.51 for $c_h(0) = 0.1$ M and $c_h(d) = 0.1$ M (*a*), 0.05 M (*b*), 0.01 M (*c*), and for $c_h(d) = 0.1$ M and $c_h(0) = 0.05$ M (*d*), 0.01 M (*e*).

the condition $\exp(z_h f \Delta\phi) \gg c_h(0)/c_h(d)$ and $\gg 1$, the current density j_h is again proportional to $\Delta\phi$, in accordance with Ohm's law; moreover, it is proportional to the ion concentration in the internal aqueous solution:

$$j_h = \frac{P_h z_h^2 F^2}{RT} c_h(d)\Delta\phi \quad \text{for } z_h > 0 \text{ and for } e^{z_h f \Delta\phi} \gg c_h(0)/c_h(d) \text{ and } \gg 1. \tag{4.54}$$

Analogously, if the ion is positive and $\Delta\phi$ is sufficiently negative to satisfy the condition $\exp(z_h f \Delta\phi) \ll c_h(0)/c_h(d)$ and $\ll 1$, the current density j_h is proportional to $\Delta\phi$, in agreement with Ohm's law, and is also proportional to the concentration of the ion in the external aqueous solution:

$$j_h = \frac{P_h z_h^2 F^2}{RT} c_h(0)\Delta\phi \quad \text{for } z_h > 0 \text{ and for } e^{z_h f \Delta\phi} \ll c_h(0)/c_h(d) \text{ and } \ll 1. \tag{4.55}$$

From Fig. 4.10, in the proximity of the Nernst potential, the curve of j_h against $\Delta\phi$, expressed by Eq. 4.51 for $z_h = +1$, shows a curvature when $c_h(0)$ is different from $c_h(d)$, since the slopes of the rectilinear segments expressed by Eqs. 4.54 and 4.55, for $\Delta\phi$ values sufficiently far from $\Delta\phi_h$, are different, being proportional to $c_h(0)$ and $c_h(d)$, respectively. The proportionality constant between j_h and $\Delta\phi$, which has the dimension of a specific conductance, is greater for $\Delta\phi$ values that move the ions from the membrane side where their concentration is higher to that where they are lower. In other words, the membrane opposes greater resistance to the flow of an ion when it is directed from a lower to a higher concentration, than when it moves in the opposite direction. This asymmetric behavior is referred to as *rectification*.

If the membrane contains more than one ion-selective channel, the GHK equation 4.51 is generalized by assuming that each ion-selective channel induces the flow of its own ion, independent of the flows of other ions; the total current is, therefore, the sum of the currents induced by the different ion channels. Confining ourselves to considering only monovalent ions, for simplicity, in view of Eq. 4.51 the total current density j is given by

$$j = \frac{F^2}{RT}\sum_c P_+\Delta\phi\frac{c_+(d)e^{f\Delta\phi} - c_+(0)}{e^{f\Delta\phi} - 1} + \frac{F^2}{RT}\sum_a P_-\Delta\phi\frac{c_-(d)e^{-f\Delta\phi} - c_-(0)}{e^{-f\Delta\phi} - 1}. \tag{4.56}$$

In this equation, the subscripts + and − denote a generic cation and anion, respectively; the summation extended over all cations is denoted by the subscript c, that extended over all anions by the subscript a. This equation is commonly applied to the case in which the total current density j equals zero, in order to determine the corresponding transmembrane potential. Equating j to zero and multiplying numerator and denominator of the terms relative to anions by $\exp(f\Delta\phi)$, we have

$$\begin{aligned} j &= \frac{F^2}{RT}\Delta\phi\left[\sum_c P_+\frac{c_+(d)e^{f\Delta\phi} - c_+(0)}{e^{f\Delta\phi} - 1} + \sum_a P_-\frac{c_-(d) - c_-(0)e^{f\Delta\phi}}{1 - e^{f\Delta\phi}}\right] \\ &= \frac{F^2}{RT}\frac{\Delta\phi}{e^{f\Delta\phi} - 1} \\ &\quad\times\left\{\sum_c P_+\left[c_+(d)e^{f\Delta\phi} - c_+(0)\right] + \sum_a P_-\left[-c_-(d) + c_-(0)e^{f\Delta\phi}\right]\right\} = 0. \end{aligned} \tag{4.57}$$

Rearranging terms, we obtain

$$e^{f\Delta\phi}\left[\sum_c P_+ c_+(d) + \sum_a P_-c_-(0)\right] - \left[\sum_c P_+ c_+(0) + \sum_a P_-c_-(d)\right] = 0 \rightarrow$$

$$\Delta\phi \equiv E_{rev} = \frac{RT}{F}\ln\frac{\sum_c P_+ c_+(0) + \sum_a P_-c_-(d)}{\sum_c P_+ c_+(d) + \sum_a P_-c_-(0)}. \tag{4.58}$$

This equation is referred to as the GHK equation for the *reversal potential* E_{rev}, or *zero-current potential*. Equation 4.58 allows an estimate of *permeability ratios* (but not absolute permeabilities) from measurements of reversal potentials, provided the ionic concentrations in the two solutions that bath the membrane are known. Obviously, in the presence of a single ion-selective channel, this equation reduces to the Nernst potential for the corresponding permeant ion. The reversal or zero-current potential expressed by the GHK equation is conceptually identical with the *resting potential* expressed by Eq. 4.4, which is based on the single-channel conductance, γ_h, of the different permeating ions and on their probability, p_h, of being in the open state. Both equations refer to a steady-state condition, and not to an equilibrium condition. The latter condition would be attained only if the concentration of each permeant ion were the same on both sides of the membrane. In Section 4.6, we show how γ_h, determined from single-channel currents by the patch-clamp technique, is related to the permeability P_h obtained from macroscopic current measurements using Eq. 4.58.

In many biological systems, the predominant permeant monovalent cations and anions consist of Na^+, K^+, and Cl^-. In this case, Eq. 4.58 takes the following form:

$$E_{rev} = \frac{RT}{F}\ln\frac{P_{Na}c_{Na}(0) + P_K c_K(0) + P_{Cl}c_{Cl}(d)}{P_{Na}c_{Na}(d) + P_K c_K(d) + P_{Cl}c_{Cl}(0)}. \tag{4.59}$$

An example of application of this equation is provided by the analysis of the selectivity of the sodium channel in the plasma membrane of the axon (the *axolemma*), where sodium and potassium channels predominate. In this experiment, the internal solution is left unchanged, while the external one, which does not contain K^+ ions, is varied each time. When the external solution contains a known concentration, $c_{Na}(0)$, of sodium ions, the reversal potential assumes a value E'_{rev}. The external solution is then exchanged for that of a different monovalent cation, X, of equal concentration, and a new reversal potential E''_{rev} is measured. In view of Eq. 4.59, with $P_{Cl} = 0$ and $c_K(0) = 0$, the difference between the two measured reversal potentials is given by

$$\begin{aligned} E'_{rev} - E''_{rev} &= \frac{RT}{F}\ln\frac{P_{Na}c_{Na}(0)}{P_{Na}c_{Na}(d) + P_K c_K(d)} - \frac{RT}{F}\ln\frac{P_X c_X(0)}{P_{Na}c_{Na}(d) + P_K c_K(d)} \\ &= \frac{RT}{F}\ln\frac{P_{Na}c_{Na}(0)}{P_X c_X(0)} \rightarrow \frac{P_X}{P_{Na}} = \frac{c_{Na}(0)}{c_X(0)}\exp\left[\frac{F}{RT}(E''_{rev} - E'_{rev})\right]. \end{aligned} \tag{4.60}$$

Upon replacing Na^+ by Li^+ in the external solution, the reversal potential varies by only −1.6 mV. This yields a P_{Li}/P_{Na} ratio of 0.93, indicating that Li^+ permeability is only slightly less than the Na^+ one. Conversely, replacement of Na^+ by K^+ yields a change in reversal potential of −59 mV, yielding a P_K/P_{Na} ratio of 0.086. The permeability of the sodium channel to potassium ions is, therefore, about one-twelfth that to sodium ions. Its permeability to rubidium, cesium, magnesium, and calcium ions is even smaller, in the manner that it is not experimentally detectable.

4.6 Open Probability and Gating Charge of Ion Channels

Let us now derive an expression of the current flowing along a single ion-selective, voltage-gated channel by applying the solution of the Nernst–Planck equation of Eq. 4.51. To this end, it is first necessary to estimate the probability of the ion channel being open. Let us denote by N_o and N_c the number of ion channels selective toward a given ion that are open and closed, respectively. Let us extend the concept of electrochemical potential to ion channels. Under equilibrium conditions, the electrochemical potential, $\tilde{\mu}_o$, of open channels is equal to that, $\tilde{\mu}_c$, of closed channels, for any value of the transmembrane potential $\Delta\phi$. In this connection, it must be noted that, for an ion channel to be voltage gated, an appropriate change of the transmembrane potential in a given direction must determine its opening, while an opposite change must induce its closing. For such an event to occur, a small charged portion of the channel is expected to undergo a shift, under the action of a change in the transmembrane potential. This mobile charge may block or unblock the access to the channel directly, or also indirectly, by inducing a mild conformational change of the protein. Consequently, the electrochemical potentials of the channel in the open and closed state, $\tilde{\mu}_o$ and $\tilde{\mu}_c$, may differ not only by their electrical contribution, due to the different position of the mobile charge within the channel, but also by their chemical contribution, due to any conformational change induced by the charge movement. We can, therefore, write

$$\tilde{\mu}_o = \mu_o^\varnothing + kT\ln N_o + q'\phi_o = \tilde{\mu}_c = \mu_c^\varnothing + kT\ln N_c + q'\phi_c. \tag{4.61}$$

Here, q' is the charge that, in passing from the open to the closed state, undergoes a shift from a position where the electric potential, measured relative to the external solution, equals ϕ_o, to a position where it equals ϕ_c. The difference, $\mu_c^\varnothing - \mu_o^\varnothing$, between the standard chemical potentials of the channel in the closed and open state accounts for the purely chemical contribution due to a possible conformational change. Rearranging terms, we obtain

$$\frac{N_o}{N_c} = \exp\left[\frac{\Delta\mu^\varnothing + q'(\phi_c - \phi_o)}{kT}\right] \quad \text{with } \Delta\mu^\varnothing = \mu_c^\varnothing - \mu_o^\varnothing. \tag{4.62}$$

Upon assuming for simplicity that the membrane spanned by the ion channel is homogeneous and that the electric field within it is constant, the electric potential varies linearly with distance, as shown in Fig. 4.11. Denoting by x_c and x_o the distances of the charge q' from the external solution of the membrane in the closed and open state, ϕ_c is given by $x_c\Delta\phi/d$, and ϕ_o by $x_o\Delta\phi/d$. Substituting into Eq. 4.62, we obtain

$$\frac{N_o}{N_c} = \exp\left[\frac{\Delta\mu^\varnothing + q\Delta\phi}{kT}\right] \quad \text{with } q = q'\frac{(x_c - x_o)}{d}. \tag{4.63}$$

The charge q is called *gating charge*, and the current arising from its movement within the membrane is called *gating current*. The probability p of the channel

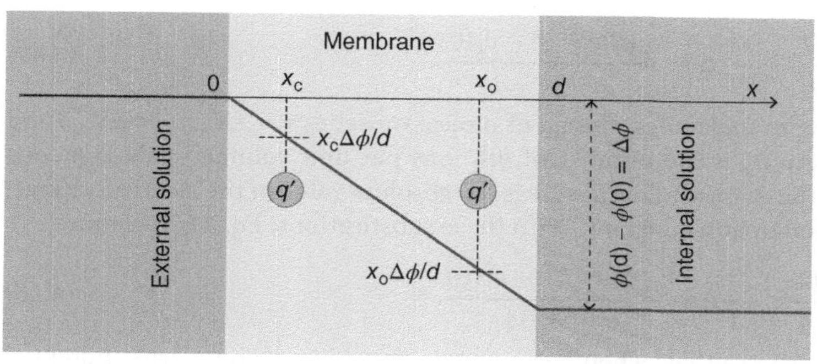

Figure 4.11 Electric potential profile (thick line) across a bilayer lipid membrane, showing the position of the gating charge q' in the closed and open states.

being open is given by the ratio of the number N_o of open channels to the total number of channels, $N_o + N_c$:

$$p = \frac{N_o}{N_o + N_c}. \tag{4.64}$$

The probability of the channel being closed is the complement of p to unity, since the channel may only be either open or closed:

$$1 - p = \frac{N_c}{N_o + N_c}. \tag{4.65}$$

From Eqs. 4.63–4.65, it follows that

$$\frac{N_o}{N_c} = \frac{p}{1-p} = \exp\left[\frac{\Delta\mu^\circ + q\Delta\phi}{kT}\right] \rightarrow$$
$$p = \frac{1}{1 + e^{-(\Delta\mu^\circ + q\Delta\phi)/kT}} = \frac{1}{1 + e^{-q(\Delta\phi - \Delta\phi_{1/2})/kT}}. \tag{4.66}$$

In writing the last member of this equation, it has been taken into account that the transmembrane potential at which the probability takes the value 1/2 (denoted by $\Delta\phi_{1/2}$) is that at which $\Delta\phi$ equals $-\Delta\mu^\circ/q$. The open probability p in Eq. 4.66 is sometimes referred to as the *one-sided Boltzmann equation*.

We can now derive the expression of the current flowing along a voltage-gated single channel. Let us denote this current by i, when the channel is open. In this connection, we use the GHK equation 4.51. In doing so, we implicitly assume that electrodiffusion, which in Section 4.5 was hypothesized to occur across a membrane considered as a homogenous medium, now occurs along a single channel. To this end, we remove the partition coefficient β_h between the membrane and the aqueous phase from Eq. 4.51, since the ion channel is filled by water molecules in the absence of permeating ions. Therefore, the inflow of an ion into the channel and its outflow from it do not involve the passage from one phase to another. Moreover, Eq. 4.51 expresses a current density, namely a current per unit area of the membrane surface. Passing to the current i along a single channel, we must multiply Eq. 4.51 by the cross-sectional area, α, of the channel. We then have

$$i_h = \alpha \frac{D_h}{d} z_h^2 \frac{F^2}{RT} \Delta\phi \; \frac{c_h(d) e^{z_h f \Delta\phi} - c_h(0)}{e^{z_h f \Delta\phi} - 1}. \tag{4.67}$$

Upon referring to molecules instead of moles, we set $c_h = n_h / N_{Av}$, $F = eN_{Av}$, and $R = kN_{Av}$, where n_h is the number of hth ions per unit volume of the aqueous phase, N_{Av} is the Avogadro number, e is the absolute value of the electron charge, and k is the Boltzmann constant. With these substitutions, Eq. 4.67 becomes

$$i_h = \alpha \frac{D_h}{d} z_h^2 \frac{e^2}{kT} \Delta\phi \frac{n_h(d) e^{z_h f \Delta\phi} - n_h(0)}{e^{z_h f \Delta\phi} - 1}. \tag{4.68}$$

This equation shows that, when $\Delta\phi$ assumes the value of the Nernst potential for the permeant ion, $\Delta\phi_h = (kT/e)\ln[n_h(0)/n_h(d)]$, the current i_h vanishes, as expected. However, as already pointed out in connection with the expression of the macroscopic current density j_h in Eq.4.51, the current i_h in the neighborhood of $\Delta\phi_h$ is not exactly proportional to the transmembrane potential measured relative to the Nernst potential, that is, $\Delta\phi - \Delta\phi_h$, except in the case of $n_h(0) = n_h(d)$. Hence, Eq. 4.2 holds strictly only in this case, while it holds approximately in close proximity of $\Delta\phi = \Delta\phi_h$ for $n_h(0) \neq n_h(d)$. By considerations analogous to those yielding Eqs. 4.54 and 4.55 for the macroscopic current density j_h, we may note that, for $\Delta\phi$ values sufficiently apart from $\Delta\phi_h$ in both directions, the single-channel current i_h is proportional to the transmembrane potential $\Delta\phi$, and not to that, $\Delta\phi - \Delta\phi_h$, measured relative to the Nernst potential:

$$i_h = \alpha \frac{D_h}{d} z_h^2 \frac{e^2}{kT} \Delta\phi n_h(d) \quad \text{for } z_h > 0 \text{ and for } e^{z_h f \Delta\phi} \gg n_h(0)/n_h(d) \text{ and } \gg 1$$
$$i_h = \alpha \frac{D_h}{d} z_h^2 \frac{e^2}{kT} \Delta\phi n_h(0) \quad \text{for } z_h > 0 \text{ and for } e^{z_h f \Delta\phi} \ll n_h(0)/n_h(d) \text{ and } \ll 1. \tag{4.69}$$

Comparison of this equation with Eq. 4.2, which introduces the single-channel-specific conductance γ, shows that, for a cation, γ is given by

$$\gamma = \alpha \frac{D_h}{d} z_h^2 \frac{e^2}{kT} n_h, \tag{4.70}$$

where n_h equals $n_h(d)$ or $n_h(0)$, depending on whether $\exp(z_h f \Delta\phi)$ is $\gg 1$ or $\ll 1$. It must be stressed that the single-channel conductance γ, responsible for the flow of the permeant ion in a given direction, is proportional to the concentration of that ion on the membrane side from which it moves to the other side. Hence, γ depends not only on the nature of the channel, but also on the concentration of the permeant ion. In patch-clamp measurements of γ, when such a concentration is not specified, one is usually referring to the physiological concentration.

The single-channel mean current, $\langle i_h \rangle$, is given by the product of i_h by the probability of the channel being open, expressed by Eq. 4.66

$$\langle i_h \rangle = \alpha \frac{D_h}{d} z_h^2 \frac{e^2}{kT} \Delta\phi \frac{n_h(d) e^{z_h f \Delta\phi} - n_h(0)}{e^{z_h f \Delta\phi} - 1} \frac{1}{1 + e^{-q(\Delta\phi - \Delta\phi_{1/2})/kT}}. \tag{4.71}$$

In view of Eq. 4.70 and of the expression for $\Delta\phi_h$, Eq. 4.71 for $z_h = +1$ takes the following form:

$$\langle i_h \rangle = \gamma_h \Delta\phi \frac{e^{-f\Delta\phi_h} e^{f\Delta\phi} - 1}{e^{f\Delta\phi} - 1} \frac{1}{1 + e^{-q(\Delta\phi - \Delta\phi_{1/2})/kT}} \quad \text{with } \gamma_h \equiv \alpha \frac{D_h}{d} \frac{e^2}{kT} n_h(0). \tag{4.72}$$

The probability p has a notable influence on the behavior of the single-channel mean current $\langle i \rangle$. Let us consider, for instance, the Nernst transmembrane potential, $\Delta\phi_h$, for the sodium channel in an axon membrane, where $c_{\text{Na}}(0) = 200$ mM and $c_{\text{Na}}(d) = 10$ mM (cf. Eq. 4.9):

$$\Delta\phi_{\text{Na}} = (kT/e) \ln[c_{\text{Na}}(0)/c_{\text{Na}}(d)] = +75 \text{ mV}. \tag{4.73}$$

Let the transmembrane potential, $\Delta\phi_{1/2}$, at which the open probability of the sodium channel is 0.5, be equal to -40 mV, with $q/(kT) = 1\ \text{mV}^{-1}$. Substituting

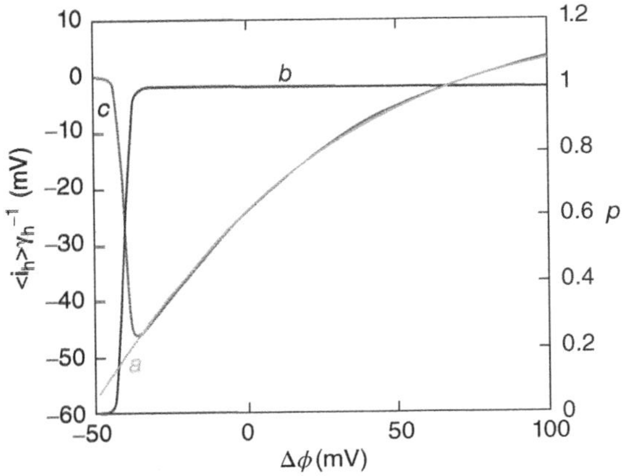

Figure 4.12 Plot of $\langle i_h \rangle/\gamma_h$ (c) and of the probability p (b) against $\Delta\phi$ for a Na channel, calculated as described in the text. Curve a, which coincides with curve c for $\Delta\phi > -40$ mV, is the $\langle i_h \rangle/\gamma_h$ value without the inclusion of the probability p.

these values in Eq. 4.72, the ratio of the mean current $\langle i_h \rangle$ to the single-channel conductance γ_h varies with $\Delta\phi$ as shown in Fig. 4.12.

Note the similarity of the shape of the current versus potential curve in Fig. 4.12 with that of the current versus time curves in Fig. 4.8A, both relative to a depolarization of the sodium channel. This similarity is not fortuitous, since both curves refer to an action potential, which is a regular transmembrane potential "wave"; this moves along the axis (z) of an axon at a constant velocity $v = dz/dt$. Hence, the shape of the action potential along the time axis, at a constant position of the z-axis, parallels that along the transmembrane potential, at constant t.

4.6.1 The Gap Junction

A peculiar example of ion channel connecting two different cells, which requires the use of two one-sided Boltzmann equations, that is, a *double-sided Boltzmann equation*, is offered by gap junctions. A *gap junction* is a specialized intercellular connection between two animal cells composed of two *connexons* (or hemichannels), one per cell, which directly link the cytoplasm of the two cells, allowing various molecules to pass freely between them (cf. Section 1.4.5). When the connexons in the plasma membranes of two adjacent cells are aligned, they form a continuous aqueous channel that joins the two cell interiors. Each connexon is formed from six channel-forming transmembrane proteins called *connexins*, which span the membrane four times. Each connexin of one cell membrane is aligned with a corresponding connexin of the membrane of its partner cell. For convenience, each pair of aligned connexins is referred to as a gate. Hence, a gap junction consists of six gates, as shown schematically in Fig. 4.13. Opening of a connexon requires its six connexins to tilt with respect to the normal to the membrane planes.

Each gate is open when both its connexins are open, and the gap junction is fully open when all its six gates are open. We denote the open state of a connexin by O, and its closed state by C. Each of the 12 connexins that form the gap junction

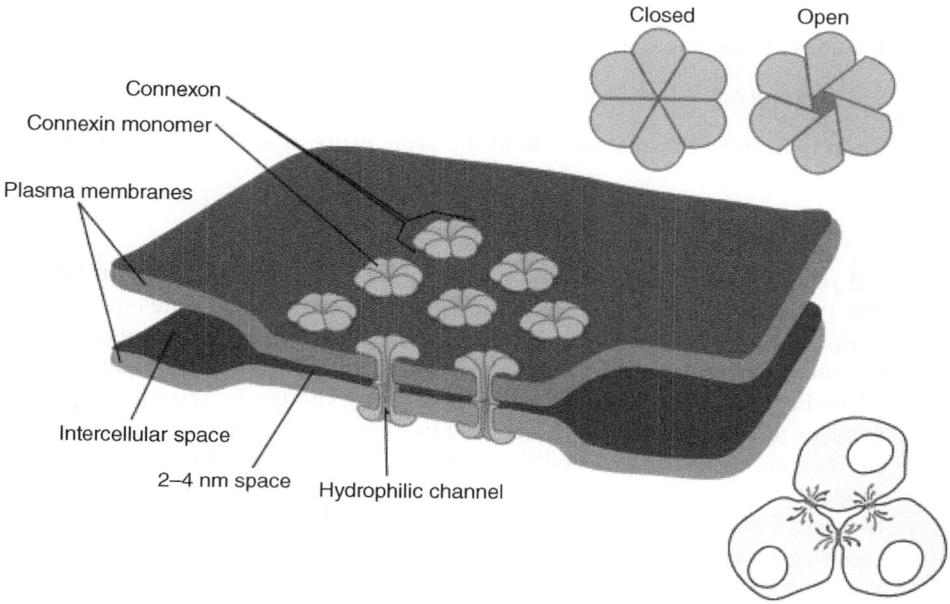

Figure 4.13 Scheme of a gap junction between two plasma membranes. The connexons decrease the intercellular space between the membranes and are open when their connexins are inclined with respect to the normal to the membrane planes. Source: https://commons.wikimedia.org/wiki/File:Adherens_Junctions_structural_proteins-fr.svg

can undergo the reaction

$$\mathrm{C} \underset{b}{\overset{a}{\rightleftharpoons}} \mathrm{O}, \tag{4.74}$$

where a and b are the rate constants governing the transitions between closed and open states, which are functions of the transmembrane potential $\Delta\phi$. Under conditions of dynamic equilibrium ($\mathrm{d[O]}/\mathrm{d}t = \mathrm{d[C]}/\mathrm{d}t = 0$), we have

$$a[\mathrm{C}] = b[\mathrm{O}] \quad \rightarrow \quad [\mathrm{O}]/([\mathrm{O}] + [\mathrm{C}]) = a/(a + b). \tag{4.75}$$

The ratio, $[\mathrm{O}]/([\mathrm{O}]+[\mathrm{C}])$, of the concentration of the open state to the sum of the concentrations of both states measures the probability, p, of the connexin being open. Hence,

$$p = a/(a + b) \quad \rightarrow \quad a/b = [p/(1 - p)]. \tag{4.76}$$

It is intuitive that the higher the rate constant b for the passage from O to C, the lower the mean time over which the channel is in the open state; hence, the mean open time is just equal to $1/b$. Analogously, the mean closed time equals $1/a$.

The dynamic equilibrium can be formally expressed by the familiar Nernst equation for a redox reaction $\mathrm{Ox} + ze = \mathrm{Red}$: $E = E^{\circ} + (\mathrm{k}T/ze)\ln(c_{\mathrm{Ox}}/c_{\mathrm{Red}})$. In the present case, the charge ze becomes the gating charge q involved in the passage from O to C, the applied potential E becomes $\Delta\phi$, and the formal potential E^{ϕ} becomes the $\Delta\phi$ value at which the [O]/[R] ratio equals unity, namely $\Delta\phi_{1/2}$:

$$\exp[q(\Delta\phi - \Delta\phi_{1/2})/kT] = [\mathrm{O}]/[\mathrm{C}] = p/(1-p). \tag{4.77}$$

Upon rearranging terms, we obtain the same expression of p as derived in Eq. 4.66:

$$p = 1/(1 + \mathrm{e}^{-q(\Delta\phi-\Delta\phi_{1/2})/kT}). \tag{4.78}$$

If p is the probability that one connexin is in the open state, the probability that two aligned connexins are in the open state at the same time equals p^2. In so doing, we must consider that, when applying a potential difference $\Delta\phi$ across the whole gap junction (the *transjunctional potential*) from left to right, the two aligned connexins will have to move their gating charges in opposite directions in order to open; this requires a change in the sign of the gating charge q in the expressions of the two probabilities. Moreover, the transmembrane potential at a single cell is conventionally taken as zero at its extracellular side; however, when two cells form a gap junction, they turn the extracellular side of their membranes toward each other. Hence, once the direction of the transjunctional potential has been chosen, its sign must change in passing from one probability p to the other; this is accomplished by changing the sign of their respective $\Delta\phi_{1/2}$ values. By so doing, the probability for one connexin tends to unity with increasing $\Delta\phi$, while that for its gate partner tends to zero, and the $\Delta\phi$ values at which the two probabilities attain the ½ value are located symmetrically with respect to $\Delta\phi = 0$. Hence, p^2 is given by

$$p^2 = 1/(1 + \mathrm{e}^{-q(\Delta\phi-\Delta\phi_{1/2})/kT}) \times 1/(1 + \mathrm{e}^{q(\Delta\phi+\Delta\phi_{1/2})/kT}). \tag{4.79}$$

This *double-sided Boltzmann equation* yields a peak with a rounded maximum when plotted against $\Delta\phi$, as shown in Fig. 4.14.

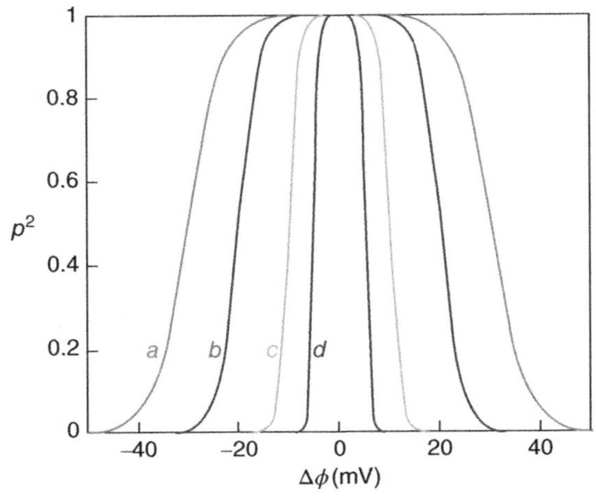

Figure 4.14 Plots of p^2 against $\Delta\phi$ for $q/kT = 0.3$ (*a*), 0.5 (*b*), 1 (*c*), 2 (*d*) and for $\Delta\phi_{1/2} = -30$ mV (*a*), -20 mV (*b*), -10 mV (*c*), -5 mV (*d*). The positive slope of the left sigmoidal branch of the curve increases with increasing q/kT, just as the absolute value of the negative slope of the right branch.

If p^2 is the probability that one gate is open, then p^{12} is the probability that the whole gap junction is fully open. As a matter of fact, this is only true if the six gates of the gap junction operate independently. In actual fact, when a connexin opens, it influences the neighboring ones to open. This mutual interaction can be expressed by the following equation (De Felice, 1997):

$$\mathrm{d}[\mathrm{O}_k]/\mathrm{d}t = a[\mathrm{C}_k] - b[\mathrm{O}_k] + \xi[\mathrm{O}_{k-1}][\mathrm{C}_k] + \xi[\mathrm{O}_{k+1}][\mathrm{C}_k] \quad \text{with } k = 1\text{–}6. \tag{4.80}$$

Here ξ is an interaction parameter, O_k and C_k are the open and closed states of the kth connexin of a given connexon, and the subscripts $(k-1)$ and $(k+1)$ denote the two nearest neighboring connexins. This expression accounts for the fact that the number of connexins in the open state O grows not only in proportion to the number of connexins in the closed state, but also in proportion to the number of neighboring connexins that are in the open state. The approximate relations in Eq. 4.80 serve to illustrate the problem of cooperativity.

4.7 Rate Theory of Membrane Transport

According to the absolute rate theory, a membrane can be viewed as a series of potential energy barriers, each one depicted as a peak interposed between two valleys. If, for example, the outer membrane surface is located on the left, a particle traversing the membrane from its outer ($x = 0$) to its inner ($x = d$) surface must overcome each barrier, passing from the bottom of the adjacent valley on its left side to the bottom of the adjacent one on its right side. The vertical distance between the bottom of the adjacent valley on the left (right) side of the given barrier and the barrier top is the minimum amount of energy that the particle must possess to overcome the barrier when moving from $x = 0$ to $\mathrm{x} = d$ (from $x = d$ to $x = 0$).

The rate constant for the movement of a particle from the j valley to the k valley, over the jk barrier, is derived from the absolute reaction rate theory:

$$k_{jk} = \kappa(kT/h)\exp(-\Delta G^{\neq}_{jk}/kT). \tag{4.81}$$

Here, κ is the *transmission coefficient*, which expresses the probability that the *transition state* at the top of the barrier will actually proceed toward the other side of the barrier jk, and is usually assumed to be equal to unity as a first approximation; k is the Boltzmann constant, h Planck's constant, and T the absolute temperature. $\Delta G^{\neq}_{jk}$ is the Gibbs energy change involved in the transition of the particle from the stable equilibrium position at the bottom of the j valley to the top of the barrier. Equation 4.81 is obtained by assuming that the particle in the transition state (i.e., the *activated state* at the top of the barrier) is in equilibrium with that at the bottom of the j valley. Hence, its concentration $c^{\neq}$ at the top of the barrier is related to its concentration c_j in the j valley by the equilibrium constant:

$$c^{\neq}/c_j = K^{\neq}_{jk}. \tag{4.82}$$

One of the vibrations of the particle in the transition state is considered as the translational degree of freedom that allows the particle to overcome the energy barrier when its energy, $h\nu$, equals the thermal energy kT, yielding $\nu = kT/h$. The rate at which the particle in the activated state overcomes the energy barrier, in molecules per second, is given by the product of its concentration $c^{\neq}$ by the number ν of times it is formed per unit time: $c^{\neq}\nu$. This is also equal to $k_{jk}c_j$, where k_{jk} is the rate constant in passing from the bottom of the j valley to the top of the jk barrier. By equating these two expressions, we immediately have

$$k_{jk} = c^{\neq}\nu/c_j = K^{\neq}_{jk}kT/h = (kT/h)\exp(-\Delta G^{\neq}_{jk}/kT). \tag{4.83}$$

In writing the last member of this equation, account has been taken of the fact that $K^{\neq}$ is an equilibrium constant and, as such, it is related to the Gibbs energy change $\Delta G^{\neq}_{jk}$ in passing from the bottom of the j valley to the top of the jk barrier by the well-known relation $kT\ln K^{\neq}_{jk} = -\Delta G^{\neq}_{jk}$. By analogous arguments, the rate constant k_{kj} in passing from the bottom of the k valley to the top of the jk barrier is given by

$$k_{kj} = c^{\neq}\nu/c_k = K^{\neq}_{kj}kT/h = (kT/h)\exp(-\Delta G^{\neq}_{kj}/kT), \tag{4.84}$$

where c_k is the particle concentration at the bottom of the k valley.

For simplicity, we confine ourselves to considering three barriers: a high central barrier located across the hydrocarbon tail region of the membrane, and two small barriers located across the two polar head regions. We further assume that the two latter barriers are small enough to be overcome by ions under quasi-equilibrium conditions, such as to satisfy Eq. 4.49. In this case, the electric potential difference between the two valleys adjacent to the high central barrier is approximately equal to the transmembrane potential $\Delta\phi$. Application of this potential difference adds an electrostatic potential energy contribution to the Gibbs energy changes in Eqs. 4.83 and 4.84, which now assume the following form:

$$\Delta\widetilde{G}^{\neq}_{0d} = \Delta G^{\neq}_{0d} + \alpha z_h e\Delta\phi; \quad \Delta\widetilde{G}^{\neq}_{d0} = \Delta G^{\neq}_{d0} - (1-\alpha)z_h e\Delta\phi. \tag{4.85}$$

for an hth ion of charge number z_h. Here, the generic j and k valleys have been identified with the valleys on the extracellular and intracellular side of the central barrier, denoted by 0 and d, respectively. Moreover, the fraction of $\Delta\phi$ between the top of the barrier and the bottom of the 0 valley is denoted by $\alpha\Delta\phi$, whereas that between the bottom of the d valley and the top of the barrier is clearly given by $(1-\alpha)\Delta\phi$. The quantity α is commonly called *charge transfer coefficient* and equals 0.5 for a symmetrical potential energy barrier. As usual, $\Delta\phi$ is measured with respect to its value at $x = 0$, taken as zero. The signs in front of the electrostatic potential energy terms in Eq. 4.85 account for the fact that a positive $\Delta\phi$ value hinders the movement of a positive ion from the 0 valley to the top of the barrier, whereas it favors its movement from the d valley to the top of the barrier. From Eqs. 4.83–4.85, it follows that the overall current density j_h for ion movement across the potential energy barrier is given by

$$\begin{aligned} j_h &= -z_h F[k_{0d}c(0) - k_{d0}c(d)] \\ &= z_h F\{k_b c(d)\exp[(1-\alpha)z_h f\Delta\phi] - k_f c(0)\exp(-\alpha z_h f\Delta\phi)\} \end{aligned} \tag{4.86}$$

with

$$\begin{aligned} &k_f \equiv (kT/h)\exp(-\Delta G^{\neq}_{0d}/kT); \quad k_b \equiv (kT/h)\exp(-\Delta G^{\neq}_{d0}/kT); \\ &f \equiv e/kT = F/RT. \end{aligned} \tag{4.87}$$

If the potential energy barrier for $\Delta\phi = 0$ is symmetrical, $\Delta G^{\neq}_{0d}$ equals $\Delta G^{\neq}_{d0}$ and k_f equals k_b. Equation 4.86 is identical with the equation commonly used in interfacial electrochemistry to express the faradaic current across a metal/solution interface and referred to as the *Butler–Volmer equation.*

The current expressed by Eq. 4.86 is based on the assumption of a single potential energy barrier located in the middle of the hydrocarbon tail region. Conversely, the GHK equation 4.51 assumes electrodiffusion of ions across the membrane, envisaged as a series of diffusional jumps above a high number of consecutive potential energy barriers of approximately the same height, similar to those in a hurdle race. The two equations may predict an appreciably different dependence of the current upon $\Delta\phi$.

4.8 Action Potential Revisited

In Section 4.3, the mechanism of action potential was outlined in simple terms. It is now convenient to provide a deeper insight into this mechanism, starting from Hodgkin and Huxley's pioneering and fascinating work, long before the existence of ion channels could be conclusively proved.

The action potential was recorded for the first time by Hodgkin and Huxley (1939, 1945) on the giant axon (0.1–1 mm in diameter) of a squid. A glass capillary, about 100 μm in diameter, was filled with sea water, inserted through the cut end of the axon and pushed along it, to allow its tip to reach an undamaged part of the axon. The transmembrane potential was measured between an electrode connected to the glass capillary and an electrode immersed in the external seawater. The axon was stimulated electrically via a pair of electrodes with square wave current pulses, whose duration and height could be varied.

Under the action of the current pulse, the axon membrane behaves to a first approximation as a parallel combination of a resistance R (the resistance opposing the ion flux through the membrane) and a capacitance C (due to the accumulation of ions of opposite sign on the two sides of the membrane). The time dependence of the transmembrane potential $\Delta\phi$ generated by a square wave current pulse across this parallel combination is expressed by the following equation:

$$\Delta\phi = IR(1 - e^{-t/RC}), \tag{4.88}$$

where I is the pulse amplitude and the product RC is the *time constant*. The lower the time constant, the shorter the time at which $\Delta\phi$ reaches its maximum limiting value, IR, expressed by Ohm's law. When a current pulse is applied, the transmembrane potential $\Delta\phi$ increases, tending asymptotically to the IR value; if this is lower than a threshold potential, no action potential is fired by the neuron. Neuron firing is possible only if the amplitude I of the current pulse is such that IR is greater than the threshold potential. Moreover, for given values of R and C, the time required to reach the threshold potential and to fire the action potential is shorter, the higher I is. The time elapsed between the application of the current pulse and the ensuing action potential is called the *latency* of the response.

We saw in Section 4.3 that, during the action potential, the probability of the sodium and potassium channels being in the open state varies with varying the transmembrane potential, which in turn varies with time. The detailed shape of the action potential started to be understood when in 1952 Hodgkin and Huxley used the *voltage clamp* technique to investigate the electrical behavior of squid axons (Hodgkin and Huxley, 1952a). This technique, which might be defined as *potentiostatic* in electrochemical jargon, consists in keeping the transmembrane potential constant at a prefixed value by monitoring it with two electrodes, one inside the axon (the recording electrode) and the other outside of it (the reference electrode), and in allowing the current to flow through the membrane between a current-passing electrode inside the axon and the same reference electrode outside of it (Fig. 4.15). As soon as the transmembrane potential deviates, by an extremely small amount, from the prefixed value, a feedback amplifier causes this deviation to be opposed through a change in the membrane current. In other words, while the transmembrane potential is kept constant at a prefixed value, the current required to maintain it constant flows through the membrane and

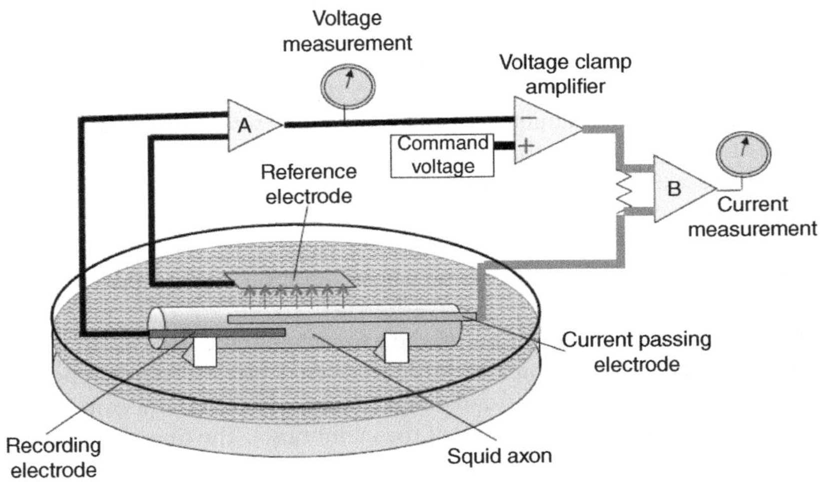

Figure 4.15 Scheme of the voltage clamp technique for studying the potentiostatic membrane current of a squid axon.

is recorded as a function of time. More precisely, the recorded current is that flowing at a prefixed transmembrane potential $\Delta\phi$ capable of causing membrane depolarization, as a consequence of a potential jump from the resting potential to $\Delta\phi$ (Purves et al., 2004).

The heart of the voltage clamp is the *operational amplifier* (OA), symbolized by an equilateral triangle in Fig. 4.15. The voltage inputs of an OA (one positive and the other negative) are marked on the left-hand side, the output on the right-hand side. A positive signal brought to the positive input causes a positive signal at the output. A signal brought to the negative input leads to an output voltage of opposite sign; for this reason it is also called the *inverting input*. If two different signals are brought to the two inputs, the output signal is the opposite of the voltage difference between the negative and the positive input, multiplied by the *gain* value, which is of the order of 10^4. An OA has a high input impedance and a low output impedance (less than 1 kΩ).

A vital part of an OA circuit is the *negative feedback loop,* which connects the OA output to its negative input. In Fig. 4.15, the feedback loop of the *voltage clamp amplifier* is represented by the gray metal wire connected to the current-passing electrode, the two black wires that connect the reference electrode and the recording electrode to the two inputs of the OA A, and the black wire that connects the output of A to the negative input of the voltage clamp amplifier. The output of A provides the potential difference between the recording electrode and the reference electrode, that is, the transmembrane potential, which is fed to the negative input of the voltage clamp amplifier. On the other hand, the positive input of the voltage clamp amplifier is kept at a constant voltage (the command voltage in the figure), which is the potential difference that the system intends to impose across the axon membrane. This goal is achieved because the output of the voltage clamp amplifier provides the opposite of the difference between the measured value of the transmembrane potential and the command voltage. When the output of the voltage clamp amplifier is fed back to its input, the OA subtracts it from the input, thus maintaining the transmembrane potential at the desired command voltage value. The OA B measures the potential difference across the resistor located along the gray wire. From the known resistance of the resistor, application of Ohm's first law provides the instantaneous value of the current across the axon membrane that the voltage clamp amplifier maintains at a constant transmembrane potential, that is, the potentiostatic current.

To determine the expression of this potentiostatic membrane current, let us first return to the expression of the action potential in Eq. 4.12, which considered a single sodium channel and a single potassium channel, an let us generalize it by the inclusion of generic number densities, N_{Na} and N_{K}, of sodium and potassium channels:

$$\begin{aligned} I &= I_{\mathrm{c}} + I_{\mathrm{K}} + I_{\mathrm{Na}} + I_{\mathrm{L}} \\ &= C(\mathrm{d}\Delta\phi/\mathrm{d}t) + N_{\mathrm{K}}\gamma_{\mathrm{K}}\,p_{\mathrm{K}}(\Delta\phi - \Delta\phi_{\mathrm{K}}) + N_{\mathrm{Na}}\gamma_{\mathrm{Na}}\,p_{\mathrm{Na}}(\Delta\phi - \Delta\phi_{\mathrm{Na}}) \\ &\quad + \gamma_{\mathrm{L}}(\Delta\phi - \Delta\phi_{\mathrm{L}}) = 0. \end{aligned} \tag{4.89}$$

Under the potentiostatic conditions of the voltage clamp technique, the capacitive current, $I_c = C\, \mathrm{d}\Delta\phi/\mathrm{d}t$, in this equation vanishes because $\Delta\phi$ is kept constant. Moreover, the total current I is no longer equal to zero, because it is forced to vary in order to maintain the transmembrane potential constant. Equation 4.89 is therefore modified as follows:

$$I = I_{\mathrm{K}} + I_{\mathrm{Na}} + I_{\mathrm{L}} = G_{\mathrm{K}}(\Delta\phi - \Delta\phi_{\mathrm{K}}) + G_{\mathrm{Na}}(\Delta\phi - \Delta\phi_{\mathrm{Na}}) + \gamma_{\mathrm{L}}(\Delta\phi - \Delta\phi_{\mathrm{L}}), \tag{4.90}$$

where $G_{\mathrm{K}} \equiv N_{\mathrm{K}}\gamma_{\mathrm{K}}p_{\mathrm{K}}$ and $G_{\mathrm{Na}} \equiv N_{\mathrm{Na}}\gamma_{\mathrm{Na}}p_{\mathrm{Na}}$ are the specific conductances of the potassium and sodium channels. Under these conditions, it is possible to separate the sodium current I_{Na} from the potassium current I_{K} in different ways.

The procedure adopted by Hodgkin and Huxley (1952a) consisted in adjusting the Na^+ concentration outside the membrane with respect to that inside it, in such a way as to satisfy the Nernst equation for the sodium ions exactly at the prefixed potential $\Delta\phi$, prior to the potential jump from the resting potential to $\Delta\phi$. With $\Delta\phi = \Delta\phi_{\mathrm{Na}}$, Eq. 4.90 becomes

$$I = I_{\mathrm{K}} + I_{\mathrm{L}} = G_{\mathrm{K}}(\Delta\phi_{\mathrm{Na}} - \Delta\phi_{\mathrm{K}}) + \gamma_{\mathrm{L}}(\Delta\phi_{\mathrm{Na}} - \Delta\phi_{\mathrm{L}}). \tag{4.91}$$

By disregarding the small contribution from the leak channel, the current I, once divided by the difference, $\Delta\phi_{\mathrm{Na}} - \Delta\phi_{\mathrm{K}}$, between the two Nernst potentials, yields directly the specific conductance G_{K} of potassium channels as a function of time. If we then subtract the current in Eq. 4.91 from that in Eq. 4.90, recorded at the same prefixed potential, but for an external Na^+ concentration different from that satisfying the Nernst equation at such a potential, we obtain the quantity $G_{\mathrm{Na}}(\Delta\phi - \Delta\phi_{\mathrm{Na}})$, and hence G_{Na}, as a function of time. An example of this procedure is shown in Fig. 4.16, where curve B corresponds to Eq. 4.91, curve A to Eq. 4.90 and curve C to the difference between the two. According to usage, the potential of +56 mV is the *depolarization potential* measured with respect to the rest potential, estimated at −60 mV; hence, $\Delta\phi$ corresponds to a transmembrane potential of −4 mV.

An alternative method to separate sodium from potassium conductance makes use of specific inhibitors. Thus, for example, tetrodotoxin, a virulent nerve poison found in the tissues of the Japanese puffer fish *Spheroides rubripes*, blocks sodium conductance if added to the external solution, while leaving potassium conductance unaltered. Conversely, it has no effect on sodium conductance if injected in the liquid inside the axon (the *axoplasm*). This indicates that its site of action is near the external surface of the axon membrane. When added to the external solution, tetrodotoxin allows the recoding of the sole current due to potassium channels. Vice versa, tetraethylammonium ions block potassium channels if injected in the axoplasm, but have no effect on sodium conductance; they are ineffective toward potassium channels if added to the external solution.

Figure 4.17 shows the sodium and potassium conductances in the squid axon as a function of time for different depolarization potentials. It is evident that depolarization causes a rapid increase in sodium conductance, followed by a slow

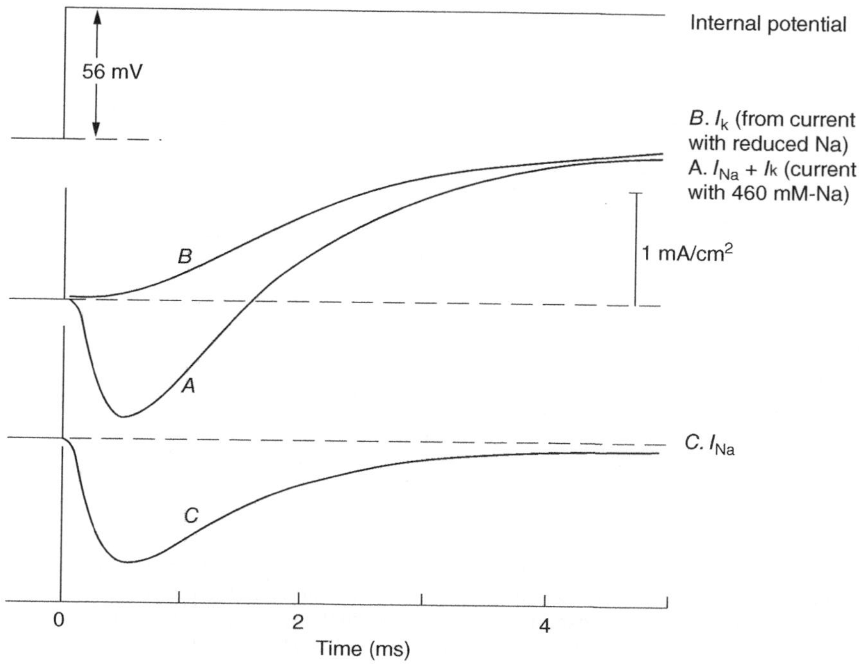

Figure 4.16 Ionic current in a squid axon in response to a depolarization of 56 mV, with the axon in sea water (*A*) and in a solution comprising 10% sea water and 90% isotonic choline chloride (*B*), which makes the selected transmembrane potential equal to the sodium Nernst potential, $\Delta\phi_{Na}$. Curve *C* is the difference between curves *A* and *B*. Source: Hodgkin (1958). Reproduced with permission of Royal Society of Chemistry.

decrease due to inactivation of the sodium channel, while it causes a slow increase in potassium conductance. The extent of these changes in conductance with time increases with increasing depolarization, attaining a maximum saturating level at 100 mV and above. In terms of voltage gating, this indicates that depolarization determines a rapid opening of sodium channels, followed by their slow inactivation, down to complete inhibition of Na^+ inflow. Depolarization also opens potassium channels, albeit more slowly; in the time scale of Fig. 4.17 this process is not yet followed by their inactivation, differently from other potassium channels that undergo a fast inactivation, such as the Shaker potassium channel.

It is quite reasonable to assume that, during the flow of the I_K and I_{Na} currents, the number densities, N_K and N_{Na}, of potassium and sodium channels do not vary, just as their single-channel conductances, γ_K and γ_{Na} (i.e., the conductances that they would exhibit if they were always open). We may therefore conclude that the G_K and G_{Na} specific conductances are proportional to the probabilities, p_K and p_{Na}, of the corresponding channels being in the open state. To estimate these probabilities at any given time and depolarization potential it is just sufficient to divide the relative specific conductances, G_K and G_{Na}, by their maximum limiting

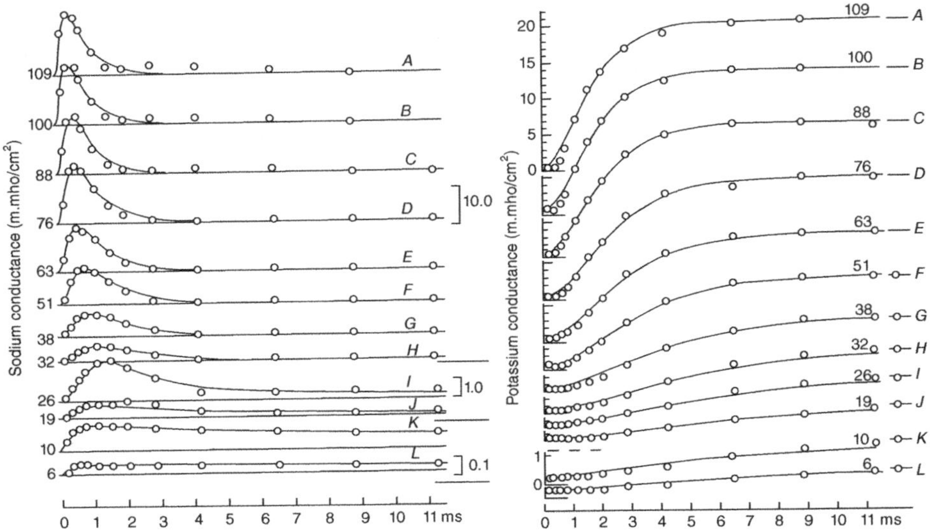

Figure 4.17 Curves of the specific conductances, G_{Na} and G_K, of sodium and potassium against time, obtained with a squid giant axon at different depolarization potentials reported near each curve. The open circles are experimental values, while the solid curves were calculated from Eqs. 4.90 and 4.92–4.97. Source: Hodgkin and Huxley (1952b). Reproduced with permission of Wiley.

values attained at depolarization potentials ≥100 mV, when the open probabilities p_K and p_{Na} reach the unit value.

Hodgkin and Huxley (1952*b*) expressed quantitatively the probabilities so determined on the basis of the following empirical expressions (see also Hodgkin, 1958):

$$p_K = n^4; \quad p_{Na} = m^3 h, \tag{4.92}$$

where the parameters n, m, and h are obviously less than unity. The rate of change of n with time is given by the following equation:

$$dn/dt = \alpha_n(1-n) + \beta_n n, \tag{4.93}$$

where α_n and β_n are rate constants that, at 6 °C, depend on the transmembrane potential $\Delta\phi$ according to the empirical relationships:

$$\alpha_n = \frac{0.01(\Delta\phi + 10)}{\exp[(\Delta\phi + 10)/10] - 1}; \quad \beta_n = 0.125\exp(\Delta\phi/80). \tag{4.94}$$

Analogously, the rates of change of m and h with time are expressed by the following equations:

$$dm/dt = \alpha_m(1-m) - \beta_m m; \quad dh/dt = \alpha_h(1-h) - \beta_h h \tag{4.95}$$

with

$$\alpha_m = \frac{0.1(\Delta\phi + 2.5)}{\exp[(\Delta\phi + 25)/10] - 1}; \quad \beta_m = 4\exp(\Delta\phi/18); \tag{4.96}$$

$$\alpha_h = 0.07\exp(\Delta\phi/20); \quad \beta_h = \frac{1}{\exp[(\Delta\phi + 30)/10] + 1}. \tag{4.97}$$

In practice, the dependence of n, m, and h upon the transmembrane potential is sigmoidal in shape (Fig. 4.18), and, hence, similar to that expressed by the equation of the open probability for an idealized voltage-gated channel (cf. the p curve in Fig. 4.12).

It should be noted from Fig. 4.18 that, while n and m increase with increasing $\Delta\phi$, h increases in the opposite direction. This causes the sodium channel to close at the same depolarization potentials at which it initially opens (inactivation). In this respect, the probability $p_{Na} = m^3h$ vaguely resembles the narrow peak d calculated from the double-sided Boltzmann expression in Fig. 4.14.

Hodgkin and Huxley's expressions for p_{Na} and p_K account exactly for the experimental behavior of G_{Na} and G_K (cf. the solid curves in Fig. 4.17) and, albeit empirically, underline a particular behavior of these channels pointed out by these authors. Thus, the opening probability $p_K = n^4$ suggests that the formation of the potassium channel might occur through the aggregation of four charged particles, where n expresses the probability of each single particle being in the appropriate position for forming the channel. This hypothesis was subsequently supported by the known structure of the potassium channel, which consists of four subunits. Analogously, the probability $p_{Na} = m^3h$ suggests that the opening of the sodium channel might be the result of the probabilities of three distinct particles being in the appropriate position for forming the channel, while h expresses the probability of a fourth particle occupying a position capable of inactivating the channel. This hypothesis is consistent with the structure of the sodium channel, which disposes of a special gate in charge of its inactivation.

Figure 4.18 Curves of *n*, *m*, and *h* against the transmembrane potential.

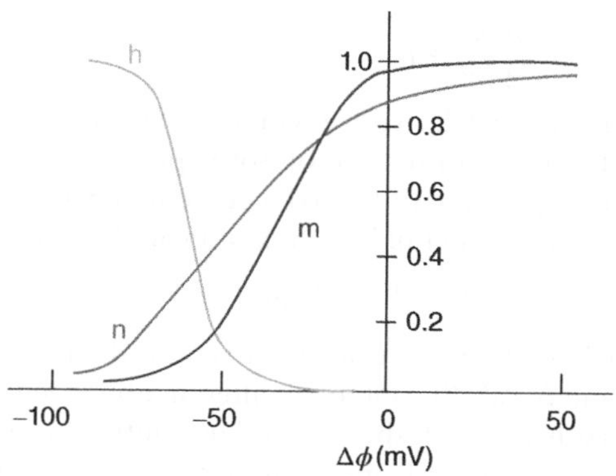

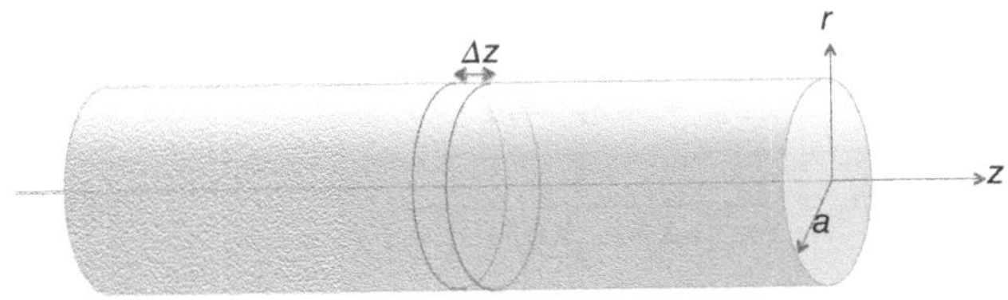

Figure 4.19 Schematic representation of the section of an axon.

4.8.1 The Shape of the Action Potential

By using the preceding expressions for p_{Na} and p_K, Hodgkin and Huxley (1952*a*) were able to calculate the shape of the action potential with high accuracy. In order to predict the propagation rate of the action potential along the axon, it was necessary to consider not only the ionic current density flowing through the axon membrane (*radial current density* j_r), but also that flowing inside the axon, approximated as a long cylinder (*axial current density* j_z). Let us denote by r an axis normal to the axon membrane and by z the axis of the cylindrical axon, as shown in Fig. 4.19. Let us consider a very short segment of the axon, of length Δz, and let us denote by a the radius of the cross section of the axon.

Under steady-state conditions, the axial current flowing out of this segment through the front cross section minus that flowing into it through the rear cross section is necessarily equal to the current that enters into the segment through its lateral surface. The lateral surface area of the segment equals $2\pi a\Delta z$, while its cross-sectional area equals πa^2. Recalling that a cationic current entering a cell across its membrane is conventionally taken as negative, we have

$$\pi a^2 \Delta j_z = -2\pi a \Delta z j_r, \tag{4.98}$$

where Δj_z is the difference between the axial current density leaving the segment and that entering it. Passing from finite to infinitesimal increments and rearranging terms, we have

$$\frac{a}{2}\frac{\partial j_z}{\partial z} + j_r = 0. \tag{4.99}$$

The partial derivative symbol is introduced to account for the fact that the current density is a function of both space z and time t. If we now identify j_z with the average value of the axial current density that flows in a given instant along our axon segment of length Δz and apply Ohm's first law, we obtain

$$\pi a^2 j_z \times R = -\Delta(\Delta\phi), \tag{4.100}$$

where $\pi a^2 j_z$ is the axial current, R the resistance of the solution inside the axon, and $\Delta(\Delta\phi)$ the potential difference between the front and rear cross sections of the axon segment; the minus sign is introduced to account for the fact that a positive current flows from points where the electric potential is higher to those where it is lower. In view of second Ohm's law, the resistance R is directly proportional to the length Δz of the segment and inversely proportional to its cross-sectional area πa^2, according to a proportionality constant ρ called *resistivity* or *specific resistance*. Upon replacing R by $\rho\Delta z/\pi a^2$ in Eq. (4.100) and passing from finite to infinitesimal increments, we have

$$\pi a^2 j_z \rho \frac{\Delta z}{\pi a^2} = -\Delta(\Delta\phi) \rightarrow j_z = -\frac{1}{\rho}\frac{\partial \Delta\phi}{\partial z}. \tag{4.101}$$

Replacement of this expression of j_z into Eq. 4.99 yields

$$j_r = \frac{a}{2\rho}\frac{\partial^2 \Delta\phi}{\partial z^2}. \tag{4.102}$$

To pass from the space dependence of $\Delta\phi$ to its time dependence, we must now consider that the action potential propagates along the axon as a regular wave, with a constant propagation rate $v = \mathrm{d}z/\mathrm{d}t$. Therefore, we can write

$$\frac{\partial^2 \Delta\phi}{\partial z^2} = \frac{\partial^2 \Delta\phi}{v^2 \partial t^2}. \tag{4.103}$$

Upon replacing $\partial^2\Delta\phi/\partial z^2$ from Eq. 4.103 into Eq. 4.102, we obtain the following expression of the radial current density j_r as a function of $\Delta\phi$:

$$j_r = \frac{a}{2\rho v^2}\frac{\partial^2 \Delta\phi}{\partial t^2}. \tag{4.104}$$

The equation of the action potential derived in Section 4.3 (cf. Eq. 4.12) considers the current across the membrane (i.e., the radial current j_r) to be practically zero, due to a compensation between its different components, in that it ignores the existence of an axial current i_z, responsible for the propagation of the nerve impulse. More correctly, the radial current density expressed by Eq. 4.104 must be equated to its expression in Eq. 4.89, which relates it to the first time derivative of $\Delta\phi$. We then obtain the following differential equation:

$$\begin{aligned}\frac{a}{2\rho v^2}\frac{\partial^2 \Delta\phi}{\partial t^2} = {} & C\frac{\mathrm{d}\Delta\phi}{\mathrm{d}t} + \bar{g}_\mathrm{K} n^4(\Delta\phi - \Delta\phi_\mathrm{K}) + \bar{g}_\mathrm{Na} m^3 h(\Delta\phi - \Delta\phi_\mathrm{Na}) \\ & + \gamma_\mathrm{L}(\Delta\phi - \Delta\phi_\mathrm{L}) \\ & \text{with } \bar{g}_\mathrm{Na} \equiv N_\mathrm{Na}\gamma_\mathrm{Na} \text{ and } \bar{g}_\mathrm{K} \equiv N_\mathrm{K}\gamma_\mathrm{K}.\end{aligned} \tag{4.105}$$

In writing this equation, p_Na and p_K were replaced by the empirical expressions in Eq. 4.92 derived by Hodgkin and Huxley (1952*a*), and their symbolism was used. This differential equation can be solved numerically and gives a finite waveform for only a particular value of the propagation rate v. By using the experimental values of C, ρ, $\Delta\phi_\mathrm{Na}$, and $\Delta\phi_\mathrm{K}$, Hodgkin and Huxley calculated a shape of the action potential very close to the experimental one (Fig. 4.20) and a propagation rate of 18.8 m s^{-1}, very close to the experimental value of 21.2 m s^{-1}. This has confirmed in an indisputable way the correctness of Hodgkin and Huxley's approach.

4.8.2 The Gating Current of the Potassium Channel

In Section 4.6, the concepts of gating current and gating charge were introduced to justify the voltage dependence of ion channels, such as sodium, potassium, and

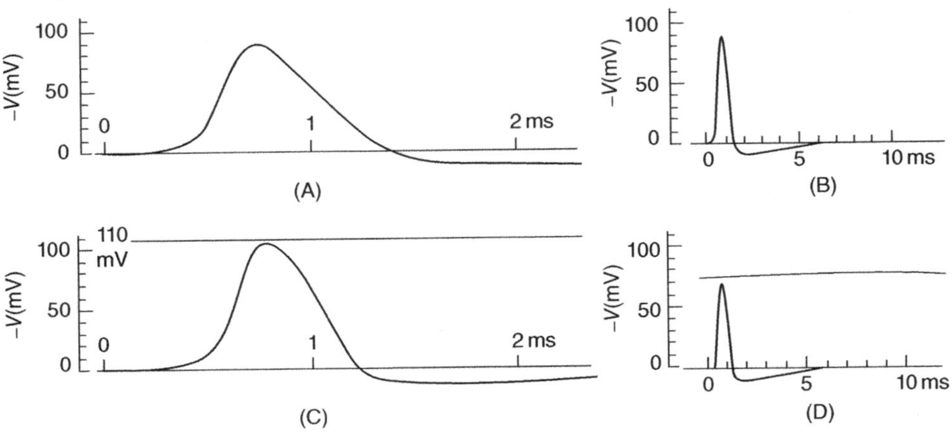

Figure 4.20 Calculated (A and B) and experimental (C and D) propagated action potentials in a squid axon at 18.5 °C. Source: Hodgkin and Huxley (1952b). Reproduced with permission of Wiley.

calcium channels. Due to the extremely low values of gating currents, only in the early 1970s was it made possible to prove their existence. To this end, the almost complete independence of the membrane capacitance C from the transmembrane potential was exploited. Thus, if one carries out two potential steps of the same amplitude in opposite directions starting from the resting potential, the capacitive currents, $I_c = C(\mathrm{d}\Delta\phi/\mathrm{d}t)$, flowing as a consequence of these two steps are expected to be equal in magnitude, but opposite in sign. If, however, one of these two steps also moves the gating charge within the voltage-gated channel, then the absolute value of the current following this step will be greater than the absolute value of the exclusively capacitive current following the opposite potential step.

This is schematically shown in Fig. 4.21, which depicts a positive gating charge located in proximity to the intracellular side of the membrane at the resting potential, where this side is more negative than the extracellular one by about −60 to −80 mV. By carrying out a potential step toward more negative values (*hyperpolarization*), the positive charge does not move, because it has no free space to move into. Conversely, by carrying out a potential step of equal amplitude toward more positive values (*depolarization*), the positive charge is able to move to the extracellular side of the membrane, originating a gating current that adds to the normal capacitive current. The sign of the exclusively capacitive current produced by hyperpolarization is negative, in agreement with the usage according to which a current is negative if involves a flow of positive charge from the extracellular to the intracellular side of the membrane. In the present case, this negative capacitive current is not due to a physical flow of ions across the membrane, but rather due to a movement of positive ions from the external bulk solution toward the external surface of the membrane and a concomitant movement of negative ions from the internal bulk solution toward the internal surface of the membrane. The

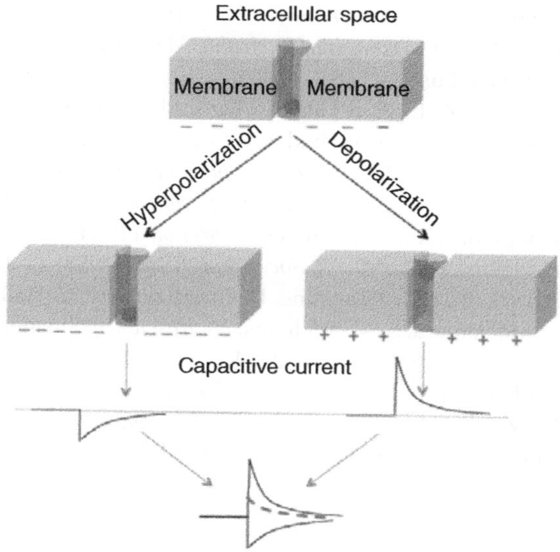

Figure 4.21 Scheme of the position of a positive gating charge in a membrane, before and after its hyperpolarization and depolarization, together with the corresponding currents. The sum of the two currents (dashed curve) yields the gating current.

sum of the capacitive and gating currents following depolarization is positive. The positive gating current is, then, obtained by summing the negative capacitive current of hyperpolarization to the total positive current of depolarization (dashed curve in Fig. 4.21). It should be noted that analogous conclusions could be drawn by hypothesizing a negative gating charge located in close proximity to the extracellular side of the membrane at the resting potential.

To measure the gating current of the potassium channel in the squid giant axon, it was necessary to eliminate the I_K and I_{Na} currents due to potassium and sodium flows along the corresponding ion channels, since they are much greater than the gating current. This goal was achieved by blocking the sodium channels with tetrodotoxin and removing potassium ions from the solutions internal and external to the axon (Gilly and Armstrong, 1980). By so doing, the potassium channel retains its voltage dependence and can move its gating charge without giving rise to a flow of potassium ions while opening.

References

Aidley, D. J., *The Physiology of Excitable Cells*, Cambridge University Press, Cambridge, U.K., 1998.

Catteral, W. A., *Trends Neurosci.* 9 (1986) 7–10.

De Felice, L. J., *Electrical Properties of Cells. Patch Clamp for Biologists*, Plenum, New York, 1997.

Einstein, A., *Ann. der Physik* 322 (1905) 549–560.

Fick, A., *Ann. der Physik* 94 (1855) 59–86.

Gilly, W. F., C. M. Armstrong, *Biophys. J.* 29 (1980) 485–492.

Goldman, D. E., *J. Gen. Physiol.* 27 (1943) 37–60.
Hille, B., *Ion Channels of Excitable Membranes*, Sinauer Associates, Sunderland, Massachusetts, U.S.A., 2001.
Hodgkin, A. L., A. F. Huxley, *Nature* 140 (1939) 710–711.
Hodgkin, A. L., A. F. Huxley, *J. Physiol.* 104 (1945) 176–195.
Hodgkin, A. L., B. Katz, *J. Physiol.* (*London*) 108 (1949) 37–77.
Hodgkin, A. L., A. F. Huxley, *J. Physiol.* 116 (1952a) 449–472; 473–496; 497–506.
Hodgkin, A. L., A. F. Huxley, *J. Physiol.* 117 (1952b) 500–544.
Hodgkin, A. L., *Proc. R. Soc. Lond. B* 148 (1958) 1–37.
Purves, D., G. J. Augustine, D. Fitzpatrick, W. C. Hall, A.-S. LaMantia, J. O. McNamara, S. M. Williams, *Neuroscience*, Sinauer Associates, Sunderland, Massachusetts, U.S.A., 2004.
Schlögl, R., *Z. Physik. Chem.* 1 (1954) 305–320.
Sigworth, F. J., *J. Physiol.* (*Lond.*) 307 (1980) 97–129.
Zhou, Y., J. H. Morais-Cabral, A. Kaufman, R. MacKinnon, *Nature* 414 (2001) 43–48.

5

Active Transport

5.1 The Ion Pumps

Only a limited number of integral proteins, called ion pumps, have the ability to couple an exergonic chemical or photochemical reaction to a flow of ions across a membrane against their electrochemical potential gradient. This coupling of a scalar process (the chemical or photochemical reaction) to a vectorial process (the ion translocation) can only take place in a heterogeneous system provided by a biological membrane interposed between its interior and exterior solutions. Only some inorganic ions, such as Na^+, K^+, Ca^{2+}, and H^+, are translocated specifically by the ion pumps. Ion pumps can be subdivided into three classes: (i) ATPases; (ii) enzymes of the electron-transport chain, which translocate protons; and (iii) light-driven proton pumps, such as bacteriorhodopsin (Läuger, 1991).

ATPases are integral membrane proteins that couple the exergonic reaction of hydrolysis of ATP into ADP and free phosphate ion (P_i) to the endergonic transport of an ion across a membrane against its electrochemical potential gradient. This coupling is widely used in all known forms of life. There are different types of ATPases, which differ in function (ATP synthesis and/or hydrolysis), structure, and type of transported ion. They can be classified into P-type, F-type, V-type, and A-type ATPases.

P-type ATPases, also known as E_1–E_2 ATPases, are a large group of evolutionarily related ion pumps that are found in bacteria, archaea, and eukaryotes. They are referred to as P-type ATPases because they catalyze phosphorylation of a key conserved aspartate residue within the pump by ATP. Vanadate, a transition-state analog of phosphate, inhibits this reaction, since it competes with phosphate for the specific binding site in the ion pump, preventing its phosphorylation. Magnesium ion is necessary for the biological activity of P-type ATPases; it forms a complex with ATP and binds to a regulatory site of the phosphorylated enzyme, from which it is released during dephosphorylation. In the course of the pumping cycle, P-type ATPases assume two principal conformations, E' (also denoted by E_1) and E'' (also denoted by E_2), with inward- and outward-facing ion-binding sites. Most members of this transporter family are specific for the pumping of

Bioelectrochemistry of Biomembranes and Biomimetic Membranes, First Edition. Rolando Guidelli.

a large number of cations (H^+, Na^+, K^+, Mg^{2+}, Ca^{2+}, Ag^+ and Ag^{2+}, Zn^{2+}, Co^{2+}, Pb^{2+}, Ni^{2+}, Cd^{2+}, Cu^+, and Cu^{2+}). One subfamily is also involved in flipping phospholipids, to maintain the asymmetric nature of the biomembrane. Prominent examples of P-type ATPases are the sodium–potassium pump (Na^+, K^+-ATPase), the plasma membrane proton pump (H^+-ATPase), the proton–potassium pump (H^+, K^+-ATPase), and the calcium pump (Ca^{2+}-ATPase). P-type ATPases have a comparatively simple structure. Thus, H^+-ATPase and Ca^{2+}-ATPase consist of only a single polypeptide chain, about 100 kD in molecular weight.

F-type (or F_0F_1) ATPases are *ATP synthetases* (also called *ATP synthases*) that couple the flow of protons down their electrochemical potential gradient to the synthesis of ATP from ADP and phosphate ion. ATP synthetases can also function in the reverse direction, by using the energy released by ATP hydrolysis to pump protons against their electrochemical potential gradient. F-type ATPases are present in the inner mitochondrial membrane of eukaryotic cells, in the thylakoid membrane of plant cells and in the plasma membrane of bacteria. They have a complex structure consisting of a hydrophilic F_1 component, which protrudes from the membrane and contains the catalytic site, and a hydrophobic F_0 component, which spans the membrane and acts as a channel for protons. The F_0 and F_1 components are linked together by a central stalk and a peripheral stalk (Freeman, 2005). F_0 consists of a subunit c-ring (probably comprising eight copies) and one copy each of subunits a, b, d, F_6, and OSCP (Fig. 5.1). Subunits b, d, F_6, and OSCP form the peripheral stalk, which lies on one side of the complex. F_1 is composed of three copies of each of subunits α and β, and one each of subunits γ, δ, and ε. F_1 subunits γ, δ, and ε constitute the central stalk. The energy released by proton flow down its electrochemical potential gradient causes rotation of the ring of c subunits in F_0 (relative to subunit a), along with subunits γ, δ,

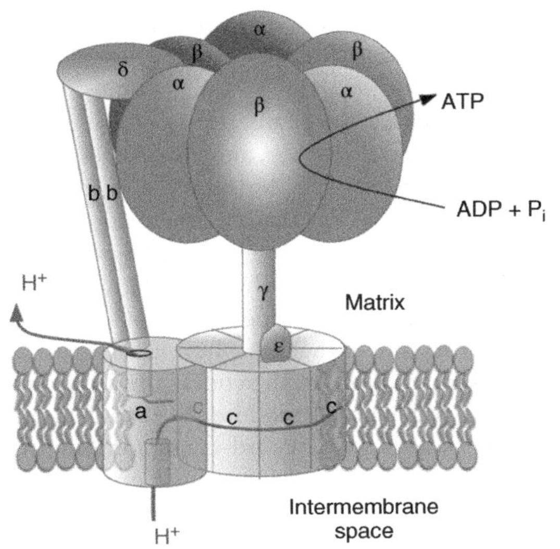

Figure 5.1 Scheme of ATP synthase, including the H^+ channel through which protons flow down into the matrix, resulting in ATP production.

and ε in F_1, to which it is attached. ATP synthetases can, therefore, be mechanically divided into "rotor" (c-ring, γ, δ, and ε) and "stator" ($\alpha_3\beta_3$, a, b, d, F_6, and OSCP) components. Protons enter F_0 via a half-channel contained in subunit *a* and bind consecutively to the aspartate residues of the different sections composing the c-ring, determining both its rotation and that of the γ subunit, which is tightly attached to the c-ring. Protons are then transferred from the c ring to the other half-channel contained in the a subunit and ultimately released into the matrix. The mechanical energy of the rotation is transmitted to the F_1 $\alpha_3\beta_3$ hexamer, where it is converted into energy for ATP synthesis. This rotary catalysis can be explained by a "binding-change" mechanism. This mechanism predicts ATP synthesis and ATP hydrolysis at catalytic sites located in each of the three β subunits, at the interface with an adjacent α subunit. In the case of ATP synthesis, each site switches cooperatively through conformations in which ADP and P_i bind, ATP is formed, and then released. ATP hydrolysis uses the same pathway, but in reverse. These transitions are caused by rotation of the γ subunit. F-type ATPases differ from P-type ATPases, in that they are not sensitive to vanadate and lack a stable phosphorylated intermediate. Conversely, they are sensitive to *N,N′*-dicyclohexylcarbodiimide (DCCD), which reacts with carboxyl groups and typically inhibits enzymes that translocate protons.

V-type ATPases are proton pumps that are found in cellular organelles, such as lysosomes and vacuoles of fungi and higher plants, and couple ATP hydrolysis to proton transport from the cytosol to the inside of these organelles, where they decrease pH. They are highly conserved evolutionarily ancient enzymes with remarkably diversified functions in eukaryotic organisms. V-type ATPases resemble F-type ATPases in their multimeric structure, by their insensitivity to vanadate and by the lack of a phosphorylated intermediate. A-type ATPases function similar to F-type ATPases and are found in a group of single-celled microorganisms called *archaea*.

Electron-transport chains are particular proton pumps used for extracting energy from redox reactions, such as the oxidation of nutrients (respiratory chain), and also from sunlight in photosynthesis. They consist of a sequence of redox couples incorporated in integral or peripheral proteins of the inner mitochondrial membrane and of the thylakoid membrane of plants, and characterized by progressively increasing reduction potentials; each redox couple accepts electrons from the redox couple that precedes it in the series, and donates them to the redox couple that follows (cf. Section 3.4.2). In so doing, each of these redox reactions translocates a number of protons. The resulting proton electrochemical gradient is used to generate chemical energy in the form of ATP, via ATP-synthase.

A very simplified mechanism of a light-driven proton pump is shown in Fig. 5.2 (Läuger, 1991).

All ion pumps are characterized by two main conformational states, a state E′ with the binding site exposed to the inside of the membrane, and a state E″ with the binding site exposed to its outside. The electrochemical potential of protons on the internal side of the membrane is lower than on its external side.

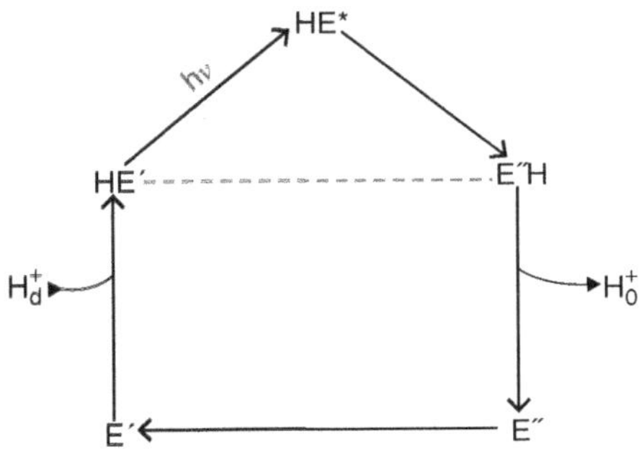

Figure 5.2 Enzymatic cycle of a light-driven proton pump. The intracellular space is on the left side, the extracellular space on the right side. The dashed step is kinetically inhibited.

A light-driven proton pump is in charge of translocating protons from its internal to its external side, that is, against their electrochemical potential gradient. Absorption of a light quantum in state HE′, formed by proton binding to the E′ state, determines a primary excited state HE*, which rapidly decays to the more stable conformational state E″H, where the proton is exposed to the outside. The intermediate E″H has a higher Gibbs energy than E′H, and hence a much lower concentration in the absence of illumination. The main effect of light absorption consists in creating a high nonequilibrium concentration of states HE* and E″H. At this point, the direct transition from state E″H to state HE′ is thermodynamically feasible, since it is exergonic, but the corresponding decrease in Gibbs energy would be completely dissipated into heat. For an effective coupling of light absorption to proton translocation, the direct transition E″H → HE′ must be kinetically inhibited, so that state E″H returns to the initial state mainly along the pathway E″H → E″ → E′ → HE′. In the course of this sequence of steps, a proton is released to the outside of the membrane and another one is taken up from its inside.

Figure 5.3 shows a schematic picture of the enzymatic process of *bacteriorhodopsin*, the simplest proton pump known so far. It is found in the *Halobacterium*, a prokaryotic microorganism of the domain *Archaea* that lives in high-salt environments (Henderson et al., 1990). This bacterium has purple membrane patches that act as tiny solar panels. The purple membranes contain a tightly packed array of bacteriorhodopsin, which uses the energy from sunlight to pump protons out of the cell. Each bacteriorhodopsin molecule contains seven helices that traverse the membrane.

Upon intense illumination, bacteriorhodopsin undergoes a cyclic sequence of transitions between photochemical intermediate states, denoted by J, K, L, M, N, and O, which differ by their spectral properties. During the first two very fast transitions, the retinal chromophore, attached to the amino group of the lysine 216 residue as a protonated Schiff base, isomerizes from the *all-trans* to the *13-cis* form upon absorbing a light quantum, and brings the proton close to the carboxyl group of the Asp-85 residue. During the L to M transition, the proton of the Schiff base is transferred to Asp-85. The protonation of Asp-85 facilitates the deprotonation of a group of neighboring protonated residues, among which Glu-204, Glu-194, and Arg-82, called the "protonation release group" (PRG), which releases a proton to the extracellular side of the purple membrane. During the M to N

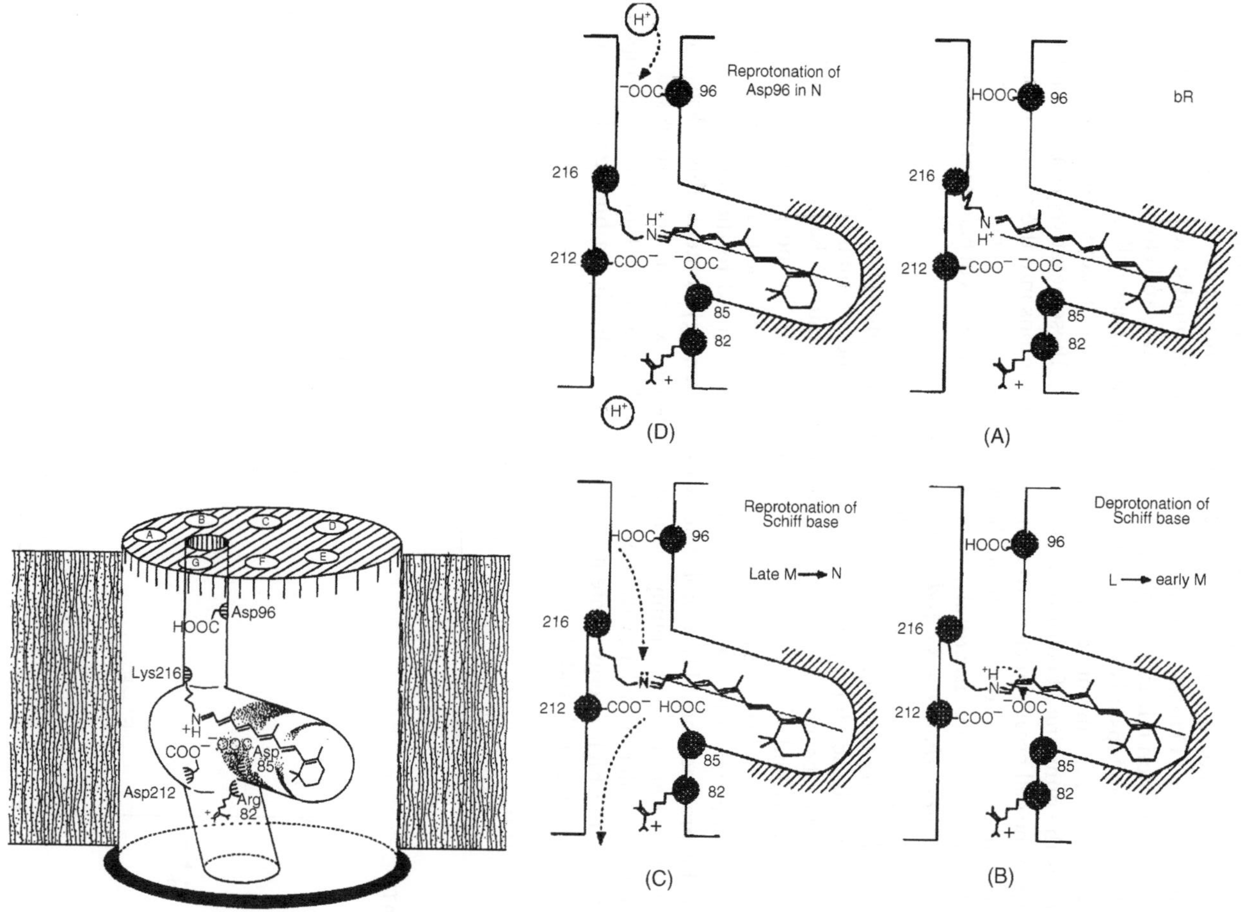

Figure 5.3 Scheme of bacteriorhodopsin and its enzymatic cycle. Source: Henderson et al. (1990). Reproduced with permission of Elsevier.

transition, the Schiff base is reprotonated by Asp-96, which is closer to the cytoplasmic side. During the N state Asp-96 regains a proton from the cytoplasmic side, while in the immediately subsequent N to O transition the retinal chromophore isomerizes back from the *13-cis* to the *all-trans* form. Finally, during the transition from O to bacteriorhodopsin (bR) in the ground state, Asp-85 reprotonates the proton release group restoring the initial state and closing the cycle.

Bacteriorhodopsin is homologous to another molecule, called *halorhodopsin*, also found in the membrane of *Halobacterium*. In halorhodopsin, however, light drives the pumping of chloride ions into the cell. These negatively charged ions inside the cell create an attractive force to draw protons into the cell through an ATPsynthase, thus contributing to the proton-motive force generated by bacteriorhodopsin. *Halobacterium* also possesses other homologs of bacteriorhodopsin, called *sensory rhodopsins I and II* (SRI and SRII).

The reaction scheme of P-type proton ATPases is more complex, because it involves the binding of a molecule of inorganic phosphate (P_i) and its subsequent release during an enzymatic cycle. Under equilibrium conditions and in the absence of ATP, the conformational state E′ has a high affinity for protons and is mainly present as HE′ (Fig. 5.4). In the presence of free phosphate ions, HE′ has a very low tendency to bind them according to the equilibrium $HE' + P_i \Leftrightarrow HE' - P$, which is notably shifted to the left. However, in the presence of ATP, a molecule with high energy content, this releases a phosphate group to the pump via a spontaneous exergonic reaction, resulting in the transfer of a large fraction of the ATP energy content to the pump. Due to the high energy content of state HE′ – P, this has a high tendency to dissociate into HE′ and P_i. This final, low-energy state can be attained following

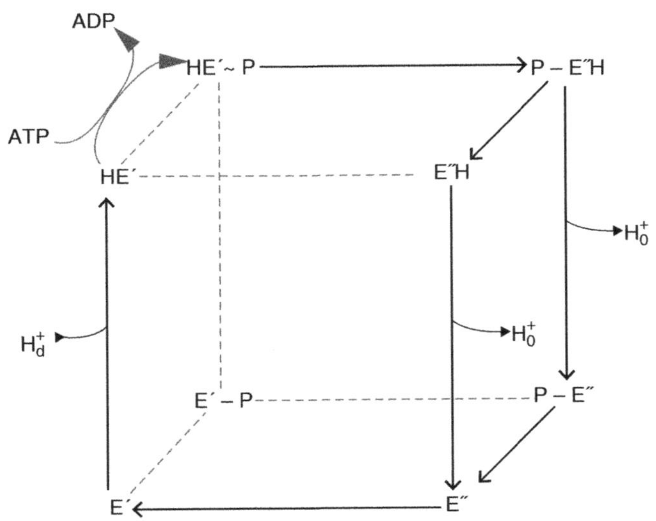

Figure 5.4 Enzymatic cycle of a P-type proton ATPase. The intracellular space is on the left side, the extracellular space on the right side. The dashed steps are kinetically inhibited.

different pathways. For an efficient coupling, the return to the HE′ state must occur via the reaction sequence HE′ − P → P − E″H → PE″ → E″ → E′ → HE′ or HE′ − P → P − E″H → E″H → E″ → E′ → HE′, which leads to the translocation of a proton against its electrochemical potential gradient.

The most direct pathway consists of the single step HE′ − P → HE′; a longer sequence leading to the same final result is HE′ − P → E′ − P → E′ → HE′. These two reaction sequences, referred to as *slippages*, are normally very slow, and hence kinetically inhibited, albeit thermodynamically feasible. Nonetheless, in real ion pumps, these sequences may sometimes occur at a detectable rate and are called *intrinsic uncouplings*. Their net result is a complete dissipation of the Gibbs energy of ATP hydrolysis into heat. In some cases, slippages may have functional roles beneficial to the organism. One possible role is to act as a buffer between proton transport and processes that exploit the resulting proton-motive force, so that these different processes can be independently regulated to some extent. Some organisms may also need to vary the heat/(proton-motive force) ratio produced from the energy source, for example, in order to regulate Gibbs energy production independently of heat production requirement. Slippage may occur in all ion pumps, including proton-pumping electron-transport chains. An uncoupled downhill proton translocation may take place via the exergonic sequence E″ → E″H → HE′ → E′ → E″, whose net result is the translocation of a proton from the outside to the inside of the membrane, down its electrochemical potential gradient, without ATP hydrolysis. In this case, the pump acts as an ion carrier. This represents a further example of intrinsic uncoupling. To prevent this pathway, the step HE′ ⇔ E″H is normally very slow.

These general features of proton pumping are also valid for other ion pumps and can be summarized as follows: (i) The pump performs a cycle of reactions $E_1 \rightarrow E_2 \rightarrow \cdots \rightarrow E_n \rightarrow E_1$ that involves transitions between various conformational states, with binding sites exposed either to the inside or to the outside of the membrane. (ii) One particular reaction of the pumping cycle, $E_1 \rightarrow E_2$, which is notably shifted to the left, is driven to the right by the energy input from a suitable external source, thus creating a high nonequilibrium concentration of state E_2. (iii) The direct back-reaction, $E_2 \rightarrow E_1$, is kinetically inhibited. The return to the initial state E_1 is only possible through the sequence $E_1 \rightarrow E_2 \rightarrow \cdots \rightarrow E_n \rightarrow E_1$, during which one or more ions are translocated across the membrane against their electrochemical potential gradients.

In what follows, the group of conformations exposing the binding site to the inside of the membrane is marked with a single prime (E′), those exposing the binding site to the outside with a double prime (E″). The primary active transport governed by the aforementioned rules does not necessarily require different binding affinities of the two E′ and E″ groups of conformations of the enzyme for the translocating ion. Nonetheless, a high binding affinity on one side of the membrane, with an easy ion uptake, and a low binding affinity on the other side, with an easy ion release, are encountered with the majority of ion pumps. This feature favors a high *turnover rate*, that is, the reciprocal of the time spent by the pump to perform a single enzymatic cycle.

5.2 Electromotive Force and Inversion Potential of Ion Pumps

An ion pump converts a large fraction of an exergonic reaction into the electrochemical potential gradient of the translocated ion (or ions). It should be noted that the *driving force* of an exergonic reaction, as expressed by Eqs. 3.33 and 3.34, is positive, and is equal to the difference between the overall electrochemical potential of reactants and that of products. This quantity can also be loosely referred to as the *Gibbs energy stored* in the exergonic reaction. Conversely, the *Gibbs energy change* ΔG that accompanies this reaction is unavoidably negative, and is equal in magnitude but opposite in sign to its driving force. The driving force measures the maximum useful work that may be provided by the exergonic reaction; it contributes to the entropy production of Eq. 3.33 to an extent that depends on its possible coupling with an endergonic process.

In the case of an ATPase, a single enzymatic cycle involves the hydrolysis of an ATP molecule into ADP and P_i and an increase in the electrochemical potential gradient of the ions translocated by the pump. The Gibbs energy decrease ΔG_{ATP} that accompanies ATP hydrolysis is given by the difference between the electrochemical potentials of products and those of reactants, each multiplied by its own stoichiometric coefficient:

$$\tilde{\mu}_{ADP} + \tilde{\mu}_{P_i} - \tilde{\mu}_{ATP} = \left(\mu^{\circ}_{ADP} + RT\ln c_{ADP} - 3F\phi\right) + \left(\mu^{\circ}_{P_i} + RT\ln c_{P_i} - F\phi\right) - \left(\mu^{\circ}_{ATP} + RT\ln c_{ATP} - 4F\phi\right), \tag{5.1}$$

where ϕ is the electric potential of the solution containing the three ionic species. The phosphate ion, P_i, is mainly present in the monovalent form $H_2PO_4^-$ at physiological pH. Obviously, within a single homogeneous phase, the electric potential has no effect on the driving force of a chemical reaction, since the latter is electronically balanced. Equation 5.1 may be written in a more concise form:

$$\Delta G_{ATP} = \Delta G^{\circ}_{ATP} + RT\ln\left(c_{ADP}c_{P_i}/c_{ATP}\right) \quad \text{with } \Delta G^{\circ}_{ATP} = \mu^{\circ}_{ADP} + \mu^{\circ}_{P_i} - \mu^{\circ}_{ATP}, \tag{5.2}$$

where ΔG°_{ATP} is the standard Gibbs energy of the reaction.

Let $\sum_h \nu_h \Delta\tilde{\mu}_h$ denote the sum of the changes in the electrochemical potentials of the ions translocated during a single enzymatic cycle of the pump, where ν_h is the number of translocated ions of the hth species. For a proton pump, this sum is given by

$$\Delta\tilde{\mu}_H = \tilde{\mu}''_H - \tilde{\mu}'_H = RT\ln\left(c_{H,0}/c_{H,d}\right) + F\left(\phi_0 - \phi_d\right), \tag{5.3}$$

where a single prime denotes a species on the internal side of the membrane, while a double prime denotes a species on its external side. Maintaining the symbolism adopted in Chapter 2, concentrations and electric potentials on the external side of the membrane are denoted by the subscript 0, those on its internal side by the subscript d. In the case of the Na^+, K^+-ATPase, the sum is written as

$$3\left(\widetilde{\mu}''_{\mathrm{Na}} - \widetilde{\mu}'_{\mathrm{Na}}\right) + 2\left(\widetilde{\mu}'_{\mathrm{K}} - \widetilde{\mu}''_{\mathrm{K}}\right) = 3RT\ln\left(c_{\mathrm{Na,0}}/c_{\mathrm{Na,d}}\right) + 2RT\ln\left(c_{\mathrm{K,d}}/c_{\mathrm{K,0}}\right) + F\left(\phi_0 - \phi_{\mathrm{d}}\right). \tag{5.4}$$

The sums in Eqs. 5.3 and 5.4 are written in such a way as to be positive during the normal function of the pump. In this way, the sum expresses the increase in the osmotic and electric Gibbs energies produced by the pump.

Imagine starting from a situation in which the driving force of the ion flows, $\sum_h \nu_h \Delta\widetilde{\mu}_h$, equals zero. According to the thermodynamics of irreversible processes, an ion flow occurring when its driving force is zero is referred to as a *level flow*. In the presence of ATP, the pump builds up an electrochemical potential gradient of the translocating ion (or ions). If the degree of coupling equals −1 (maximum coupling) and if leakage paths are negligible, the system attains an equilibrium situation in which the electrochemical potential gradient exactly counterbalances the chemical driving force, $-\Delta G_{\mathrm{ATP}}$, and the ion flow ceases. In this ideal situation, the whole driving force of ATP hydrolysis is converted into osmotic and electric Gibbs energies, and the equilibrium condition states that the change in Gibbs energy accompanying the whole process equals zero:

$$\Delta G_{\mathrm{ATP}} + \sum_h \nu_h \Delta\widetilde{\mu}_h = 0. \tag{5.5}$$

This equation allows us to derive the equilibrium transmembrane potential. Thus, by combining this equation with Eq. 5.3, the equilibrium potential for a proton pump is given by

$$\begin{aligned} &\Delta G_{\mathrm{ATP}} + \Delta\widetilde{\mu}_{\mathrm{H}} = \Delta G_{\mathrm{ATP}} + RT\ln\left(c_{\mathrm{H,0}}/c_{\mathrm{H,d}}\right) + F\left(\phi_0 - \phi_{\mathrm{d}}\right) = 0 \\ &\rightarrow \Delta\phi \equiv \left(\phi_{\mathrm{d}} - \phi_0\right) = \Delta G_{\mathrm{ATP}}/F + \frac{RT}{F}\ln\left(c_{\mathrm{H,0}}/c_{\mathrm{H,d}}\right), \end{aligned} \tag{5.6}$$

where $(RT/F)\ln(c_{\mathrm{H,0}}/c_{\mathrm{H,d}}) = \Delta\phi_{\mathrm{H}}$ is the Nernst potential of protons, referred, as usual, to the electric potential outside the membrane, taken conventionally equal to zero. Analogously, in the case of Na^+, K^+-ATPase, the equilibrium potential is given by

$$\begin{aligned} &\Delta G_{\mathrm{ATP}} + 3RT\ln\left(c_{\mathrm{Na,0}}/c_{\mathrm{Na,d}}\right) - 2RT\ln\left(c_{\mathrm{K,0}}/c_{\mathrm{K,d}}\right) + F\left(\phi_0 - \phi_{\mathrm{d}}\right) = 0 \\ &\rightarrow \Delta\phi \equiv \phi_{\mathrm{d}} - \phi_0 = \Delta G_{\mathrm{ATP}}/F + 3\Delta\phi_{\mathrm{Na}} - 2\Delta\phi_{\mathrm{K}}, \end{aligned} \tag{5.7}$$

where $\Delta\phi_{\mathrm{Na}}$ and $\Delta\phi_{\mathrm{K}}$ are the Nernst potentials of sodium and potassium ions.

The equilibrium potential $\Delta\phi$ is sometimes called *electromotive force of the pump*. A real pump never converts completely the chemical energy into osmotic and electric energies, due to intrinsic uncoupling and/or leakage paths. Even in this case, a transmembrane potential at which the ion flow vanishes exists. However, it no longer corresponds to an equilibrium situation, since the suppression of the ion flow is maintained at the expense of a certain rate of ATP hydrolysis; in the thermodynamics of irreversible processes, this state is referred to as a *static head*. This transmembrane potential is called *inversion potential*; its absolute value is less than the electromotive force of the pump, for obvious thermodynamic reasons.

In the most unfavorable situation of total slippage, the pump operates under conditions of level flow, namely without creation of electrochemical potential gradients, with complete dissipation of the Gibbs energy stored in ATP hydrolysis. In a real pump, the Gibbs energy change accompanying an enzymatic cycle of the pump takes the value

$$\Delta G_{\text{cycle}} = \Delta G_{\text{ATP}} + \sum_h \nu_h \Delta\tilde{\mu}_h = -\Delta G_{\text{dissip}} < 0. \tag{5.8}$$

This change in Gibbs energy must be necessarily negative for ion pumping to proceed spontaneously. Its opposite measures the Gibbs energy, ΔG_{dissip}, dissipated during the cycle and contributing to the entropy production. The limiting case of a rigorously coupled pump corresponds to $\Delta G_{\text{cycle}} = 0$, while the limiting case of total slippage corresponds to $\Delta G_{\text{cycle}} = \Delta G_{\text{ATP}}$. It should be clear that a certain dissipation of the Gibbs energy stored in ATP hydrolysis is a necessary requisite for any pump, and it would be present even in the hypothetical case of a rigorously coupled pump. In fact, a positive dissipation energy, albeit small, is required to sustain a finite turnover rate.

5.3 Energy Levels of the Enzymatic Cycle of Ion Pumps

The energy transduction performed by ion pumps is not the result of a single step of its enzymatic cycle, but rather of the cycle as a whole. In spite of this, some steps of the cycle are more important than others. To understand this feature, it is import to gain insight into the energy levels of the intermediate states of the cycle. The introduction of these energy levels draws its justification from the consideration that these macromolecular states are relatively long-lived, so that they may be considered as in equilibrium with respect to the internal degrees of freedom of the protein (e.g., vibrations of the peptide backbone, rotations of amino acid side chains). Accordingly, long-lived conformational states of an ionic pump can be treated as distinct chemical species, with a well-defined chemical potential. The number of these conformational states cannot be considered as settled, and further investigations of ion pumps may well reveal additional states so far unknown (Läuger, 1991).

Let us consider, for example, the enzymatic cycle of a P-type proton pump. If x_i is the mole fraction of the pump in state E_i, the chemical potential $\mu(E_i)$ of the *quasi-species* E_i is expressed by the usual relation:

$$\mu(E_i) = \mu^{\circ}(E_i) + RT\ln x_i. \tag{5.9}$$

The mole fractions x_i are defined by $x_i = N_i/N$, where N_i is the average number of pumps in state E_i and N is the total number of pumps. The standard chemical potential $\mu^{\circ}(E_i)$ expresses the Gibbs energy per mole of the pump, when all the N pumps are exclusively in state E_i (i.e., when $x_i = 1$). While $\mu(E_i)$ depends on the kinetic parameters of the whole cycle via x_i, the standard chemical potential $\mu^{\circ}(E_i)$ is a rigorously molecular property of state E_i. In Eq. 5.9, it is implicitly assumed that the single pump molecules do not interact with each other within the membrane.

When the pump operates under turnover conditions, that is, through a continuous series of enzymatic cycles, a situation in which all N ion pumps are in the same state E_i is never realized. To determine $\mu^{\circ}(E_i)$, it is, therefore, necessary to investigate the kinetics of the pump, in an attempt to single out the various steps connecting each state to the subsequent one and to determine the forward and backward rate constants of these steps. Consider, for example, the ith step connecting state E_i to state E_{i+1}:

$$E_i \underset{b_i}{\overset{f_i}{\rightleftharpoons}} E_{i+1}, \tag{5.10}$$

where f_i and b_i are the forward and backward rate constants of this step, respectively. The rate constants of the various steps may be determined from a kinetic investigation carried out under *steady-state conditions*, when all steps proceed at the same rate, or else under *pre-steady-state conditions*, that is, before attaining the steady state (e.g., by perturbing a system in steady state with an abrupt jump of temperature, electric potential, or concentration of an activating substance, and by following the system while it relaxes toward a new steady state).

If we imagine a hypothetical equilibrium condition of the pump, the rates of all steps composing its enzymatic cycle must vanish, in view of the *principle of microscopic reversibility*. This principle is based on the consideration that equilibrium has a dynamical nature and, as such, it is the result of the balance between the forward and backward reaction rates. In other words, the forward and backward rates of each single elementary step of a given process must be equal at equilibrium. Thus, for the ith step, we can write

$$v_i = f_i x_i - b_i x_{i+1} = 0, \tag{5.11}$$

where x_i and x_{i+1} are the mole fractions of the pump in states E_i and E_{i+1}. From chemical thermodynamics, we also know that, at equilibrium, the chemical potentials of these two states are equal:

$$\mu_i^{\circ} + RT\ln x_i = \mu_{i+1}^{\circ} + RT\ln x_{i+1} \quad \text{with } \mu_i^{\circ} \equiv \mu^{\circ}(E_i)\,; \;\; \mu_{i+1}^{\circ} \equiv \mu^{\circ}(E_{i+1})\,. \tag{5.12}$$

From Eqs. 5.11 and 5.12, it follows that

$$f_i/b_i = x_{i+1}/x_i = \exp\left[-\left(\mu_{i+1}^{\circ} - \mu_i^{\circ}\right)/RT\right] \equiv K_i, \tag{5.13}$$

where K_i is the equilibrium constant for the ith step. This equation allows the determination of the standard chemical potentials of all steps, except for one, from the forward and backward rate constants, provided kinetic measurements are carried out not too far from equilibrium.

Let us assume that the enzymatic cycle of the pump consists of the following consecutive steps, starting from state E_1 and returning to this state at the end of the cycle:

$$E_1 \underset{b_1}{\overset{f_1}{\rightleftharpoons}} E_2 \underset{b_2}{\overset{f_2}{\rightleftharpoons}} \cdots E_i \underset{b_i}{\overset{f_i}{\rightleftharpoons}} E_{i+1} \cdots E_{n-1} \underset{b_{n-1}}{\overset{f_{n-1}}{\rightleftharpoons}} E_n \underset{b_n}{\overset{f_n}{\rightleftharpoons}} E_1. \quad (5.14)$$

The sum of the mole fractions of all states of this cycle is clearly equal to unity:

$$x_1 + x_2 + \cdots + x_{n-1} + x_n = \sum_{j=1}^{n} x_j = 1. \quad (5.15)$$

The product of the expressions in Eq. 5.13 relative to all the steps of the cycle included between the ith step and any further jth step is given by

$$\frac{x_{i+1}}{x_i}\frac{x_{i+2}}{x_{i+1}}\cdots\frac{x_{j-1}}{x_{j-2}}\frac{x_j}{x_{j-1}} = \frac{x_j}{x_i} = \exp\left(-\frac{\mu_j^{\circ} - \mu_i^{\circ}}{RT}\right) = \frac{e^{-\mu_j^{\circ}/RT}}{e^{-\mu_i^{\circ}/RT}}. \quad (5.16)$$

Substituting x_j from Eq. 5.16 into Eq. 5.15, we obtain

$$\sum_{j=1}^{n} x_j = 1 = \frac{x_i}{e^{-\mu_i^{\circ}/RT}}\sum_{j=1}^{n} e^{-\mu_j^{\circ}/RT} \quad \rightarrow \quad x_i = e^{-\mu_i^{\circ}/RT} \Big/ \sum_{j=1}^{n} e^{-\mu_j^{\circ}/RT}. \quad (5.17)$$

This equation shows that only states with low standard chemical potentials μ_i° are populated under equilibrium conditions (i.e., for $\Delta G_{cycle} = 0$).

By multiplying by themselves the expressions of Eq. 5.13 relative to all the steps of the cycle of Eq. 5.14, we obtain

$$\frac{f_1 f_2 \cdots f_{n-1} f_n}{b_1 b_2 \cdots b_{n-1} b_n} = \frac{x_2 x_3 \cdots x_n x_1}{x_1 x_2 \cdots x_{n-1} x_n} = 1 = \exp\left(-\frac{\mu_1^{\circ} - \mu_1^{\circ}}{RT}\right). \quad (5.18)$$

This relation is based on the principle of microscopic reversibility and is referred to as the *principle of detailed balance*. In this equation, $(\mu_1^{\circ} - \mu_1^{\circ})$ expresses the Gibbs energy change, $\Delta G_{cycle} = 0$, which accompanies an enzymatic cycle when the pump is rigorously coupled, that is, when it operates under equilibrium conditions and the driving forces of ion flow and of ATP hydrolysis exactly compensate each other. It could be demonstrated that the relationship of Eq. 5.18 holds even under steady-state conditions, when $\Delta G_{cycle} < 0$:

$$\frac{f_1 f_2 \cdots f_{n-1} f_n}{b_1 b_2 \cdots b_{n-1} b_n} = \exp\left(-\frac{\Delta G_{cycle}}{RT}\right). \quad (5.19)$$

This equation shows that, of the $2n$ rate constants f_i and b_i, only $(2n - 1)$ are actually independent.

The elementary steps considered so far are monomolecular. Identical conclusions can be extended to pseudomonomolecular steps, that is, steps involving a ligand L whose concentration during pumping remains substantially constant; the ligand may be an ATP molecule or a translocating ion, such as H^+, K^+, Na^+, or Ca^{2+}. To this end, it is just sufficient to include the electrochemical potential of the ligand L in the chemical potential of the state that is expected to bind the ligand. Thus, for example, the standard chemical potential of an ith state E_i involved in a step that foresees its binding to the ligand L,

$$E_i + L \underset{b_i}{\overset{f_i = f_i^* c_L}{\rightleftharpoons}} E_i - L, \tag{5.20}$$

is defined as

$$\mu_i^{\varnothing} = \mu_i^{\varnothing}(E_i) + \tilde{\mu}_L^{\varnothing}. \tag{5.21}$$

Obviously, the forward rate constant of the corresponding step, f_i, is proportional to the concentration c_L of the ligand, according to a proportionality constant f_i^*, which is a concentration-independent quantity expressed in $M^{-1}\,s^{-1}$. Analogously, the equilibrium constant, $K_i = K_i^* c_L$, of the given step is proportional to the ligand concentration.

Let us apply the aforementioned concepts to the proton ATPase in Fig. 5.4; many of these concepts can be easily extended to other P-type ATPases. The conformational states of the enzymatic cycle of the pump can be defined in different ways, depending on the way in which the ligand molecules are ascribed to the states of the pump. Here, the standard chemical potentials of the different conformational states are defined as follows:

$$\mu_1^{\varnothing} = \mu^{\varnothing}(HE') + \mu_P;\ \mu_2^{\varnothing} = \mu^{\varnothing}(HE'P);\ \mu_3^{\varnothing} = \mu^{\varnothing}(PE''H);\ \mu_4^{\varnothing} = \mu^{\varnothing}(PE'') + \tilde{\mu}_H'';$$
$$\mu_5^{\varnothing} = \mu^{\varnothing}(E'') + \tilde{\mu}_H'' + \mu_P;\ \mu_6^{\varnothing} = \mu^{\varnothing}(E') + \tilde{\mu}_H'' + \mu_P. \tag{5.22}$$

The left side of the diagram in Fig. 5.5B reports the driving force provided by ATP hydrolysis, $-\Delta G_{ATP} = \mu_{ATP} - \mu_{ADP} - \mu_P$. If the pump starts from state $HE' + P_i$ and moves clockwise along the cycle depicted in Fig. 5.5A, it returns to state HE′ at the end of the cycle. However, this new state, $(HE' + P_i + H_0^+ - H_d^+)$, differs from the initial state, $HE' + P_i$, by the translocation of one proton against its electrochemical potential gradient, $\Delta\tilde{\mu}_H$:

$$\left(\mu_1^{\varnothing}\right)^* = \mu_1^{\varnothing} + \Delta\tilde{\mu}_H = \mu^{\varnothing}\left(HE'\right) + \mu_P + \tilde{\mu}_H'' - \tilde{\mu}_H'. \tag{5.23}$$

The levels of the standard Gibbs energies in Fig. 5.5B depict qualitatively the energetic properties typical of all P-type ATPases. The major change in standard Gibbs energy takes place in the $HE' + P_i \rightarrow HE' \sim P$ step. The acylphosphate bond in state $HE' \sim P$ is highly energetic and is herein denoted by the tilde symbol ~; its bound phosphate has an energy comparable with that of the terminal phosphate residue in the ATP molecule.

It should be noted that a highly energetic bond is unstable, since it tends to release a part of its high Gibbs energy; conversely, a stable bond is characterized by a low Gibbs energy. Thus, the $ME' \sim P$ (M = H, Na, Ca) reacts readily with ADP to yield ATP. The high energy level of $HE' \sim P$ (and also of $Ca_2E' \sim P$ or $Na_3E' \sim P$) is an important property of the enzymatic cycle, because it allows the pump to store a large fraction of the Gibbs energy of the ATP hydrolysis reaction. In step $HE' \sim P \rightarrow P - E''H$, the pump undergoes a conformational transition that causes the binding sites to be exposed on the opposite side of the membrane. As a rule,

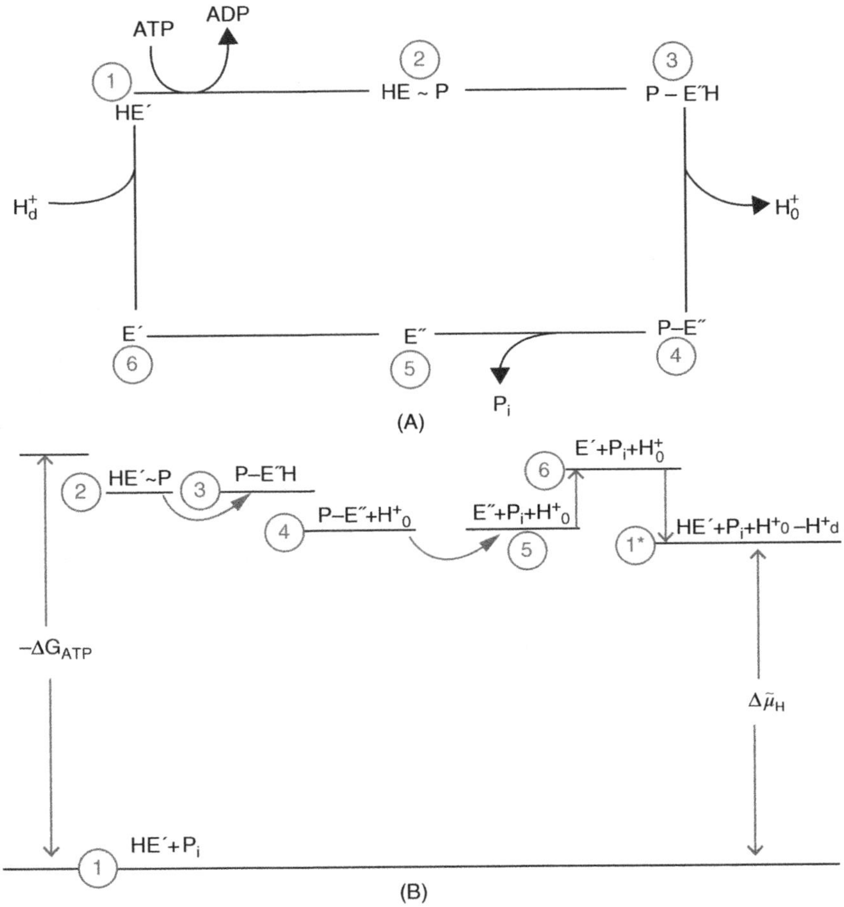

Figure 5.5 Enzymatic cycle of a P-type proton ATPase (A) and the corresponding Gibbs energy levels (B).

this conformational change is accompanied by a notable change in the affinity of the binding site for the translocating ion. In the case of Ca^{2+}-ATPase of the sarcoplasmic reticulum, for example, the affinity for calcium ion is from three to four orders of magnitude higher in state E′ than in state E″. The kinetic advantage of this affinity change is evident, because calcium ion is taken up in state E′ from the cytoplasm at submicromolar concentrations and released into the sarcoplasmic reticulum at millimolar concentrations.

The change in ion affinity is accompanied by a transition of the acylphosphate bond from a high to a low energy level. This statement is supported by the observation that the protein in conformation E″ can be phosphorylated at the active site by free phosphate ion, P_i. This reaction, called *backdoor phosphorylation*, is possible when the bond of the phosphate residue in E″ – P is stabilized by additional noncovalent interactions with the protein. Therefore, in spite of appreciable variations in the binding affinities of phosphate, the standard Gibbs energies of $ME' \sim P$ and $ME'' - P$ (with $M = H, Na, Ca$) are usually comparable. In other words, the decrease in Gibbs energy of the acylphosphate bond in passing from the $ME' \sim P$ to the $ME'' - P$ conformational state is to a large extent compensated for by an increase in the Gibbs energy of the bond of the translocating ion to the pump, which facilitates its release. Taking the Ca^{2+}-ATPase of the sarcoplasmic reticulum as an example, the equilibrium constant of the reaction $Ca_2E' \sim P \rightarrow Ca_2E'' - P$ is close to unity. Consequently, the standard Gibbs energy levels of states $HE' \sim P$ and $HE'' - P$ have been set approximately equal in Fig. 5.5B. Both reaction $P - E''H \rightarrow P - E'' + H_0^+$ and reaction $E' + H_d^+ \rightarrow HE'$ are considered as exergonic in the figure; for $\widetilde{\mu}''_H > \widetilde{\mu}'_H$, this implies a large change in the affinity of the bond, as already stated. Naturally, the difference in Gibbs energy between $P - E''H$ and $P - E''$ and between HE' and E' depends on the actual values of the proton concentration and on the transmembrane potential. The only reaction that is considered as endergonic in the enzymatic cycle model of Fig. 5.5B, apart from the first step $HE' + P_i \rightarrow HE' \sim P$, is the conformational transition $E'' \rightarrow E'$.

Diagrams of standard Gibbs energy levels, such as that in Fig. 5.5B, allow us to depict optimal conditions for energy transduction. A particularly favorable situation is encountered when the difference in standard Gibbs energy between states $HE' \sim P$ and $HE' + P_i$ is close to the opposite, $-\Delta G_{ATP}$, of the standard Gibbs energy change for ATP hydrolysis, that is, its standard driving force. If this difference is much less than $-\Delta G_{ATP}$, the energy stored by the ion pump is small, and the driving force that moves the subsequent steps is also small. On the other hand, if the difference between the standard Gibbs energies of states $HE' \sim P$ and $HE' + P_i$ is greater than $-\Delta G_{ATP}$, state $HE' \sim P$ is rarely formed and the turnover rate is too low. For analogous reasons, the steps of the enzymatic cycle are particularly advantageous when they are approximately isoenergetic. Thus, a highly exergonic step involves a strong energy dissipation. On the other hand, a step $i \rightarrow j$ that is endergonic under standard conditions tends to proceed at a very low rate. In this connection, it should be noted that such a step may well take place, provided that the ratio x_i/x_j of the mole fractions of states i and j under steady-state conditions is high enough to make the difference $\mu_j - \mu_i = \mu_j^\circ - \mu_i^\circ + RT\ln(x_j/x_i)$ negative, and hence the step exergonic, even if $(\mu_j^\circ - \mu_i^\circ) > 0$.

It is a notable fact that ion pumps may use different energy sources, such as light, redox energy, or ATP hydrolysis. The elementary process of absorption of a light quantum by bacteriorhodopsin is fundamentally different from an electron exchange in an electron-transport chain fueled by redox reactions, or from the transfer of a phosphate residue from ATP to an aspartyl group of an ATPase. In spite of this, all these different ion pumps are capable of coupling a highly exergonic reaction to an endergonic ion translocation. It is, therefore, legitimate to wonder whether the function of these different types of ion pumps relies on common basic principles.

In summary, some properties are undoubtedly common to all known ion pumps. Thus, an ion pump may assume a certain number of states, which differ by the presence or absence of ligands, such as ions or phosphate residues, and by the inward and outward configurations of the binding sites of the translocating ions. A state E′ of low Gibbs energy and a state E″ of high Gibbs energy are energetically important and present in all ion pumps. In the absence of a reaction capable of providing energy, the ion pump is mainly present in state E′, while state E″ is populated only to a minor extent. In the presence of an external energy source, the concentration of state E″ increases far above its equilibrium level. In particular, the high-energy state E″ of an ion pump activated by light is populated under illumination via an excited state E*, while that of a P-type ATPase is phosphorylated via state E′ ~ P. An efficient use of the supplied Gibbs energy requires a kinetic selection among possible reaction pathways. A reaction connecting directly state E″ to state E′ must be avoided; instead, the ion pump must follow a sequence of reactions capable of determining ion translocation from one side of the membrane to the opposite side.

5.4 Kinetics of Ion Pumps Under Steady-State Conditions

Let us now consider the kinetics of a perfectly coupled pump operating under steady-state conditions according to a cyclic sequence of steps through states $E_1, E_2, \ldots, E_n, E_1$, as shown in Fig. 5.6. During a single enzymatic cycle, ν ions of charge number z are translocated from one side of the membrane to the other. Consequently, the Gibbs energy change associated with a single cycle is given by

$$\Delta G_{\text{cycle}} = \Delta G + \nu\Delta\tilde{\mu} = \Delta G + \nu\Delta\mu + \nu zF\left(\phi_0 - \phi_{\text{d}}\right). \tag{5.24}$$

Here, ΔG is the Gibbs energy change undergone by the external energy source and $\Delta\tilde{\mu} \equiv \tilde{\mu}'' - \tilde{\mu}'$ is the electrochemical potential difference that accompanies the translocation of the ion. In this reaction scheme, the rate constants of the different steps in the forward and backward directions are denoted by f_i and b_i, respectively;

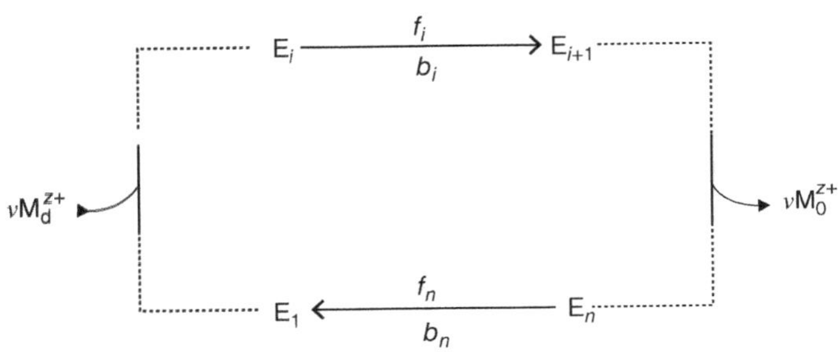

Figure 5.6 Enzymatic cycle in which only the initial and final states and a generic intermediate step of the cycle are shown.

f_i and b_i are monomolecular or pseudomonomolecular rate constants (expressed in s^{-1}), which may incorporate the concentrations of the translocating ion or of an activating substance and may depend upon the transmembrane potential. Thus, for example, the rate constant for the phosphorylation of an ATPase may be written $f_i = f_i^* c_{ATP}$, where f_i^* is a concentration-independent bimolecular rate constant, expressed in $M^{-1}\,s^{-1}$.

The distribution of the states of the cycle may be described by their mole fractions $x_i = N_i/N$, where N_i is the number of pumps in state E_i and N is the total number of pumps. Obviously, one has

$$\sum_{i=1}^{n} x_i = 1. \tag{5.25}$$

Under steady-state conditions, all steps proceed at the same rate, such that the mole fractions in the various states are time independent. The rate at which a given state E_i is generated from the immediately preceding state, E_{i-1}, is equal to the rate at which the given state generates the immediately subsequent state, E_{i+1}. Hence, we can write

$$\begin{aligned} dx_i/dt &= f_{i-1}x_{i-1} - b_{i-1}x_i - \left(f_i x_i - b_i x_{i+1}\right) \\ &= f_{i-1}x_{i-1} + b_i x_{i+1} - \left(f_i + b_{i-1}\right) x_i = 0 \quad \text{with } i = 1, \dots, n. \end{aligned} \tag{5.26}$$

In view of the constraint of Eq. 5.25, only $(n-1)$ out of the n algebraic equations 5.26 are independent.

It should be noted that each electrogenic step of the cycle, $E_i \rightarrow E_{i+1}$, involves a charge equal to $\alpha_i F$, where α_i is its *dielectric coefficient* (for the significance of dielectric coefficient, cf. Section 5.5). Clearly, the sum of the charges involved in all steps of the cycle is equal to the overall charge vzF translocated during the cycle: $F\sum_i \alpha_i = vzF$. Single dielectric coefficients appear in the expressions of the rate, $f_i x_i - b_i x_{i+1}$, of each electrogenic step of the cycle. In particular, if an electrogenic step is rate determining, its rate depends on the transmembrane potential $\Delta\phi$ and can be expressed on the basis of the rate theory of ion transport across membranes (Schultz, 1980) (see Section 4.7 and Eq. 4.86). The electric potential difference involved in such a step is the fraction, $\alpha_i \Delta\phi$, of the whole transmembrane potential $\Delta\phi$ across the portion of the dielectric medium covered by the charge that moves during the given step. Denoting by α the charge transfer coefficient expressing the symmetry of the potential energy barrier, the rate of the given step is expressed by the following equation:

$$\begin{aligned} f_i x_i - b_i x_{i+1} &= f_i^0 x_i \exp\left[(1-\alpha) f \alpha_i \Delta\phi\right] - b_i^0 x_{i+1} \exp\left(-\alpha f \alpha_i \Delta\phi\right) \\ &\text{with } f \equiv F/(RT). \end{aligned} \tag{5.27}$$

Here, f_i^0 and b_i^0 are the forward and backward rate constants at zero transmembrane potential, $\Delta\phi = 0$. In applying Eq. (4.86), the charge transfer coefficient α is usually equated to 0.5, upon assuming a symmetric potential energy barrier. Naturally, for the nonelectrogenic steps of the enzymatic cycle, the dielectric coefficient α_i is set equal to zero in Eq. 5.27. Upon solving the set of $n-1$ algebraic equations expressed by Eq. 5.27 so as to derive the various x_i values, one obtains the turnover rate ρ, which under steady-state conditions is equal to the rate of any step, $E_i \rightarrow E_{i+1}$, of the cycle. The pump current I is just given by the turnover rate ρ, multiplied by the charge, vzF, of the ions translocated in a single enzymatic cycle and by the mole number, N, of the pump, in order to convert mole fractions into moles:

$$I = vzF\rho = vzFN\left(f_i x_i - b_i x_{i+1}\right) \quad \text{with } i = 1, 2, \ldots n. \tag{5.28}$$

The $(n-1)$ relationships in Eq. 5.27 can be solved analytically or numerically. An analytical expression of I can be obtained using the so-called diagram method (Läuger, 1991). A particularly simple expression of I is obtained when the forward rate constant, f_i, of each single step is much greater than the corresponding backward rate constant, b_i, to such an extent as to permit us to ignore the backward rates of all steps with respect to the corresponding forward rates; these steps are said to be *irreversible*. This situation is encountered when the Gibbs energy dissipated during the cycle, ΔG_{dissip}, is high. In this case, I is given by

$$I = vzFN\left(\sum_i 1/f_i\right)^{-1}. \tag{5.29}$$

This equation shows that the steps that make the major contribution to the ion flux are those with the lower rate constants, as expected. These steps are improperly called *slow steps*, in spite of the fact that all steps proceed at the same rate under steady-state conditions.

When the number of steps in the cycle is higher than five or six, the analytical expression provided by the diagram method is particularly complex, and it is preferable to solve the set of $n-1$ relations in Eq. 5.27 numerically. Some rate constants in these equations are bimolecular and, hence, depend upon the concentration of the translocating ion or of a substrate (e.g., ATP or ADP); some others depend upon the transmembrane potential. This allows experimental changes in turnover rate with varying the concentration of the translocating ion or of a substrate, or also with varying the transmembrane potential, to be compared with those predicted on the basis of the solution of Eqs. 5.28. By so doing, the set of rate constants providing the best fit of the calculated changes in turnover rate to the experimental ones can be determined.

As a rule, the forward and backward rates of several steps are much greater than their difference, which expresses the net rate. In this case, the forward and backward rates in Eq. 5.27 can be set equal to each other to a good approximation, yielding a pseudo-Nernst equation:

$$\left(x_{i+1}/x_i\right) = \left(f_i^0/b_i^0\right) \exp\left(f\alpha_i \Delta\phi\right). \tag{5.30}$$

In this expression, the turnover rate is exclusively affected by the ratio of the forward and backward rate constants, that is, the corresponding equilibrium constant.

Usually, these steady-state measurements are carried out by the patch-clamp technique in the whole-cell configuration, using a wide-tipped pipette in combination with a device for exchanging the solution inside the pipette. Possible ion channels in the membrane patch are blocked by specific inhibitors. The ion pump current is obtained upon subtracting the residual current recorded after blocking the pump with one of its specific inhibitors. This system also allows the potential difference between an electrode inside the pipette and an identical electrode in the external solution to be varied.

5.5 Electrogenicity of the Ion Pumps

Before examining the procedures adopted to investigate the kinetics of an enzymatic cycle under pre-steady-state conditions, it is convenient to consider how it is possible to determine the charge involved in a single step of the cycle. In fact, the majority of ion pumps are electrogenic, that is, they translocate a net charge across the membrane. A typical example of electrogenic ion pump is provided by the Na^+, K^+-ATPase of mammalian cells, which translocates three sodium ions from the intracellular to the extracellular side of the plasma membrane and two potassium ions in the opposite direction, thus generating an electric current across the membrane. In this case, the ion pump acts as a current source and contributes to altering the transmembrane potential. An example of nonelectrogenic ion pump is provided by H^+, K^+-ATPase, a P-type ATPase that transports two protons from the cytoplasm of the parietal cells, located in the gastric mucosa, in exchange for two potassium ions retrieved from the gastric lumen, per each hydrolyzed ATP molecule. Despite the fact that in this case the overall ion translocation process is electrically silent, some steps of the enzymatic cycle of the pump induce a charge movement within the membrane, thus producing a temporary change in the transmembrane potential.

Let us consider an electrogenic step $E_i \rightarrow E_{i+1}$ that translocates a charge density Nq_i across a given distance within a membrane of thickness d (Fig. 5.7). Here, q_i is the charge moved by a single pump and N is the number density of the ion pumps.

Let us assume that the charge q_i is located at a distance x_1 from the external surface of the membrane immediately before this step, and at a distance x_2 immediately after it. A constant potential difference $\Delta\phi$ is applied between two identical reference electrodes placed in the solutions that bath the two sides of the membrane. To understand how this charge movement causes a current flow along the external circuit connecting the two reference electrodes, let us consider all the contributions to the constant potential difference $\Delta\phi$, both before and after the charge movement. Let us denote by σ the charge density located on the membrane external surface immediately before step $E_i \rightarrow E_{i+1}$, and by ε_m the dielectric constant of the membrane.

By considerations analogous to those made in Section 2.6 to determine the electric potential profile across a membrane, the potential difference across the dielectric slab enclosed between the membrane external surface, $x = 0$, and the initial location, $x = x_1$, of the charge density Nq_i, is given by $-\sigma x_1/(\varepsilon_0\varepsilon_m)$. Analogously, the potential difference across the dielectric slab enclosed between $x = x_1$

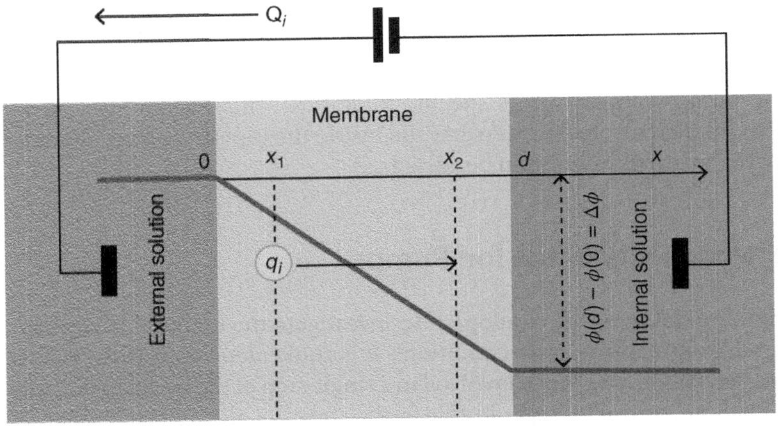

Figure 5.7 Movement of a charge q_i in a membrane over a distance $(x_2 - x_1)$ at constant transmembrane potential, as induced by an electrogenic step, and the resulting movement of a charge Q_i along the external circuit.

and the membrane internal surface, $x = d$, is given by $-(\sigma + Nq_i)(d - x_1)/(\varepsilon_0\varepsilon_m)$. The transmembrane potential $\Delta\phi$ is the sum of the two aforementioned contributions :

$$\Delta\phi = -\left(\sigma x_1/\varepsilon_0\varepsilon_m\right) - \left(\sigma + Nq_i\right)(d - x_1)/(\varepsilon_0\varepsilon_m). \tag{5.31}$$

During the movement of the charge density Nq_i from the $x = x_1$ to the $x = x_2$ location, the charge density on the membrane external surface must pass from the initial value σ to a new value σ', in order to maintain $\Delta\phi$ constant. After such a charge movement, the expression of $\Delta\phi$ is modified as follows:

$$\Delta\phi = -\left(\sigma' x_2/\varepsilon_0\varepsilon_m\right) - \left(\sigma' + Nq_i\right)(d - x_2)/(\varepsilon_0\varepsilon_m). \tag{5.32}$$

Upon equating the two expressions of $\Delta\phi$ before and after the q_i movement, one obtains

$$Q_i \equiv \sigma' - \sigma = \left[\left(x_2 - x_1\right)/d\right] Nq_i. \tag{5.33}$$

Q_i is the charge density that flows along the external circuit toward the membrane external side, to keep the transmembrane potential $\Delta\phi$ constant during the movement of the charge density Nq_i from x_1 to x_2. If q_i is positive and $x_2 > x_1$, the q_i movement would tend to make the membrane internal surface more positive, and the charge Q_i that moves toward the membrane external side is also positive to counterbalance this effect, as appears from Eq. 5.33. The experimentally measurable charge Q_i, once divided by the elementary charge (proton charge) e and by the number density N, yields a dimensionless number, α_i, which characterizes the charge movement associated with step $E_i \rightarrow E_{i+1}$, and is referred to as the *dielectric coefficient* of this step:

$$\alpha_i = \left[\left(x_2 - x_1\right)/d\right]\left(q_i/e\right). \tag{5.34}$$

The quantity Q_i is the charge density that step $E_i \rightarrow E_{i+1}$ would hypothetically move across the whole thickness d of the membrane, rather than across the unknown thickness $x_2 - x_1$. In particular, the quantity $(x_2 - x_1)/d$ expresses the ratio of two lengths only upon assuming that the membrane is a homogeneous dielectric layer of thickness d separating two conducting phases. In reality, a membrane incorporating integral proteins is an inhomogeneous dielectric medium. Henceforth, $(x_2 - x_1)$ should not be considered as a geometric distance, but rather as an effective *dielectric distance*, which may be appreciably different from the geometric distance. Moreover, the charge density Q_i does not exclusively include the charge density of a translocating ion, but also any possible charge (usually small) moved by the conformational change accompanying step $\mathrm{E}_i \rightarrow \mathrm{E}_{i+1}$.

If a single cycle of the pump involves the translocation of ν ions of charge number z, the overall amount of translocated charge equals νzF per mole of the pump. Consequently, if the enzymatic cycle of the pump is composed of n steps, the sum of the dielectric coefficients of all steps must be equal to νz:

$$\sum_{i=1}^{n} \alpha_i = \nu z. \tag{5.35}$$

Naturally, the dielectric coefficient equals zero for all steps that do not involve a charge movement.

Instead of interposing a *bilayer lipid membrane* (BLM) incorporating an ion pump between two aqueous phases, a membrane fragment containing the pump may be adsorbed on a BLM (Hartung et al., 1987) or on a mixed alkanethiol/lipid bilayer tethered to a gold electrode via the sulfhydryl group (Fig. 5.8) (Pintschovius and Fendler, 1999). Let us estimate the charge density Q_i that flows along the external circuit as a consequence of a given step $\mathrm{E}_i \rightarrow \mathrm{E}_{i+1}$. To this end, let us denote by d_b and ε_b the thickness and the dielectric constant of the bilayer supporting the membrane, namely the BLM or the alkanethiol/lipid mixed bilayer. Moreover, let us denote by σ the charge density on the BLM surface opposite to the membrane fragment (or else that on the surface of the gold electrode supporting the mixed bilayer, with the membrane fragment on top) just before the movement of the charge density Nq_i induced by step $\mathrm{E}_i \rightarrow \mathrm{E}_{i+1}$. For convenience, let us place the origin of the x axis, normal to the membrane, at the boundary between the supporting bilayer and the membrane fragment.

Prior to the movement of the charge density Nq_i from x_1 to x_2, the potential difference across the supporting bilayer is given by $-\sigma d_\mathrm{b}/(\varepsilon_0\varepsilon_\mathrm{b})$, that across the membrane section enclosed between $x=0$ and $x=x_1$ by $-\sigma x_1/(\varepsilon_0\varepsilon_\mathrm{m})$, and that across the membrane section enclosed between $x=x_1$ and $x=d$ by $-(\sigma+Nq_i)(d-x_1)/(\varepsilon_0\varepsilon_\mathrm{m})$. The potential difference $\Delta\phi$ across the whole interface, which is kept constant by the potentiostatic system, is the sum of the aforementioned three contributions:

$$\Delta\phi = -\sigma d_\mathrm{b}/(\varepsilon_0\varepsilon_\mathrm{b}) - \sigma x_1/(\varepsilon_0\varepsilon_\mathrm{m}) - (\sigma + Nq_i)(d - x_1)/\varepsilon_0\varepsilon_\mathrm{m}. \tag{5.36}$$

Incidentally, in writing this equation, the possible presence of a nonzero charge density at the boundary, $x=0$, of the supporting bilayer was ignored; nonetheless,

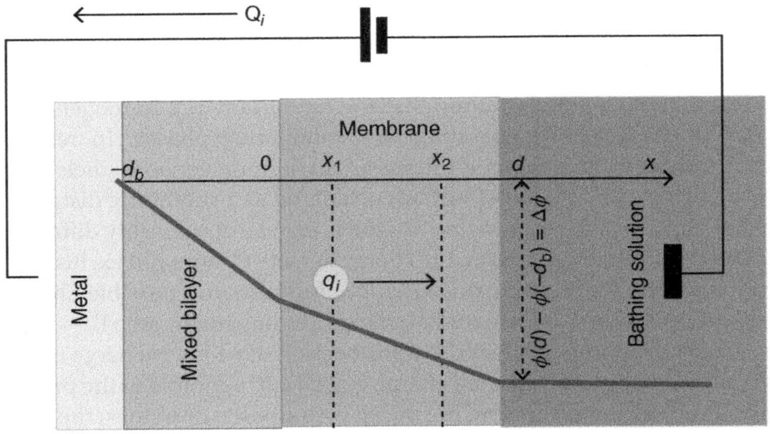

Figure 5.8 Movement of a charge q_i in a membrane fragment over a distance $(x_2 - x_1)$, as induced by an electrogenic step. The membrane fragment is adsorbed on a mixed bilayer anchored to a metal support and is inserted into an electric circuit that maintains the potential difference across the whole electrified interface constant, causing the flow of a charge Q_i along the external circuit.

provided it remains constant, it does not affect the final conclusions. Following the movement of the charge density Nq_i from x_1 to x_2, the charge density on the BLM surface opposite to the membrane fragment, or else on the surface of the gold electrode, takes a new value σ', so as to maintain $\Delta\phi$ constant. We then have

$$\Delta\phi = -\sigma' d_b/(\varepsilon_0\varepsilon_b) - \sigma' x_2/(\varepsilon_0\varepsilon_m) - (\sigma' + Nq_i)(d - x_2)/\varepsilon_0\varepsilon_m. \tag{5.37}$$

On equating the two expressions of $\Delta\phi$ and rearranging terms, we obtain

$$Q_i \equiv \sigma' - \sigma = \frac{x_2 - x_1}{d}\frac{C_b}{C_b + C_m}Nq_i \quad \text{with } C_b \equiv \frac{\varepsilon_0\varepsilon_b}{d_b};\; C_m \equiv \frac{\varepsilon_0\varepsilon_m}{d}. \tag{5.38}$$

C_b and C_m are the capacitances of the supporting bilayer and of the membrane fragment, respectively, in view of the well-known expression for the capacitance of a parallel plate capacitor (cf. Eq. 2.19). It is apparent that, when using this experimental technique, the charge Q_i that flows along the external circuit is only a fraction, $C_b/(C_b + C_m)$, of the charge that would flow if a membrane incorporating the pump were directly interposed between two aqueous solutions. Such conclusion is expressed by stating that this technique is based on a *capacitive coupling* between supporting bilayer and membrane.

5.6 Kinetics of Ion Pumps Under Pre-Steady-State Conditions

Let us now examine the kinetics of an enzymatic cycle under pre-steady-state conditions. Consider a set of identical ion pumps that are initially in the steady state. If this set is perturbed by an abrupt change in an external parameter, the set relaxes to a new steady state. In particular, a temperature jump alters the rate constants, f_i and b_i, that are temperature dependent. An electric potential jump alters the potential-dependent rate constants of the electrogenic steps of the enzymatic cycle. The concentration jump of an activating substance affects the pseudomonomolecular rate constants incorporating the concentration of that substance. The pre-steady-state current transients accompanying these relaxation processes provide useful information on the microscopic parameters of the pump cycle.

Concentration jumps consist in bringing, as rapidly as possible, a substance capable of activating the pump in contact with membrane fragments or with vesicles (*proteoliposomes*) incorporating the ion pump and adsorbed on a suitable support, while keeping the applied potential E constant. The substances that are abruptly injected into the system may be ATP, translocating ions either in the absence or presence of ATP, and many other activating species. The analysis of the current transients elicited by a concentration jump is based on the cycle of reactions already examined for the steady state. The system state before the perturbation is described by the initial values of the various rate constants and of the initial values, $x_i(0)$, of the different mole fractions. Following such a perturbation, the system evolves toward a new steady state, characterized by a new set, $x_i(\infty)$, of mole fraction values. The rate of change of the mole fraction x_i of any given state i with time is still expressed by Eq. 5.26; however, this is obviously no longer equal to zero, as in the case of steady-state conditions:

$$\begin{aligned} \mathrm{d}x_i/\mathrm{d}t &= f_{i-1}x_{i-1} - b_{i-1}x_i - \left(f_i x_i - b_i x_{i+1}\right) \\ &= f_{i-1}x_{i-1} + b_i x_{i+1} - \left(f_i - b_{i-1}\right)x_i \quad \text{with } i = 1, \ldots, n. \end{aligned} \tag{5.39}$$

The pump current I_p is expressed by the following equation:

$$I_\mathrm{p} = FN\sum_i \alpha_i \left(f_i x_i - b_i x_{i+1}\right) \quad \text{with } i = 1, \ldots, n, \tag{5.40}$$

in which the rate of each electrogenic step of the cycle is multiplied by the corresponding charge $\alpha_i F$ per mole, where α_i is the dielectric coefficient of the given step, and N is the number of moles of the ion pump. The set of differential equations in Eq. 5.39 can be solved numerically. To obtain the capacitive current*I*that flows along the external circuit and yields the current transient, a further differential equation accounting for the instantaneous activation of the pump is required.

Pre-steady-state measurements are often carried out by adsorbing membrane fragments or proteoliposomes on one side of a BLM or on a gold-supported alkanethiol/phospholipid bilayer. This system is represented by the equivalent circuit shown in Fig. 5.9. The mesh between the A and B junctions simulates the membrane fragment, while that between the B and C junctions simulates the supporting bilayer (either a BLM or a gold-supported alkanethiol/phospholipid bilayer). The membrane fragment consists of a current source generating a *pump current* I_p, with in parallel the capacitance C_p and resistance R_p of the fragment.

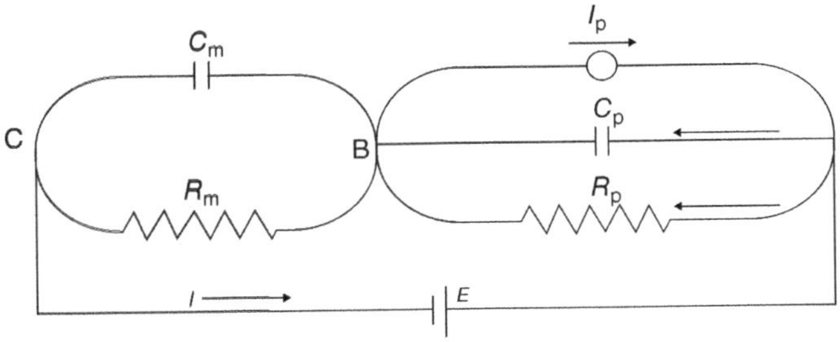

Figure 5.9 Equivalent circuit simulating a membrane fragment adsorbed on a lipid bilayer and closed with an external applied potential *E*.

The supporting bilayer is represented by a further $R_m C_m$ mesh in series with the membrane fragment. The equivalent circuit is closed with the external applied potential E. The current source is activated at time $t = 0$ and deactivated at time $t = T$ by a *gate function*, $G(t, T)$, that is, a function representing a rectangular pulse of unit height, which starts at $t = 0$ and lasts for a time T.

The analysis of this equivalent circuit yields the two differential equations (Bartolommei et al., 2008):

$$I = I_p G(t, T) + C_p \frac{d\Delta\phi_p}{dt} + \frac{\Delta\phi_p}{R_p}, \tag{5.41a}$$

$$I = \frac{\Delta\phi_m}{R_m} + C_m \frac{d\Delta\phi_m}{dt}. \tag{5.41b}$$

Equation 5.41a expresses that fact that the overall current I flowing both along the external circuit and across the membrane fragment is equal to the sum of the pump current I_p (during the activation period $0 < t < T$), the capacitive current, $C_p \, d\Delta\phi_p/dt$, and the resistive current, $\Delta\phi_p/R_p$, where $\Delta\phi_p$ is the potential difference across the membrane fragment. Equation 5.41b states that the overall current I is also equal to the sum of the capacitive current, $C_m \, d\Delta\phi_m/dt$, and the resistive current, $\Delta\phi_m/R_m$, of the supporting bilayer in series with the membrane fragment, where $\Delta\phi_m$ is the potential difference across the supporting bilayer. In practice, R_m is so high that the first term in Eq. 5.41b can be neglected to a good approximation. At constant applied potential E, we have $E = (\Delta\phi_p + \Delta\phi_m)$ and $d\Delta\phi_m/dt = -d\Delta\phi_p/dt$. Combining the latter equation with Eqs. 5.41a and 5.41b yields the further differential equation:

$$\frac{d\Delta\phi_p}{dt} = -\frac{I_p}{C_m + C_p} - \frac{\Delta\phi_p}{R_p (C_m + C_p)}. \tag{5.42}$$

The capacitance C_m and resistance R_m can be measured directly on the supporting bilayer in the absence of the adsorbed membrane fragment. As a rule, the quantities C_p and R_p can be accurately determined, so as to attain the best fit of the experimental current I by the calculated one; in particular, C_p is usually close to $1 \, \mu F \, cm^{-2}$.

The resulting system of $(n+1)$ differential equations can be readily solved numerically, for example, by the *fourth-order Runge–Kutta method*. After obtaining the mole fractions of all the states of the pump and $\Delta\phi_p$ as a function of time, the pump current I_p is obtained from Eq. 5.40. The current I is then calculated from the following equation:

$$I = -C_m \, d\Delta\phi_p/dt, \tag{5.43}$$

which is obtained from Eq. 5.41b by disregarding the $\Delta\phi_m/R_m$ term and noting that $d\Delta\phi_m/dt$ equals $-d\Delta\phi_p/dt$.

This approach is rigorous, but requires a detailed knowledge of the kinetics of all steps of the cycle, not all attainable by electrophysiological or electrochemical means. Biochemical methods may be of help in providing the kinetics of a number of steps, even though they cannot establish their electrogenicity. Upon using the available kinetic data provided by biochemical methods, the fit of experimental current transients by current transients calculated by numerical solution of the aforementioned set of differential equations may yield useful information about the electrogenicity of certain steps of the enzymatic cycle of ion pumps. For example, the black curve in Fig. 5.10 shows a typical current transient following an ATP concentration jump on proteoliposomes incorporating the *sarco/endoplasmic reticulum* Ca^{2+}*-ATPase* (SERCA) and adsorbed on an octadecanethiol/phospholipid bilayer anchored to gold, in the presence of free Ca^{2+} ions at pH 7 (Bartolommei et al., 2008). The gray curve was calculated by solving the system of differential equations in Eqs. 5.39 and 5.42, upon using kinetic parameters available in the literature and ascribing an appropriate set, $\{\alpha_i\}$, of dielectric coefficients to the steps considered to be electrogenic.

If the perturbation is small enough not to shift the system too much from its initial steady state, the solution of the set of differential equations in Eq. 5.39 takes the simplified form:

$$x_i(t) = \sum_{j=1}^{n} a_{ij} \exp\left(-t/\tau_j\right) + x_i(\infty). \tag{5.44}$$

The time constants τ_j and amplitudes a_{ij} are functions of the various rate constants f_j and b_j. From Eqs. 5.40 and 5.44, it follows that the pump current has an analogous form (Borlinghaus et al., 1987):

$$I_p(t) = \sum_{j=1}^{n} a_j \exp\left(-t/\tau_j\right) + b. \tag{5.45}$$

In practice, the dependence of I_p on time is expressed as a sum of exponentially decaying contributions plus a constant contribution b, which represents the stationary pump current. The differential equation for $\Delta\phi_p$, as obtained by substituting $I_p(t)$ from Eq. 5.45 into Eq. 5.42, can be solved by the use of Laplace transforms (Tadini Buoninsegni et al., 2004). The expression of the current I, for both $t < T$ and $> T$, is then obtained from Eq. 5.43. In addition to the time constants τ_j in Eq. 5.45, the expression for $I(t)$ contains a further exponentially decaying term

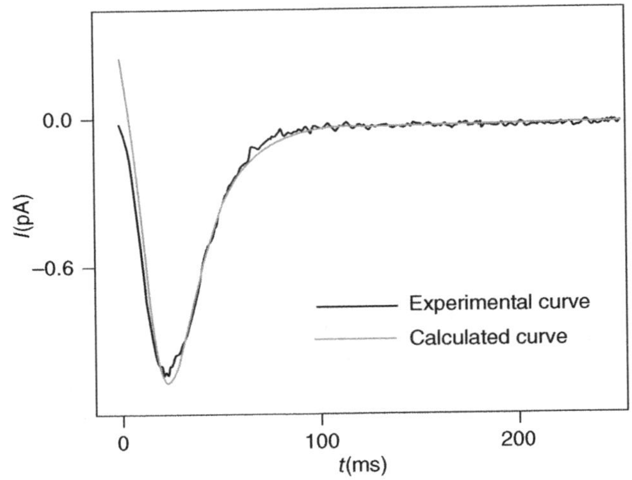

Figure 5.10 Experimental (black curve) and calculated current transient (gray curve) following a 100 μM ATP concentration jump on proteoliposomes incorporating Ca^{2+}-ATPase and adsorbed on an octadecanethiol/DOPC mixed bilayer anchored to gold, in the presence of 100 μM free Ca^{2+} at pH 7. Source: Bartolommei et al. (2008). Reproduced with permission of Elsevier.

with a time constant $\tau_c = (C_p + C_m)R_m R_p/(R_m + R_p)$, which depends exclusively on the resistive and capacitive elements of the equivalent circuit, as well as the time constant, $\tau_m = R_m C_m$, of the supporting bilayer.

As soon as the pump is activated, the pump current flows along the R_p and C_p network elements in the direction of the arrows in Fig. 5.9, toward the supporting bilayer. Under these conditions, the capacitive coupling with the $R_m C_m$ mesh causes electrons to flow along the external circuit to the supporting bilayer and, hence, the experimental current $I(t)$ to flow in the opposite direction. This is due to the potentiostatic system, which keeps the potential difference across the whole electrified interface constant. Consequently, the potential difference $\Delta\phi_p$ across the membrane fragment (positive toward the supporting membrane), built up by the ion pump, is instantaneously compensated for by an equal and opposite potential difference $\Delta\phi_m$ across the supporting bilayer, built up by the flow of electrons along the external circuit to the supporting bilayer; this corresponds to a negative capacitive current from the supporting bilayer toward the membrane fragment. The capacitance C_p remains substantially charged until the pump is inactivated, at time $t = T$.

When examining both the activation and the subsequent inactivation of an ion pump, the current transient following the activation is called *on-current* (I_{on}), whereas that following the subsequent inactivation is referred to as *off-current* (I_{off}). As soon as the pump is inactivated, C_p undergoes discharge, causing a positive capacitive current I_{off} with time constant τ_c. The inset in Fig. 5.11 shows the current transient due to activation of Ca^{2+}-ATPase by a Ca^{2+} concentration jump

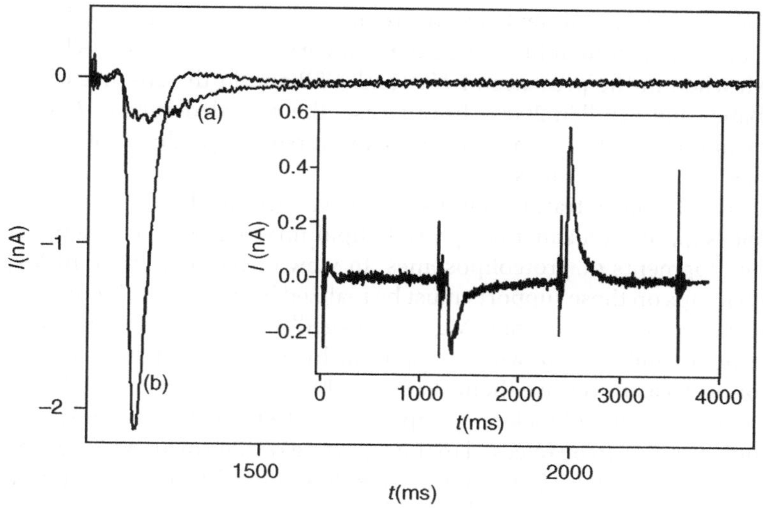

Figure 5.11 Current transients on proteoliposomes containing Ca^{2+}-ATPase and adsorbed on an octadecanethiol/DOPC bilayer anchored to gold: (a) following a 28.2 μM Ca^{2+} concentration jump in the absence of ATP; (b) following a 100 μM ATP concentration jump in the presence of 28.2 μM Ca^{2+}. The inset shows a current transient caused by a 28.2 μM Ca^{2+} concentration jump in the absence of ATP (*on-current*), as well as the subsequent *off-current* following a rapid displacement of the Ca^{2+}-containing solution by an inactivating solution differing from the activating one by the absence of Ca^{2+}. Source: Tadini Buoninsegni et al. (2004). Reproduced with permission of Elsevier.

in the absence of ATP (on-current), followed by a current transient due to inactivation of the calcium pump by removal of Ca^{2+} ion from the bathing solution. Much before the end of the activation period, $0 < t < T$, the I_{on} current vanishes, in spite of the fact that the pump current I_p maintains a stationary value b due to continuous pumping. In fact, I_{on} is sensitive to I_p only as long as the latter changes with time, so as to produce a change in $\Delta\phi_p$ and, hence, in $I_{on} = -C_m \, d\Delta\phi_p/dt$.

If the stationary pump current b is too low, it may happen that, during the activation period, the I_{on} current, after attaining a negative current peak, becomes slightly positive before vanishing. This positive current is due to a rapid decrease in I_{on}, which causes the capacitance C_p of the membrane fragment to be temporarily discharged across R_p, causing the potential difference $\Delta\phi_p$ across C_p to become negative, with a resulting inversion in the sign of the capacitive I_{on} current, called *overshoot*; a mild overshoot is exhibited by the I_{on} transient (b) in Fig. 5.11.

The current I flowing along the external circuit can be made to coincide with the pump current I_p if the ions pumped across the membrane fragments adsorbed on the supporting bilayer may also be allowed to traverse the latter bilayer. This is only possible by using a BLM as the supporting bilayer, and by making it permeable to the ion translocated by the pump upon addition of an ion carrier that can shuttle the ion across the BLM. For example, addition of monensin, a polyether ionophore complexing monovalent cations, removes the Na^+ ion accumulation within the space between membrane fragments incorporating Na^+, K^+-ATPase and the BLM, when in the presence of Na^+ and ATP and in the absence of K^+ (Fig. 5.12) (Babes and Fendler, 2000). In this case, the recorded current I coincides with the pump current I_p. Thus, it shows no overshoot and does not vanish at longer times, but maintains its stationary value b.

As already stated, concentrations jumps can be carried out by using either a BLM or a gold-supported alkanethiol/phospholipid mixed bilayer as a support for membrane fragments or proteoliposomes. In view of the fragility of BLMs, concentration jumps on these supports must be realized by using *caged* activating substances (e.g., "caged" ATP, "caged" Mg^{2+}, or "caged" Ca^{2+}), that is, substances blocked by a photolabile group, which is split off by an intense light flash in a few milliseconds, so as to release them (Läuger, 1991). For example, caged ATP is a derivative of ATP with a blocking group attached to the terminal phosphate residue, the one destined to be released to the aspartate residue of an ATPase. The substance released from its cage cannot be rapidly removed from the solution. Hence, no off-current can be recorded on a BLM support.

The limitation imposed by the availability of a "caged" activated substance is removed if the membrane fragments are adsorbed on a *solid-supported membrane* (SSM). The mechanical stability of the latter allows a jet of solution of an activating substance to be impinged on its surface. The time required by the concentration of the activating substance to attain its bulk value in direct contact with the surface of the membrane fragments or proteoliposomes is longer by this hydrodynamic procedure than by flash photolysis of a caged activating substance. This prevents this method from appreciating relaxation time constants τ_j

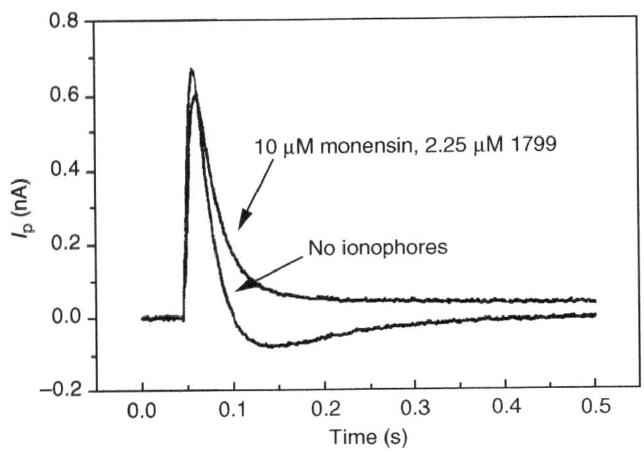

Figure 5.12 Current transients following 0.225 mM ATP concentration jumps on membrane fragments containing Na^+, K^+-ATPase and adsorbed on a BLM, in a pH 6.2 solution of 0.13 M NaCl and 3 mM $MgCl_2$, both in the absence and presence of monensin and of the protonophore 1799. Source: Babes and Fendler (2000). Reproduced with permission of Elsevier.

in Eq. 5.45 less than 200 ms (Tadini Buoninsegni et al., 2004), while flash photolysis can detect τ_j values less than 5 ms (Hartung et al., 1987). The SSM consists of a gold electrode obtained by evaporation on a glass plate, on which a monolayer of a long-chain alkanethiol is anchored via its sulfhydryl group (Pintschovius and Fendler, 1999). An unsaturated phospholipid monolayer is then self-assembled on top of the thiol monolayer, with the hydrocarbon tails of the lipid directed toward those of the thiol. The gold electrode is inserted in a two-electrode system with a reference electrode, under computer control. On the mixed thiol/lipid bilayer, membrane fragments or proteoliposomes are easily adsorbed, as shown schematically in Fig. 5.13 for the case of a proteoliposome containing Ca^{2+}-ATPase.

By directing a rapid flux of ATP solution to the coated gold electrode in the presence of Ca^{2+} ions, the calcium pumps turning the cytoplasmic side toward the external solution pump Ca^{2+} ions into the vesicle, imparting a positive charge to the solution side of the electrified interface. At constant applied potential, an electron charge of equal magnitude flows to the electrode surface along the external circuit in order to maintain the electroneutrality of the whole electrified interface. This flow is recorded as a negative capacitive current transient, namely an on-current I_{on}. After a few seconds, the activating solution is rapidly displaced by a nonactivating solution, which contains the same ingredients as the activating one, except for the activating substance (in the present case, ATP). The fast switching from the activating to the nonactivating solution marks the end of the activation period and the start of the positive off-current, I_{off}.

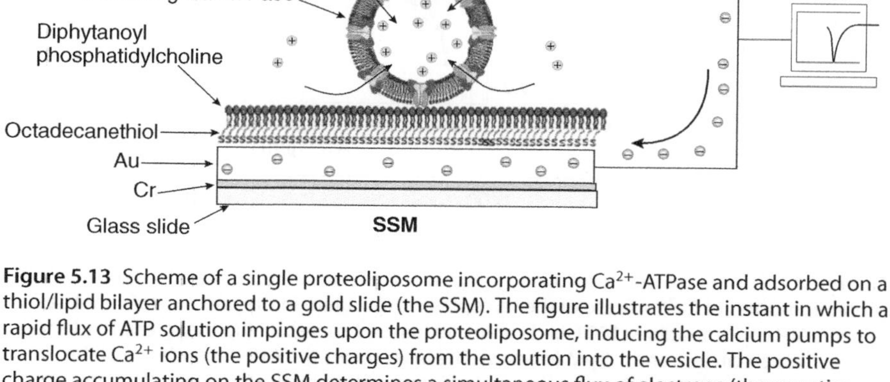

Figure 5.13 Scheme of a single proteoliposome incorporating Ca^{2+}-ATPase and adsorbed on a thiol/lipid bilayer anchored to a gold slide (the SSM). The figure illustrates the instant in which a rapid flux of ATP solution impinges upon the proteoliposome, inducing the calcium pumps to translocate Ca^{2+} ions (the positive charges) from the solution into the vesicle. The positive charge accumulating on the SSM determines a simultaneous flux of electrons (the negative charges) along the external circuit connecting the reference electrode (RE) to the SSM. A personal computer (PC) records the resulting negative on-current.

The I_{on} current provides valuable information about the binding affinity of activating substances (called *substrates*) for the ion pump and about the kinetics of the enzymatic cycle. Thus, the descending branch of the current transients, when fitted by the sum the exponentially decaying terms in Eq. 5.45, provides values of the relaxation time constants, τ_j, of different steps of the enzymatic cycle. Moreover, the binding constant of substrates to the pump can be estimated from peak currents or from the charge obtained by integration of current transients. Thus, the binding constant of ATP to a P-type ATPase can be determined by performing jumps of increasing ATP concentrations in the presence of the translocating ion, which is initially bound to the pump in its E′ conformation. For example, Fig. 5.14 shows current transients recorded by carrying out jumps of increasing ATP concentrations on proteoliposomes of Ca^{2+}-ATPase of the sarcoplasmic reticulum in the presence of 1 mM Ca^{2+} at pH 7 (Bartolommei et al. 2004).

At this high Ca^{2+} concentration, the charge obtained by integration of the current transients is practically the same for all ATP concentrations and is involved in a single enzymatic cycle; this starts from state $E'\cdot 2Ca^{2+}$ and returns to the same state, after releasing two Ca^{2+} ions to the luminal side of the pump and taking up two other Ca^{2+} ions from its cytoplasmic side. It should be recalled that the calcium pump continues cycling during the whole activation period. However, after the first enzymatic cycle, the pump current I_p attains its stationary value b, which

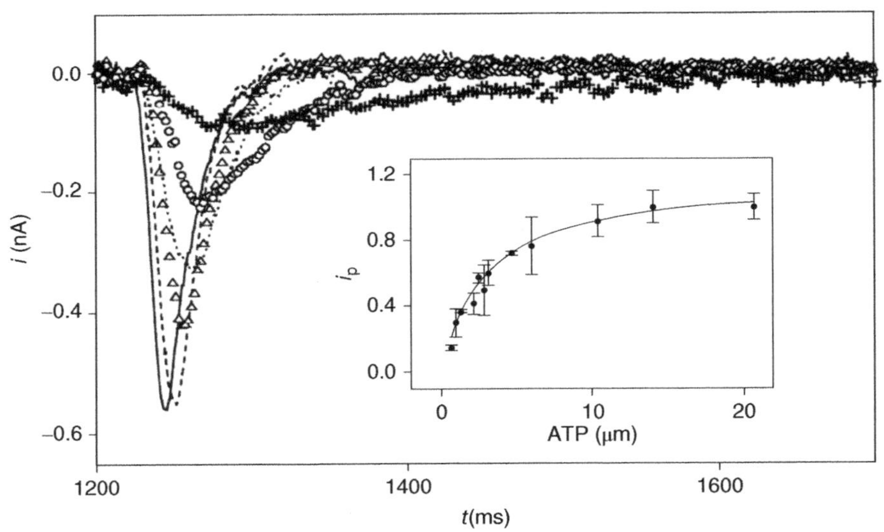

Figure 5.14 Current transients following jumps of increasing ATP concentrations on proteoliposomes of Ca^{2+}-ATPase of the sarcoplasmic reticulum in the presence of 1 mM $CaCl_2$ at pH 7. ATP concentrations: 100 (solid curve), 50 (dashed curve), 25 (up triangles), 10 (dotted curve), 5 (open circles), and 1.5 µM (crosses). The inset shows the ATP dependence of the normalized peak currents. The experimental points are fitted by the Michaelis–Menten equation, yielding a K_S value of 2.9 ± 0.3 µM. Source: Bartolommei et al. (2004). Reproduced with permission of Elsevier.

cannot be monitored under conditions of capacitive coupling, and I_{on} vanishes. On the other hand, the peak current, I_p, increases with the ATP concentration, due to the increase in the translocation rate.

In general, I_p is proportional to the ratio of the concentration of the pumps bound to a given substrate, denoted by [PS], to the overall concentration, [P] + [PS], of the pumps, where P stands for the free pump and S for the substrate. The peak current attains its maximum value, $I_{p,max}$, when all pumps are bound to the substrate, that is, for [P] = 0. Hence, we can write

$$I_p = I_{p,max}[PS]/([P] + [PS]) \rightarrow I_p/I_{p,max} = [S]/([S] + K_S) \quad \text{with } K_S \equiv [P][S]/[PS]. \tag{5.46}$$

K_S is the dissociation constant of the pump–substrate complex PS and measures the S concentration that binds one half of the pumps. Equation 5.46, which holds for S concentrations sufficiently high not to be depleted by binding to the pump, is a form of the *Michaelis–Menten equation*. It accounts for the fact that, as the substrate concentration increases, the binding sites of the ion pump tend to be saturated. The inset of Fig. 5.14 shows a plot of I_p against [ATP] obtained from Ca^{2+}-ATPase proteoliposomes, together with its fit by the Michaelis–Menten equation. The fit yields a K_S value of 2.9 μM, denoting a notable affinity of ATP for Ca^{2+}-ATPase.

The binding constant of the translocating ion to the E′ conformation of an ion pump can be estimated by carrying out jumps of increasing concentrations of this ion in the absence of ATP. In this case, it is the plot of the charge Q under each current transient against the corresponding ion concentration that must be fitted by the Michaelis–Menten equation, in order to obtain the binding constant of the ion to the pump. However, if the ion pump binds two ions, the binding of the first ion may affect the binding of the second, giving rise to a cooperative effect. In other words, the two ions are not bound independently from each other. In this case, the plot of Q against the ion concentration should be fitted by the phenomenological *Hill function*, which accounts for cooperativity via a Hill coefficient. In the absence of cooperativity, the Hill coefficient equals unity and the Hill function becomes identical with the Michaelis–Menten equation. In the case of the calcium pump, Q measures the amount of ionic charge moved by the pump in the passage of two Ca^{2+} ions from the cytoplasmic side of the pump to the $E' \cdot 2Ca^{2+}$ state. The fit by the Hill function yields a half-saturating Ca^{2+} concentration of 1.5 μM and a small Hill coefficient of 1.1 (Tadini Buoninsegni et al., 2004).

The binding constant of the translocating ion to the pump may also be determined by carrying out jumps of a high, saturating ATP concentration in the presence of different concentrations of the translocating ion bound to the E′ conformation. Even in this case, the plot of the charge Q under each current transient against the corresponding ion concentration provides the binding constant of the ion to the pump, upon fitting this plot by the appropriate binding isotherm. However, as a rule, the charge Q moved by a saturating ATP concentration jump in the presence of a given concentration of the translocating ion turns out to be higher than that moved by a jump of the same concentration of the translocating ion in the absence of ATP. This behavior is exemplified by the two current transients in Fig. 5.11, obtained by the two different procedures on proteoliposomes of Ca^{2+}-ATPase for the same Ca^{2+} concentration. Here, the charge Q obtained by the Ca^{2+} concentration jump is less than that obtained by the ATP concentration jump by a factor of 0.63. This is due to the fact that the Ca^{2+} ion is moved deeper inside the proteoliposome membrane during the ATP concentration jump than during the Ca^{2+} concentration jump. This difference in charge can be used to make an approximate estimate of the distance of the Ca^{2+}-binding site on the cytoplasmic side of the pump from that on its luminal side, with respect to the pump thickness (Tadini Buoninsegni et al., 2004).

Concentration jumps at gold-supported alkanethiol/phospholipid bilayers can also be profitably employed for the investigation of the mechanism of integral proteins responsible for electrogenic secondary active transport, such as the ATP/ADP translocase (cf. Section 3.4.2), which couples the flow of ATP from the mitochondrial matrix to the intermembrane space to the flow of ADP in the opposite direction (Gropp et al., 1999).

A particularly simple case is encountered when all steps of the cycle are irreversible and the rate of each step is much less than that of the preceding step. In this case, Eq. 5.39 is notably simplified, in that the backward rate constants, $b_{i-1}x_i$ and $b_i x_{i+1}$, are negligible. Moreover, the concentration x_{i-1} of state E_{i-1}, which immediately precedes the state E_i under consideration, is practically zero, since the rate constant f_{i-1} that consumes it is much greater than the rate constant f_i that consumes x_i. Equation 5.39 is simplified as follows:

$$\mathrm{d}x_i/\mathrm{d}t = -f_i x_i \quad \text{with } i = 1, \ldots, n. \tag{5.47}$$

These equations are readily solved by separation of variables, yielding the following expressions:

$$\int_{x_i(0)}^{x_i(t)} \frac{\mathrm{d}x_i}{x_i} = -\int_0^t f_i\,\mathrm{d}t \rightarrow \ln\frac{x_i(t)}{x_i(0)} = -f_i t \rightarrow x_i(t) = x_i(0)\,\mathrm{e}^{-f_i t} \quad \text{with } i = 1, \ldots, n. \tag{5.48}$$

A pre-steady-state situation can also be elicited by a potential step across a BLM, on which membrane fragments incorporating an ion pump are adsorbed, or across the membrane of a cell attached to the tip of a pipette in the whole-cell configuration. A potential step causes the ion pump to relax toward a new equilibrium state, yielding a current transient that decays exponentially to zero and can be integrated over time giving a charge Q. If the ion translocated by the ion pump is present on both sides of the membrane, a series of steps from the initial inversion potential to both more positive and more negative transmembrane potentials $\Delta\phi$ yields a charge Q that saturates at both extreme $\Delta\phi$ values. The plot of Q against $\Delta\phi$ has a sigmoidal shape (Wuddel and Apell, 1995).

The enzymatic cycle of ion pumps may also be investigated in aqueous suspensions of membrane fragments incorporating the pump by using the amphiphilic electrochromic styryl dye RH 421, which is thought to penetrate into lipid domains of the membrane fragments (Heyse et al., 1994). RH 421 does not respond to conformational transitions, but detects changes of electric field strength inside the membrane dielectric. These changes are caused by binding or release of ions and/or by movements of charge inside the membrane, thus allowing a discrimination between differently charged states of the ion pump. Steady-state fluorescence measurements are carried out with a fluorescence spectrophotometer, by following the relative fluorescence changes of RH 421-labeled membrane fragments upon addition of increasing concentrations of an activating substance, for example, the translocating ion. Measurements of transient fluorescent signals are carried out immediately after photochemical release of ATP from caged ATP excited by light.

To get a better view of the enzymatic cycle of P-type ATPases, as investigated by the procedures previously described, we now examine in some detail the cycle of the two most extensively investigated P-type ATPases, namely Ca^{2+}-ATPase of the sarcoplasmic reticulum and Na^+, K^+-ATPase of the mammalian plasma membrane.

5.6.1 Ca^{2+}-ATPase of the Sarcoplasmic Reticulum

Ca^{2+}-ATPase of the sarcoplasmic reticulum (SERCA) is present in different forms, called *isoforms*, which have slightly different functions and are distributed in various tissues, where they are codified. The most known isoform is SERCA1a, which is expressed in the skeletal fast striated muscles. SERCA pumps calcium ions from the cytoplasm into the lumen of the sarcoplasmic reticulum, a network of closed pockets that acts as a calcium reservoir and where calcium concentration is from three to four orders of magnitude higher than in the cytoplasm. By so doing, SERCA promotes muscle relaxation.

A simplified reaction scheme of the enzymatic cycle of SERCA is shown in Fig. 5.15. According to this scheme, Ca-ATPase exists in two different groups of conformational states, denoted by E_1 (or E′) and E_2 (or E″). The conformational states of group E_1 have a high affinity for calcium ions and their binding sites face the cytoplasmic side of the membrane. Conversely, the conformational states of group E_2 have a much lower affinity for calcium ions and their binding sites face the luminal side of the membrane. Starting from the conformational

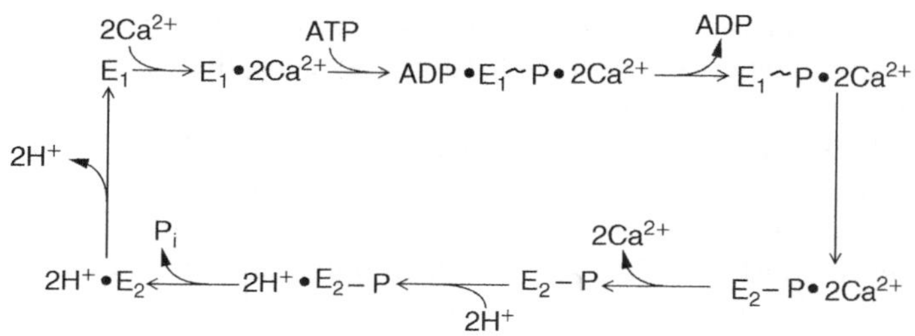

Figure 5.15 Eight-state enzymatic cycle of SERCA.

state E_1, the cycle in Fig. 5.15 consists of the following steps: (i) the sequential binding of two calcium ions, which occurs even in the absence of ATP; (ii) the binding of ATP, which leads to the formation of the acylphosphate bond, while ADP is still bound to the pump; the formation of this bond takes place only after two calcium ions are bound to the pump; (iii) the release of ADP and occlusion of the two calcium ions; (iv) the conformational change from E_1 to E_2; (v) the release of the two calcium ions to the lumen of the sarcoplasmic reticulum; (vi) the uptake of two protons from the lumen of the sarcoplasmic reticulum; (vii) the hydrolysis of the acylphosphate bond, with release of phosphate ion, P_i; and (viii) the return to the E_1 conformation, with release of two protons to the cytosol.

The structural features of SERCA are well known, since more than 20 crystal structures at 2.6 Å resolution have been determined (Toyoshima, 2008; 2009). SERCA is a tall molecule, about 15 nm high and 8 nm thick, and comprises a large cytoplasmic headpiece, a transmembrane domain made of 10 (M1–M10) α-helices, and short luminal loops. Four of the transmembrane helices, M2–M5, have long cytoplasmic extensions, referred to as the *stalk* region. The cytoplasmic headpiece consists of three domains, designated as A (*actuator*), N (*nucleotide binding domain*), and P (*phosphorylation domain*), as shown in Fig. 5.16. The A domain is the smallest of the three cytoplasmic domains and is highly mobile. It is connected to the M1–M3 helices, with loops of 10 residues or more, and acts as the actuator of the gates that regulate the binding and release of calcium ions. The N domain is the largest of the three cytoplasmic domains and is connected to the P domain with two strands that have a β-sheet-like hydrogen-binding pattern. It contains the binding site for the adenosine moiety of ATP. The binding of the adenine ring of ATP to the N domain is predominantly hydrophobic, devoid of hydrogen bonds. The P domain contains the residue of phosphorylation, Asp^{351}. The three cytoplasmic domains are widely split in the $E_1 \cdot 2Ca^{2+}$ state, but gather to form a compact headpiece in the other states. In fact, the N domain must approach the P domain to release the terminal phosphate group of ATP to Asp^{351}, forming the acylphosphate bond. In the transmembrane domain, the M4–M6 and M8 helices contain the residues directly coordinating the two calcium ions through seven oxygen atoms, one of which acts as a bridge between the two ions. The two Ca^{2+}-binding sites are located side by side, approximately at the center of the transmembrane domain. The amino acidic sequences are well conserved for M4–M6 within the members of closely related P-type ATPases, such as Na^+, K^+-ATPase and H^+, K^+-ATPase. All helices from M1 to M6 move considerably during the reaction cycle, whereas the M7–M10 helices keep their positions, apparently acting as a membrane anchor.

To ensure vectorial transport of calcium ions and to avoid channel-like behavior, the ion pump is expected to have two gates: one on the cytoplasmic side and the other on the luminal side. The two gates must operate in strict coordination, and the enzymatic cycle must include a step in which both gates are closed, so that the transmembrane binding sites become inaccessible from either side of the membrane (i.e., the *occluded state*). The hallmark of P-type ATPases is the

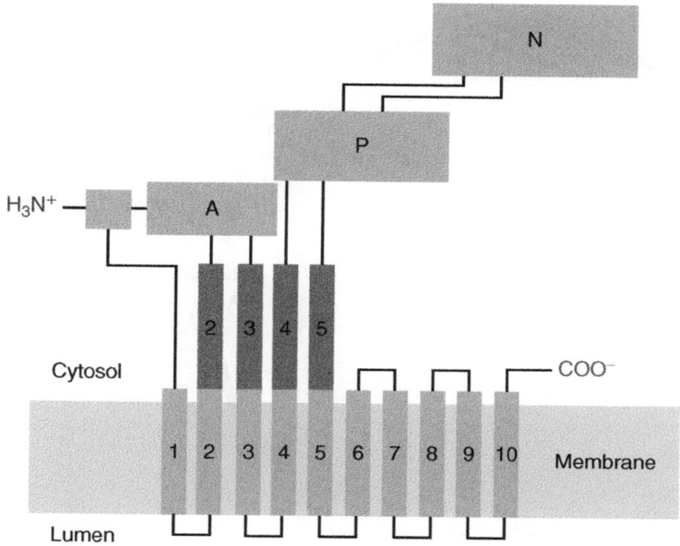

Figure 5.16 Structural model of SERCA.

autophosphorylation and dephosphorylation of the catalytic aspartate during the reaction cycle. Only after two calcium ions are bound is the cytoplasmic gate closed and locked by phosphoryl transfer to the aspartate residue, leading to the $E_1 \sim P{\cdot}2Ca^{2+}$ state. In this state, sometimes denoted by $P \sim E_1(2Ca^{2+})$, bound calcium ions are occluded. During the transition to $E_2 - P{\cdot}2Ca^{2+}$, the luminal gate opens, and the two calcium ions are released sequentially. Then, the two protons to be countertransported occupy the binding sites. Subsequently, the luminal gate is closed by the hydrolysis of the acylphosphate and locked in the $2H^+{\cdot}E_2 - P_1$ state. This is only a transient state, followed by spontaneous release of protons to the cytosol, accelerated by the presence of ATP and ADP.

A pictorial representation of opening and closing of the two gates during the enzymatic cycle is provided in Fig. 5.17, where closed gates are represented as small triangles on one or both sides of the binding site for Ca^{2+} ions. The bottom of the Ca^{2+}-binding site is depicted as sharp edged in the E_1 conformations and rounded in E_2 conformations, to denote a more difficult Ca^{2+} release in the former case. The opening and closing of these gates are due to appropriate rotations of the bundle of helices along which the calcium ions move. When the binding sites of some helices with a moderate affinity for calcium ions face the interior of the bundle, they allow Ca^{2+} passage. Conversely, when rotation of the helices causes the binding sites to face the exterior of the bundle, a potential energy barrier to Ca^{2+} passage is generated. These alternatively favorable and adverse rotations of the helices composing the bundle move along the transmembrane domain during an enzymatic cycle, involving different sections of the bundle.

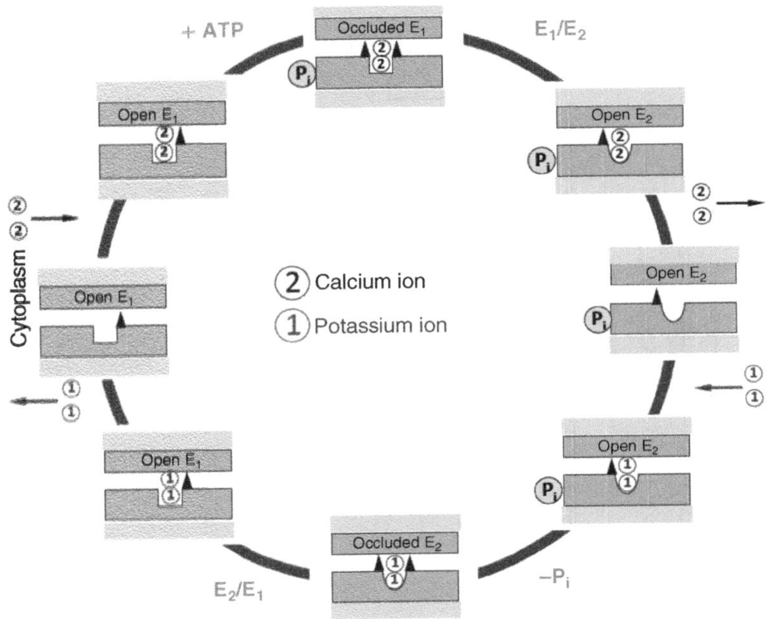

Figure 5.17 Pictorial scheme of the enzymatic cycle of SERCA.

The connection between phosphorylation and movement of transmembrane helices is ensured by the P domain, which is directly linked to the M3–M5 helices by hydrogen bonds. The top part of M5 moves together with the P domain as a single entity. M4 is also linked to the P domain, but much less rigidly compared with M5. M6 is also connected, albeit less directly, to the P domain through a cytosolic loop. Thus, the M3–M6 helices are all linked to the P domain with flexible "joints," except for M5. This allows the realization of largely different configurational states, all coordinated by the P domain. For example, if the P domain inclines, then, due to the concomitant bending of M5, all the M3–M6 helices will incline, generating movements that have components normal to the membrane surface. Important movements of the transmembrane helices that are directly related to the release of calcium ions to the sarcoplasmic reticulum lumen are a shift of M4 toward the luminal side by one turn of this α-helix and the bending of the upper part of M5 toward M4. These movements determine a profound reorganization of the Ca^{2+}-binding residues, decreasing the number of coordinating oxygen atoms.

The dependence upon pH of the current transients elicited by saturating Ca^{2+} concentration jumps in the absence of ATP and by ATP concentration jumps in the presence of Ca^{2+} ions proves unequivocally the countertransport of two protons for each SERCA enzymatic cycle (Tadini Buoninsegni et al., 2004; 2006). Figure 5.18 shows a series of current transients elicited by 100 μM ATP

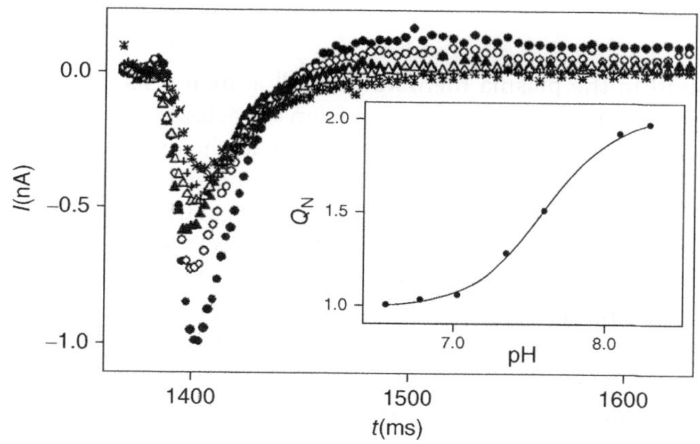

Figure 5.18 Current transients following 100 μM ATP concentration jumps on proteoliposomes of Ca^{2+}-ATPase of the sarcoplasmic reticulum in 100 mM free calcium and 0.15 M choline chloride, at different pH values: 6.55 (stars), 6.78 (crosses), 7.03 (open up-triangles), 7.35 (solid up-triangles), 7.58 (open circles), and 8.13 (solid circles). Calcium ionophore A23187 and protonophore 1799 (1.25 μM) were also used. The inset shows the dependence of the normalized charge under the peaks upon pH. The solid curve is the best fit of the experimental points to the phenomenological Hill function. Source: Tadini Buoninsegni et al. (2004). Reproduced with permission of Elsevier.

concentration jumps in the presence of Ca^{2+} ions at different pH values. These jumps activate a whole enzymatic cycle that starts from the $E_1 \cdot 2Ca^{2+}$ state and returns to the same state, after releasing two Ca^{2+} ions to the luminal side of the pump and taking up two other Ca^2 ions from its cytoplasmic side. The peak current increases with an increase in pH first slowly, and then more rapidly. In the inset of Fig. 5.18, the charge Q obtained by integration of the current transients is almost constant at pH < 7, and then practically doubles in passing from pH 7 to pH 8.2, tending to a maximum limiting value. This behavior is explained by considering that the ATP concentration jump causes the binding sites in the $E_1 \cdot 2Ca^{2+}$ state, which are initially exposed to the cytoplasmic side of the pump, to be exposed to its luminal side as a consequence of a conformational change. Here, the two calcium ions are released to the lumen. At pH 7, they are replaced by two protons because the acidic residues that bind the calcium ions on the luminal side are protonated at this pH. The enzymatic cycle is then completed through the exposition of the binding sites to the cytoplasmic side, where the protons are released and two further Ca^{2+} ions are taken up. During the whole cycle, two Ca^{2+} ions are, therefore, translocated from the cytoplasmic to the luminal side of the pump, while two protons are translocated in the opposite direction, with a net translocation of two elementary charges across the membrane. On the other hand, at pH 8.3, the acidic residues that bind the Ca^{2+} ions on the luminal side are completely deprotonated, and one enzymatic cycle involves exclusively the translocation of two Ca^{2+} ions, that is, of four elementary charges.

5.6.2 Na^+, K^+-ATPase

Na^+, K^+-ATPase is found in the plasma membrane of all animal cells and is in charge of the endergonic transport of three Na^+ ions from the cytosol to the extracellular fluid and of two K^+ ions in the opposite direction, at the expense of the exergonic hydrolysis of one molecule of ATP into ADP and inorganic phosphate P_i. In addition to the catalytic unit of approximately 1000 residues, analogous to that of SERCA and referred to as the *α-subunit*, Na^+, K^+-ATPase contains a heavily glycosylated *β-subunit* of about 300 residues, which modulates the pump function, including cation binding affinity and K^+-occlusion. Na^+, K^+-ATPase is the most important member of the P-type ATPase family, which includes SERCA and the gastric H^+, K^+-ATPase, among others. It is primarily a sodium pump, since potassium ion can be substituted equally well with other monovalent metal ions, including sodium ion itself. The Na^+, K^+-ATPase α-subunit contains a cytoplasmic headpiece consisting of the three domains A, N, and P. The β-subunit contains only one transmembrane helix that runs nearly parallel to the M7 helix of the α-subunit. Figure 5.19 shows an approximate scheme of the enzymatic cycle of Na^+, K^+-ATPase, which includes the dielectric coefficient, α_i, of each electrogenic step. Steps for which α_i is not specified are electrosilent (Apell and Karlish, 2001).

The initial uptake of two Na^+ ions in the E_1 conformation is not electrogenic, while the uptake of the third Na^+ ion is moderately electrogenic. The subsequent steps involving the occlusion of the three Na^+ ions and the transition from the E_1 to the E_2 conformation are not electrogenic. The last Na^+ ion taken up by the pump from the cytosol is the first one to be released to the extracellular fluid, with a high dielectric coefficient. Conversely, the release of the other two Na^+ ions and the subsequent uptake of two K^+ ions are only very slightly electrogenic. Finally, the occlusion of the two K^+ ions, the return to the E_1 conformation, and the release of these ions to the cytosol are not electrogenic. It should be noted that charge movements within the cytoplasmic headpiece of P-type ATPases (e.g., the transfer of P_i from the N to the P domain) are not detectable, because they are screened by the water molecules surrounding the cytoplasmic domains. The only possible electrogenic steps are those involving a movement of the translocating ions along the transmembrane domain.

Studies of the electrogenicity performed with SERCA and H^+, K^+-ATPase have shown that, in these P-type ATPases, all ion-binding and ion-release steps within the transmembrane domain are electrogenic. Before the three-dimensional structure of the SERCA with atomic resolution was known, the apparently anomalous nonelectrogenic steps of the Na^+, K^+-ATPase were explained by the presence, in the sodium pump, of large vestibula full of water and all kind of ions (Läuger, 1991). These species were assumed to screen the pathways of two of the three Na^+ ions and of the two K^+ ions to and from the corresponding binding sites. Nowadays, the knowledge of the structure of SERCA (Morth et al., 2007) down to 2.4 Å resolution (Toyoshima et al., 2011) excludes the existence of such vestibula. Electrogenicity investigations over the pH range from 6 to 8.5 performed by Apell and coworkers (Apell and Diller, 2002; Apell, 2003) have shown that the ion-binding

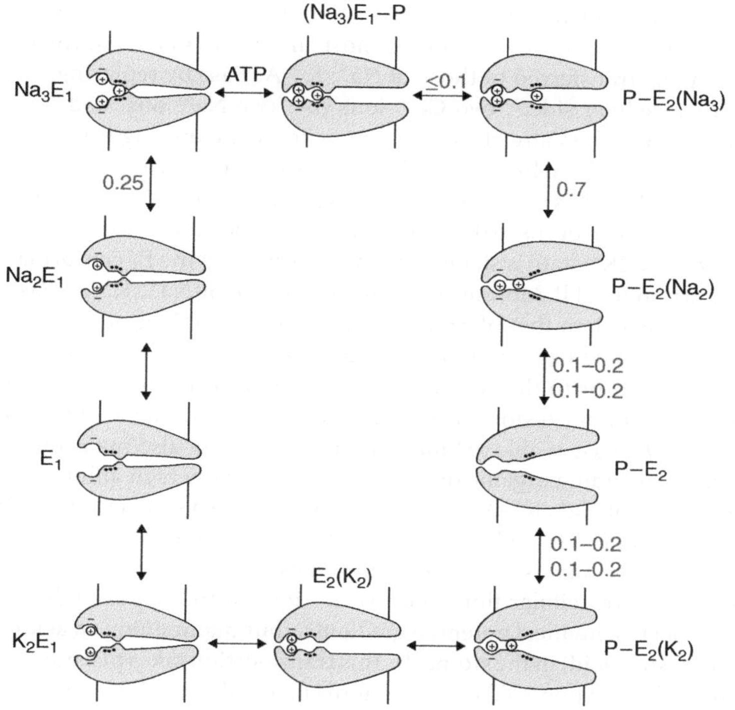

Figure 5.19 Schematic picture of the enzymatic cycle of Na^+, K^+-ATPase, with a representation of nowadays superseded large vestibula to justify nonelectrogenic ion-binding and ion-release steps. Source: Apell and Karlish (2001). Reproduced with permission of Springer.

and ion-release steps that have zero or very small dielectric coefficient at physiological pH in the scheme of Fig. 5.19 become increasingly electrogenic with increasing pH, such as to become hypothetically wholly electrogenic at pH > 9.5. This behavior is explained by a transient binding of protons to carboxylate groups of (or close to) the Na^+- and K^+-binding sites; this binding, which is fast enough as not to be resolved so far, practically compensates the positive charge of a Na^+ or K^+ ion moving in one direction by the equal charge of a H^+ ion moving almost simultaneously in the opposite direction. The amino acid residues involved in the cation binding sites of two of the three Na^+ ions in Na^+, K^+-ATPase appear similar to those involved in the binding of the two Ca^{2+} ions in Ca^{2+}-ATPase. The third Na^+ ion, which is taken up after the other two and released before them, is a peculiar feature of Na^+, K^+-ATPase. Its binding to the sodium pump and its release from it are highly electrogenic and the appreciable asymmetry of the corresponding dielectric coefficients (0.25 for the uptake and 0.7 for the release) suggests a location of the binding site at about 25% of the membrane thickness from the cytoplasmic side. This binding site is exclusively selective for Na^+ and is not significantly affected by a change in pH.

Besides having a very similar structure, the α-subunit of Na^+, K^+-ATPase and SERCA have a similar enzymatic cycle. Thus, almost all features of the enzymatic cycle of SERCA can be transferred to those of Na^+, K^+-ATPase by replacing "luminal side" by "extracellular side," "two Ca^{2+} ions" by "two Na^+ ions" and "two H^+ ions" by "two K^+ ions." As already stated, the main distinguishing feature of the sodium pump is represented by the third Na^+ ion. Another difference between Na^+, K^+-ATPase and SERCA is represented by the ion release from the E_2 conformation. The release of two protons to the cytosol by SERCA occurs spontaneously. On the contrary, the $E_2{\cdot}2K^+$ state is stable, and the transition to the E_1 conformation is very slow without ATP. Moreover, the turnover rate of Na^+, K^+-ATPase is about six times higher than that of SERCA. This difference might possibly be ascribed to the cytoplasmic headpiece of the sodium pump being less compact than that of SERCA, and/or to the coordination geometry of the bound ions in Na^+, K^+-ATPase being rather distorted compared with SERCA. This might also explain why Na^+, K^+-ATPase is able to bind not only K^+ ion, but also many other monovalent cations, including organic ones, implying a not very high affinity.

The enzymatic cycles of Na^+, K^+-ATPase and SERCA are also similar to that of the catalytic α-subunit of H^+, K^+-ATPase, present in the plasma membrane of the parietal cells of the gastric mucosa. This ion pump translocates two protons from the cytoplasm to the extracellular fluid and two K^+ ions in the opposite direction, against their electrochemical potential gradients, thus maintaining an acidic environment (the gastric acid) in the stomach. This translocation takes place at the expense of the hydrolysis of one ATP molecule at pH 6, but of two ATP molecules at $pH < 3$. Considering that the gastric acid has a pH ranging from 1.5 to 3.5, the less favorable stoichiometry tends to prevail. It should be noted that the charge translocated by an enzymatic cycle of H^+, K^+-ATPase in one direction is equal to that translocated in the opposite direction, as distinct from SERCA and Na^+, K^+-ATPase, which translocate two and one elementary charge, respectively, during a cycle at physiological pH. For this reason, H^+, K^+-ATPase is electrically silent, while the other two ion pumps are electrogenic. However, single steps of the H^+, K^+-ATPase cycle may well be electrogenic, as is actually the case.

5.7 Transporters

Ionic electrochemical potential gradients store energy packets smaller than the chemical energy stored in the ATP molecules that fuel ion pumps. However, they are more readily exploitable in case of need, and with less dissipation of energy. It is up to transporters to utilize these energy packets.

5.7.1 Cotransporters

Let us first consider cotransporters. Cotransport (sometimes referred as symport) is a mechanism whereby the concerted movement of two solutes takes place in the same direction. A cotransporter can be considered as operating according to the following cyclic scheme:

$$\begin{array}{l} T_0 \Leftrightarrow ST_0 \Leftrightarrow ST_d \Leftrightarrow T_d \\ \uparrow\underline{\hspace{9em}}| \end{array} \tag{5.49}$$

where T is the transporter and S is the substrate, namely a group of ionic and/or neutral species that must act together in order to be shuttled across the membrane (De Felice, 1997). The subscripts "0" and "d" denote the positions of the empty transporter T and of the loaded transporter ST at the outer and inner surface of the membrane, respectively. The transporter T is allowed to shuttle across the membrane unloaded: $T_0 \Leftrightarrow T_d$. Conversely, free ions and molecules are prevented from traversing the membrane; the only way for them to move is by the ST combination. For net transport to occur, empty transporters must return to the other side, where they are free to load ions and molecules for another cycle. In other words, substrate transport is allowed only through the transition $ST_0 \Leftrightarrow ST_d$. Even though the transporter concentration in the membrane is unknown, it is reasonable to assume that the total number of unloaded and loaded transporters is constant in the membrane; it is denoted by $N = [T_0] + [ST_0] + [T_d] + [ST_d]$. Now, in the steady state, the flux, Φ_{ST}, of loaded transporters, that is, the number of loaded transporters that traverse the unit area of the membrane per unit time, must be equal in magnitude but opposite in sign to the flux, Φ_T, of the empty transporters:

$$\Phi_{ST} = -\Phi_T. \tag{5.50}$$

On the other hand, the total current density across the membrane is given by

$$j = z_{ST}e\Phi_{ST} + z_Te\Phi_T = \left(z_{ST}e - z_Te\right)\Phi_{ST} \equiv Q\Phi_{ST}, \tag{5.51}$$

where z_{ST} and z_T are the charge numbers of the loaded and empty carrier, respectively, and e is the elementary charge. In Eq. 5.51, account has been taken of Eq. 5.50. $Q \equiv (z_{ST}e - z_Te)$ is the total charge transported in a single cycle. The substrate transport from the outer to the inner side of the membrane by the loaded transporter moves a charge $z_{ST}e$, while the subsequent passage of the free transporter from the inner to the outer membrane surface moves a charge z_Te in the opposite direction. Hence, Q is the net charge of all components of the substrate S bound to T, without the charge of T. Note that the current density j expressed by Eq. 5.51 may flow in any of the two possible directions, namely toward the inside or the outside of the membrane.

The current can be expressed either by the GHK equation or by the Butler–Volmer equation. If we choose to use the GHK equation, as expressed by Eq. 4.68

$$\frac{i}{\alpha} = \frac{D}{d} ze\, zf\Delta\phi \frac{n(d)\,e^{zf\Delta\phi} - n(0)}{e^{zf\Delta\phi} - 1} \quad \text{with } f \equiv \frac{e}{kT} = \frac{F}{RT}, \tag{5.52}$$

a number of changes are required to apply it to the present case. First, $n(d)$ and $n(0)$ are volume concentrations, whereas $[ST_d]$ and $[ST_0]$ are surface concentrations; hence, the latter concentrations must be divided by the membrane thickness *d*, to convert them into volume concentrations within the membrane. Second, the charge number z associated to the transmembrane potential $\Delta\phi$, both in the pre-exponential term and in the exponent of the exponential functions of Eq. 5.52, must be replaced by the charge number z_{ST} of the loaded transporter that traverses the membrane under the action of $\Delta\phi$. Third, the other charge, ze, refers to the total charge shuttled by the loaded transporter in a single cycle, and hence must be ascribed the value Q. With these changes, the current density $j = i/\alpha$ is given by

$$j = \frac{D}{d^2} Q z_{ST} f \Delta\phi \frac{[ST_d] \exp(z_{ST} f \Delta\phi) - [ST_0]}{\exp(z_{ST} f \Delta\phi) - 1}. \tag{5.53}$$

In this equation, it is assumed that ST_0 and ST_d have a unitary probability of translocating across the membrane. However, the probability, p_0, that the translocation of ST_0 occurs is not unitary; rather, it is measured by the ratio of $[ST_0]$ to the overall concentration, $[T_0] + [ST_0]$, of the transporter at the external membrane surface. Consequently, j must be multiplied by

$$\begin{aligned} p_0 &= \frac{[ST_0]}{[T_0] + [ST_0]} = \frac{1}{1 + [T_0]/[ST_0]} = \frac{[S_0]}{[S_0] + [S_0]\,[T_0]/[ST_0]} \\ &= \frac{[S_0]}{[S_0] + K_0}, \end{aligned} \tag{5.54}$$

where $K_0 \equiv [S_0][T_0]/[ST_0]$ is the dissociation constant of ST_0. It should be noted that, while $[T_0]$ and $[ST_0]$ are measured in molecules per unit surface, $[S_0]$ is measured in molecules per unit volume. By analogous considerations, j must also be multiplied by the probability, p_d, for ST_d translocation to occur, which is given by

$$p_d = \frac{[ST_d]}{[T_d] + [ST_d]} = \frac{[S_d]}{[S_d] + K_d} \quad \text{with } K_d \equiv \frac{[T_d]\,[S_d]}{[ST_d]}. \tag{5.55}$$

5.7.2 Countertransporters

The dependence of the current upon $\Delta\phi$ for countertransport is more involved than for cotransport. We confine ourselves to examining the problem in the absence of an applied potential, to determine the conditions under which the system is capable of transporting one species against its chemical potential gradient. Let S′ and S″ denote the two species transported across the membrane in opposite directions. Upon assuming that the total concentration of all forms of the countertransporter T is constant, we can write

$$N = [T_0] + [T_d] + [S'T_0] + [S'T_d] + [S''T_0] + [S''T_d], \tag{5.56}$$

where the same symbolism adopted in the preceding subsection is employed.

Under steady-state conditions, the fluxes of the bound transporters from the membrane outer surface (0) to the inner one (d) must be equal to the flux of the empty transporter in the opposite direction (Schultz, 1980). In the absence of an applied potential, the flux of any species from 0 to d is proportional to the difference between the concentration of the given species at 0 and that at d, according to a proportionality constant P. Upon assuming for simplicity that this constant is the same for all the species, we can write

$$P\left(\left[S'T_0\right]-\left[S'T_d\right]\right)+P\left(\left[S''T_0\right]-\left[S''T_d\right]\right)=P\left(\left[T_d\right]-\left[T_0\right]\right). \quad (5.57)$$

Rearranging terms, we obtain

$$\left[T_0\right]+\left[S'T_0\right]+\left[S''T_0\right]=\left[T_d\right]+\left[S'T_d\right]+\left[S''T_d\right]. \quad (5.58)$$

Let us now introduce the dissociation constants of S′T and S″T, which can be reasonably assumed to be the same on the outer and inner membrane surfaces:

$$K'=\left[S_0'\right]\left[T_0\right]/\left[S'T_0\right]=\left[S_d'\right]\left[T_d\right]/\left[S'T_d\right];$$
$$K''=\left[S_0''\right]\left[T_0\right]/\left[S''T_0\right]=\left[S_d''\right]\left[T_d\right]/\left[S''T_d\right]. \quad (5.59)$$

Combining Eqs. 5.56 and 5.58 and inserting these dissociation constants, we get

$$N=2\left(\left[T_0\right]+\left[S'T_0\right]+\left[S''T_0\right]\right)=2\left[T_0\right]\left(1+\left[S_0'\right]/K'+\left[S_0''\right]/K''\right). \quad (5.60)$$

The unidirectional flux of S′T from 0 to d, $\Phi'_{0d}=P[S'T_0]$, attains its maximum value when S′T is the only form of the transporter present at 0. From Eq.5.60, it follows that $\Phi'_{0d,max}=PN/2=\Phi'_{d0,max}\equiv\Phi'_{max}$. Therefore, we can write

$$\Phi'_{0d}=P\left[S'T_0\right]=P\frac{\left[S_0'\right]\left[T_0\right]}{K'}=\frac{2\Phi'_{0d,max}}{N}\frac{\left[S_0'\right]\left[T_0\right]}{K'}$$
$$=\frac{\Phi'_{0d,max}\left[S_0'\right]}{\left[S_0'\right]+K'\left(1+\left[S_0''\right]/K''\right)}. \quad (5.61)$$

In deriving this equation, N was removed by using Eq. 5.60. Clearly, Φ'_{0d} decreases with increasing $[S_0'']$. In view of the symmetry of the problem with respect to the two membrane surfaces, the unidirectional flux of S′T from d to 0, Φ'_{d0}, is obtained by replacing $[S_0']$ with $[S_d']$ and $[S_0'']$ with $[S_d'']$ in Eq. 5.61. Therefore, the overall flux of S′T is given by

$$\Phi'=\Phi'_{0d}-\Phi'_{d0}$$
$$=\Phi'_{max}\left(\frac{\left[S_0'\right]}{\left[S_0'\right]+K'\left(1+\left[S_0''\right]/K''\right)}-\frac{\left[S_d'\right]}{\left[S_d'\right]+K'\left(1+\left[S_d''\right]/K''\right)}\right). \quad (5.62)$$

By setting this equation equal to zero and rearranging, it can be readily shown that Φ' equals zero when

$$\left[S_d'\right]/\left[S_0'\right]=\left(K''+\left[S_d''\right]\right)/\left(K''+\left[S_0''\right]\right). \quad (5.63)$$

Now, a net flux of S′T from 0 to d causes a translocation of S′ against its concentration gradient if $[S'_d] > [S'_0]$. For this event to be possible, $[S''_d]$ must be $> [S''_0]$, allowing the countertransporter to move S″ from d to 0 along its concentration gradient. If S″ is present only in d ($[S''_0] = 0$), the transporter can move S′ from 0 to d against its concentration gradient provided that $[S''_d] > 0$, in view of Eq. 5.63. In this case, S″ in compartment d competes with S′ for the binding site of the empty countertransporter slowing down the flux of S′ from d to 0, whereas that from 0 to d is unimpeded.

References

Apell, H.-J., S. J. Karlish, *J. Membrane Biol.* 180 (2001) 1–9.

Apell, H.-J., A. Diller, *FEBS Lett.* 532 (2002) 198–202.

Apell, H.-J., *Ann. New York Acad. Sci.* 986 (2003) 133–140.

Babes, A., K. Fendler, *Biophys J.* 79 (2000) 2557–2571.

Bartolommei, G., F. Tadini Buoninsegni, M. R. Moncelli, *Bioelectrochemistry* 63 (2004) 157–160.

Bartolommei, G., F. Tadini-Buoninsegni, M. R. Moncelli, R. Guidelli, *Biochim. Biophys. Acta* 1778 (2008) 405–413.

Borlinghaus, R., H.-J. Apell, P. Läuger, *J. Membr. Biol.* 97 (1987) 161–178.

De Felice, L. J., *Electrical Properties of Cells. Patch Clamp for Biologists*, Plenum, New York, 1997.

Freeman, S., *Biological Science*, 2nd Edition, Pearson Prentice Hall, Upper Saddle River, NJ, U.S.A., 2005.

Gropp, T., N. Brustovetsky, M. Klingenberg, V. Müller, K. Fendler, E. Bamberg, *Biophys. J.* 77 (1999) 714–726.

Hartung, K., E. Grell, W. Hasselbach, E. Bamberg, *Biochim. Biophys. Acta* 900 (1987) 209–220.

Henderson, R., J. M. Baldwin, T. A. Ceska, F. Zemlin, E. Beckmann, K. H. Downing, *J. Mol. Biol.* 213 (1990) 899–929.

Heyse, S., I. Wuddel, H.-J. Apell, W. Stürmer, *J. Gen. Physiol.* 104 (1994) 197–240.

Läuger, P., *Electrogenic Ion Pumps*, Sinauer Associates, Sunderland, Massachusetts, U.S.A., 1991.

Morth, J. P., B. P. Pedersen, M. S. Toustrup-Jensen, T. L. Sørensen, J. Petersen, J. P. Andersen, B. Vilsen, P. Nissen, *Nature* 450 (2007) 1043–1049.

Pintschovius, J., K. Fendler, *Biophys. J.* 76 (1999) 814–826.

Schultz, S. G., *Basic Principles of Membrane Transport*, Cambridge University Press, London, 1980.

Tadini Buoninsegni, F., G. Bartolommei, M. R. Moncelli, G. Inesi, R. Guidelli, *Biophys. J.* 86 (2004) 3671–3686.

Tadini Buoninsegni, F., G. Bartolommei, M. R. Moncelli, R. Guidelli, G. Inesi, *J. Biol. Chem.* 281 (2006) 37720–37727.
Toyoshima, C., *Arch. Biochem. Biophys.* 476 (2008) 3–11.
Toyoshima, C., *Biochim. Biophys. Acta* 1793 (2009) 941–946.
Toyoshima, C., R. Kanai, F. Cornelius, *Structure* 19 (2011) 1732–1738.
Wuddel, I., H.-J. Apell, *Biophys. J.* 69 (1995) 909–921.

6

Biomimetic Membranes

In view of the complexity and diversity of the functions performed by the different proteins embedded in a biomembrane, incorporating single integral proteins or smaller lipophilic biomolecules into experimental models of biological membranes, so as to isolate and investigate their functions, has been found to be convenient. This serves to reduce complex membrane processes to well-defined interactions between selected proteins, lipids, and ligands. There is great potential for application of experimental models of biomembranes for the elucidation of structure – function relationships of many biologically important membrane proteins, for the fabrication of biosensors and other devices, and for the development of novel pharmaceuticals.

6.1 The Various Types of Biomimetic Membranes

The first successfully employed biomembrane model consists of a lipid bilayer interposed between two aqueous phases. The lipid bilayer may be formed across a small hole in a hydrophobic septum (bilayer lipid membrane, BLM) (Mueller et al., 1962). It may also consist of the membrane enveloping a large unilamellar vesicle (LUV), or of a plasma membrane fragment clamped at the tip of a micropipette. This is the most realistic biomimetic membrane, but suffers from low mechanical stability and low resistance to electric fields. In an effort to overcome the drawbacks of these traditional BLMs, supported BLMs have been devised. They are normally obtained by *self-assembly*, which can be driven exclusively by noncovalent, hydrophobic interactions or with the additional contribution of covalent linkages.

With only a few exceptions, supported biomimetic membranes consist of a more or less complex architecture that includes a lipid bilayer. In order of increasing complexity, they can be classified into *solid-supported BLMs* (sBLMs), *tethered BLMs* (tBLMs), *polymer-cushioned BLMs* (pBLMs), *S-layer stabilized BLMs* (ssBLMs), and *protein-tethered BLMs* (ptBLMs). Mercury-supported lipid monolayers and alkanethiol/lipid *hybrid bilayers* are particular biomimetic membranes that contain a single lipid monolayer (Guidelli and Becucci, 2011*b*; 2011*c*; 2012*b*).

Bioelectrochemistry of Biomembranes and Biomimetic Membranes, First Edition. Rolando Guidelli.

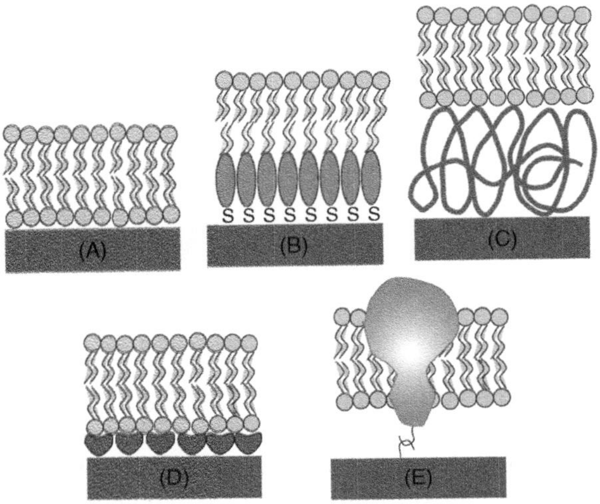

Figure 6.1 Schematic picture of different biomimetic membranes: (A) solid-supported BLM (sBLM); (B) tethered BLM (tBLM) consisting of a thiolipid monolayer with a lipid monolayer on top; (C) polymer-cushioned BLM (pBLM); (D) S-layer stabilized BLM (ssBLM) consisting of an S-layer with a lipid bilayer on top; (E) protein-tethered BLM (ptBLM). Source: Guidelli and Becucci (2011b). Reproduced with permission of Springer.

The sBLMs are lipid bilayers in direct contact with a hydrophilic surface, such as Au, Ag, or mica. In tBLMs, the lipid bilayer is separated from a metal support by a hydrophilic chain tethered to the metal via a sulfhydryl or disulfide group, in pBLMs by a soft polymeric material of typically less than 100 nm thickness, and in ssBLMs by a monomolecular crystalline array of protein subunits called S-layer (Fig. 6.1). Finally, in ptBLMs certain particular recombinant membrane proteins engineered to bear a stretch of six consecutive histidine residues are tethered to a gold electrode, and lipid molecules are allowed to self-assemble around them.

Tethered BLMs make use of large synthetic molecules called *thiolipids*. A thiolipid molecule consists of a polyethyleneoxy or oligopeptide hydrophilic chain terminated at one end with a sulfhydryl or disulfide group for anchoring to a metal support, such as gold, silver, or mercury, and covalently linked at the other end to two alkyl chains simulating the hydrocarbon tails of a lipid. Methyl-, methyloxy-, or chloride-substituted silane groups are used for tethering to glass, quartz, silica, or mica; the latter supports are nonconducting and cannot be investigated by electrochemical techniques. A tethered thiolipid monolayer exposes a hydrophobic surface to the bulk aqueous phase and provides one half of the lipid bilayer. The other half is obtained by self-assembling a lipid monolayer on top of the thiolipid monolayer, usually by vesicle fusion. The hydrophilic chain (the *spacer*) serves to separate the lipid bilayer from the metal support, to compensate for surface roughness effects, to prevent incorporated peptides or proteins from touching the support surface (thus avoiding loss of their functionality due to denaturation), and to provide an ionic reservoir underneath the lipid bilayer. A particularly convenient thiolipid, often referred to by the acronym DPTL, consists of a tetraethyleneoxy (TEO) hydrophilic chain terminated at one end with a lipoic acid residue for anchoring to the metal surface, and covalently linked at the other end to two phytanyl chains mimicking the hydrocarbon tails of a lipid (Fig. 6.2) (Schiller et al., 2003).

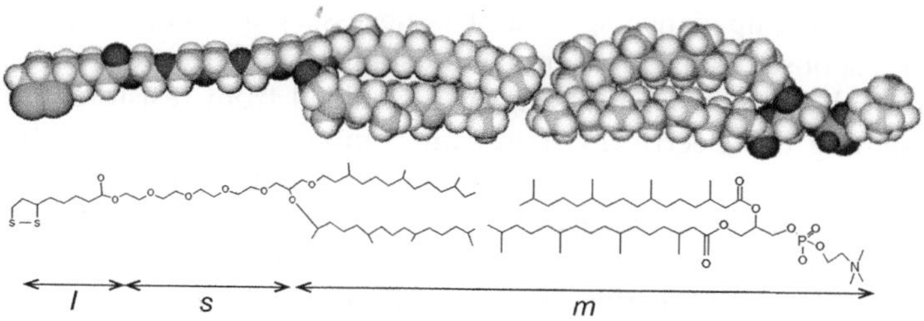

Figure 6.2 Space-filling model and structure of 2,3,di-*O*-phytanyl-*sn*-glycerol-1-tetraethylene-glycol-D,L-α lipoic acid ester lipid (DPTL) and of diphytanoylphosphatidylcholine (DPhPC), in a tail-to-tail configuration; *l*, *s*, and *m* denote the lipoic acid residue, the tetraethyleneoxy spacer, and the hydrocarbon tail region, forming the monomeric unit of a DPTL/DPhPC-tethered bilayer lipid membrane.

Before discussing the advantages and disadvantages of these biomimetic membranes, we describe in detail the electrochemical techniques that are more commonly employed for their investigation. In doing so, particular emphasis is given to their application to mercury-supported DPTL-based tBLMs.

6.2 Electrochemical Techniques for the Investigation of Biomimetic Membranes

Among the electrochemical techniques that can be used for the investigation of BLMs and metal-supported biomimetic membranes, those that have found more frequent and profitable applications are electrochemical impedance spectroscopy (EIS), potential-step amperometry, potential-step chronocoulometry, and cyclic voltammetry.

6.2.1 Electrochemical Impedance Spectroscopy

EIS is the technique of choice for discriminating the various sections of a BLM or tBLM on the basis of their different dielectric properties, and for monitoring the way in which these properties are affected by the incorporation and functional activity of peptides and membrane proteins. EIS consists in perturbing the system under investigation (the biomimetic membrane) with an electric potential generated by superimposing on a constant applied potential E an AC voltage of small amplitude (~10 mV peak to peak), whose frequency, f, is progressively varied from 10^5 to 10^{-3} Hz. For each frequency, the EIS instrument measures the amplitude, I, and the phase shift φ, of the current flowing through the electrochemical cell with the same frequency, thus generating an impedance spectrum. The current may be displayed by reporting its *in-phase component*, $I \cos \varphi$, along the abscissa of an orthogonal coordinate system (the *complex plane*) and its *quadrature component*, $I \sin \varphi$, along the ordinate (Barsoukov and Macdonald, 2005).

To interpret measured impedance spectra, it is necessary to compare them with the electrical response of an equivalent circuit simulating the system under investigation. As a rule, an equivalent circuit is assembled from resistors and capacitors, representing the electrically dominant components of biomimetic membranes. In general, a BLM or a metal-supported self-assembled mono- or multilayer can be regarded as consisting of a series of slabs with different dielectric properties. When ions flow across each slab, they give rise to an ionic current $\boldsymbol{I}_{\mathrm{ion}} = \sigma \boldsymbol{E}$, where $\boldsymbol{E}$ is the electric field and σ is the *conductivity*. Ions may also accumulate at the boundary between contiguous dielectric slabs, causing a discontinuity in the *electric displacement vector* $\boldsymbol{D} = \varepsilon_0 \varepsilon \boldsymbol{E}$, where ε is the dielectric constant (cf. Eq. 7.23), Under AC conditions, the accumulation of ions at the boundary of the dielectric slabs varies with time, and so does the electric displacement vector, giving rise to a capacitive current $\boldsymbol{I}_{\mathrm{c}} = \mathrm{d}\boldsymbol{D}/\mathrm{d}t$. The total current is, therefore, given by the sum of the ionic current and the capacitive current. In this respect, each dielectric slab can be simulated by a parallel combination of a resistance, accounting for the ionic current, and of a capacitance, accounting for the capacitive current, namely by an *RC mesh*. Accordingly, the impedance spectrum of a self-assembled layer can be simulated by a series of *RC* meshes. It should be noted that lateral heterogeneities, such as defects or microdomains, cannot be accounted for by simulating them by *RC* meshes in parallel with each other. In fact, in view of *Kirchhoff's laws* for the combination of circuit elements, the parallel connection of *RC* meshes is again reduced to a single *RC* mesh, with averaged values for resistance and capacitance.

Application of an AC voltage of amplitude V and frequency f to a pure resistor of resistance R yields a current of equal frequency f and of amplitude V/R, in phase with the voltage. Conversely, application of the AC voltage to a pure capacitor of capacitance C yields a current of frequency f and amplitude $2\pi fC$, out of phase by $-\pi/2$ with respect to the voltage, that is, in quadrature with it. This state of affairs can be expressed by stating that the *admittance* Y of a resistance element equals $1/R$, while that of a capacitance element equals $-\mathrm{i}\omega C$, where $\omega = 2\pi f$ is the *angular frequency* and i is the *imaginary unit*. More generally, in an equivalent circuit consisting of several resistances and capacitances, Y is a complex quantity, and the *impedance* Z is equal to $1/Y$, by definition. Hence, Z equals R for a resistance element, and $\mathrm{i}/\omega C$ for a capacitance element. In analogy with the resistance in DC measurements, the overall impedance of two circuit elements in series is equal to the sum of the impedances of the single circuit elements. Conversely, the overall impedance of two circuit elements in parallel is such that its reciprocal is equal to the sum of the reciprocals of the single circuit elements. Consequently, if two circuit elements have appreciably different impedances, their overall impedance is controlled by the circuit element of higher impedance if they are in series, and by the circuit element of lower impedance if they are in parallel.

Application of impedance spectroscopy to a conventional BLM may allow four regions with different dielectric properties to be distinguished, namely the polar head region, the ester linkage region, the hydrocarbon tail region, and a thin layer of aqueous solution bathing the BLM. To this end, the impedance spectrum of the

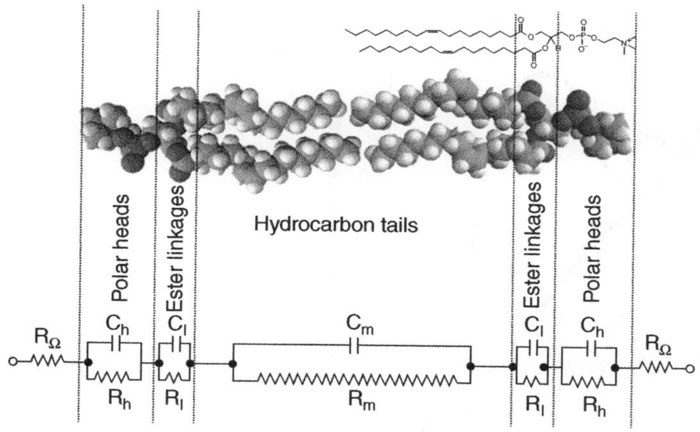

Figure 6.3 Equivalent circuit for a BLM of dioleoylphosphatidylcholine.

BLM is fitted by an equivalent circuit consisting of three *RC* meshes and a resistance simulating the aqueous solution (Ashcroft et al., 1983), as shown in Fig. 6.3. The two different polar head regions, as well as the two ester linkage regions, albeit spatially separated, cannot be distinguished by impedance spectroscopy, because they have almost identical properties.

Due to its greater complexity, as a rule, the impedance spectrum of a tBLM formed by lateral aggregation of the units in Fig. 6.2 cannot distinguish the ester linkage region of the outer lipid leaflet. It can be simulated by the equivalent circuit shown in the inset of Fig. 6.4 and consisting of four meshes in series, $(R_lC_l)(R_sC_s)(R_mC_m)(R_\Omega C_\Omega)$; here, the subscripts l, s, m, and Ω denote the lipoic acid residue, the spacer, the lipid bilayer, and the aqueous solution (Guidelli and Becucci, 2012*a*). The spectrum is often displayed on a ***Bode plot***, namely a plot of $\log|Z|$ and phase angle φ against $\log(f)$, where $|Z|$ is the magnitude of the impedance (Fig. 6.4). In the absence of exogenous species incorporated in the lipid bilayer moiety, the C_l, C_s, and C_m values for the mercury-supported tBLM in Fig. 6.4 amount to 2–5, 3–7, and 0.8–1 $\mu F\ cm^{-2}$, respectively, whereas the corresponding resistances R_l, R_s, and R_m amount to 5–20, 0.05–0.2, and 2–5 $M\Omega\ cm^2$ (Becucci et al., 2005). Moreover, at a tBLM in aqueous solution of 0.1 M KCl, C_Ω and R_Ω are approximately equal to 25 $nF\ cm^{-2}$ and 4 $\Omega\ cm^2$. Hence, the time constant, *RC*, of the lipoic acid residue is of the order of 10 s, those of the spacer and lipid bilayer moieties are comparable and of the order of 1 s, and that of the aqueous solution is of the order of 10^{-7} s. Reconstituting an ion channel such as gramicidin or melittin, or an ion carrier such as valinomycin, into the lipid bilayer moiety of a tBLM in contact with aqueous 0.1 M KCl changes the capacitances of the various slabs of the tBLM only slightly. Conversely, it decreases the resistance of the lipoic residue by about one order of magnitude, that of the spacer by about two orders of magnitude, and that of the bilayer by about three orders of magnitude. Figure 6.4 shows the Bode plot of a mercury-supported tBLM incorporating the ion carrier valinomycin,

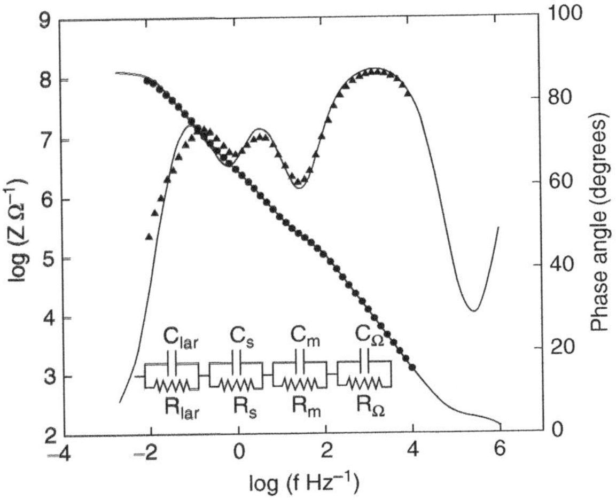

Figure 6.4 Plot of log|Z| (solid circles) and φ (solid triangles) against log(f) (Bode plot) for a mercury-supported DPTL/DPhPC bilayer incorporating valinomycin from its 0.15 μM solution in aqueous 0.1 M KCl at −0.375 V versus Ag/AgCl(0.1 M KCl). The solid curve is the best fit of the impedance spectrum by the equivalent circuit shown in the figure, with R_{lar} = 2.8 MΩ cm^2, C_{lar} = 4.2 μF cm^{-2}, R_s = 48 kΩ cm^2, C_s = 3.3 μF cm^{-2}, R_m = 3.4 kΩ cm^2, C_m = 0.9 μF cm^{-2}, R_Ω = 3.7 Ω cm^2, and C_Ω = 43 nF cm^{-2}. Source: Guidelli and Becucci (2012b). Reproduced with permission of Springer.

a hydrophobic depsipeptide that cages a desolvated potassium ion, shuttling it across the lipid bilayer.

As already stated, the impedance of circuit elements in series is determined by the element with the highest impedance; conversely, the impedance of circuit elements in parallel is determined by the element with the lowest impedance. Therefore, at the highest frequencies, $f = \omega/2\pi$, the overall impedance $|Z|$ is determined by the resistance R_Ω, which is lower than the impedance $1/(\omega C_\Omega)$ of the capacitor in parallel with it, due to the very low C_Ω value. On the other hand, R_Ω is higher than the impedance of the C_s capacitor, $1/(\omega C_s)$, which is lower than the resistance R_s of the resistor in parallel with it; the same is true for the impedance of the $R_m C_m$ mesh, which is determined by the lowest of the impedances of these two elements in parallel, that is, $1/(\omega C_m)$. At the highest frequencies, $|Z|$ is therefore controlled by R_Ω, which is independent of frequency and is characterized by a phase angle $\varphi = 0$. With decreasing frequency, $1/(\omega C_m)$ becomes greater than R_Ω, while still remaining lower than R_m, and it is also greater than $1/(\omega C_s)$, because C_s is greater than C_m. Hence, $|Z|$ coincides with $1/(\omega C_m)$, and the log$|Z|$ versus log(f) plot has a slope equal to −1, while the phase angle tends to 90°. With a further decrease in frequency, $1/(\omega C_m)$ becomes comparable with and ultimately greater than R_m, and $|Z|$ tends to become independent of frequency, which would correspond to complete control by R_m. At the same time, φ decreases tending to zero. However, much before this can occur, a further decrease in frequency makes $1/(\omega C_s)$ much greater than R_m, causing $|Z|$ to coincide with $1/(\omega C_s)$. Hence, the slope of the $\log|Z|$ versus $\log(f)$ plot becomes once again equal to -1 and φ tends to 90°. With a further decrease in frequency, R_s tends to become comparable with $1/(\omega C_s)$, causing a slight decrease in the slope of the $\log|Z|$ versus $\log(f)$ plot. A further decrease in frequency prevents this decrease in slope from becoming significant, because the capacitive impedance $1/(\omega C_l)$ of the lipoic acid residue becomes greater than R_s and the slope of the Bode plot becomes again equal to -1. Finally, at still lower frequencies, $1/(\omega C_l)$ tends to become greater than R_l, which tends to take control of the overall impedance by causing a decrease in the slope of the $\log|Z|$ versus $\log(f)$ plot and a decrease in the phase angle. The solid curve in Fig. 6.4 is the best fit of the impedance spectrum by the equivalent circuit depicted in Fig. 6.4.

To better visualize the layered structure of a tBLM, it is convenient to display the spectrum on a plot that yields a semicircle when a single *RC* mesh is considered. Two plots satisfy this requirement, namely a plot of Z'' against Z' and a plot of $\omega Z'$ against $-\omega Z''$, where Z' and Z'' are the in-phase and quadrature component of Z; the former plot is commonly called the *Nyquist plot*, whereas the latter can be referred to as an M *plot*, since $\omega Z'$ and $-\omega Z''$ are the components of the *modulus function M*.

To understand their behavior, we must consider that the impedance Z of a single *RC* mesh is given by

$$Z^{-1} = Y = R^{-1} - \mathrm{i}\omega C. \tag{6.1}$$

Writing $Z \equiv Z' + \mathrm{i}Z''$, we obtain

$$Z' = R/\left(1 + \omega^2 R^2 C^2\right), \tag{6.2a}$$

$$Z'' = Z'\omega RC. \tag{6.2b}$$

Eliminating ωRC from Eqs. 6.2a and 6.2b, we get

$$Z''^2 + Z'^2 - RZ' = 0 \quad \rightarrow \quad \left(Z' - R/2\right)^2 + Z''^2 = (R/2)^2. \tag{6.3}$$

Equation (6.3) yields a semicircle of diameter R and center of coordinates $(R/2, 0)$ on a Z'' versus Z' plot, that is the Nyquist plot. Noting that the maximum of this semicircle is characterized by the equality of Z' and Z'', from Eq. 6.2b it follows that the angular frequency ω at the maximum equals the reciprocal of the *time constant RC* of the mesh. This time constant is related to the time required by an *RC* mesh subjected to a constant current I, applied instantaneously at time $t = 0$, to reach the final electric potential $V = IR$ expressed by Ohm's law. The time dependence of V is expressed by the equation $V = IR[1 - \exp(-t/RC)]$ (cf. Eq. 4.88); hence, the lower the RC, the shorter the time t at which V approaches its asymptotic limiting value. Analogously, when interrogated by a series of AC signals of variable frequency, an *RC* mesh will "respond" at a frequency that is higher, the lower its time constant is.

In the presence of a series of *RC* meshes, their time constants may be close enough to cause the corresponding semicircles to overlap partially. In this case, if the mesh of highest time constant has also the highest resistance, R_{max}, as is often the case, then the Nyquist plot of the whole impedance spectrum exhibits a single

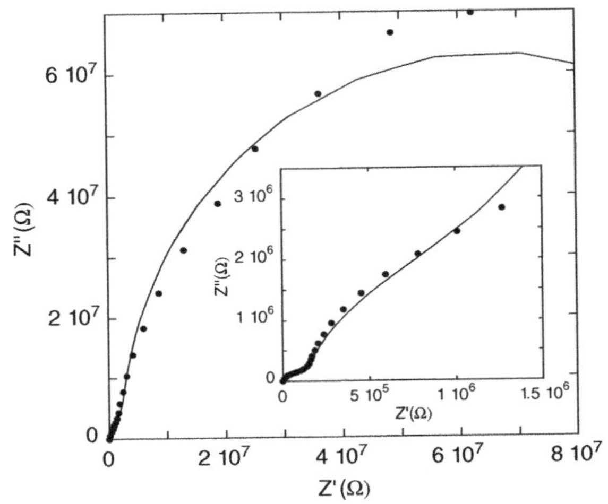

Figure 6.5 Plot of Z'' against Z' (Nyquist plot) for the same tBLM as in Fig. 6.4. The solid curve is the best fit of the impedance spectrum by the equivalent circuit shown in Fig. 6.4, obtained by using the same R and C values. The semicircle in the figure corresponds to the mesh of highest time constant and highest resistance, ascribable to the lipoic acid residue. The inset shows an enlargement of the initial portion of the Nyquist plot. Source: Guidelli and Becucci (2012b). Reproduced with permission of Springer.

well-formed semicircle, R_{max} in diameter. The semicircles of the remaining meshes are compressed in a very narrow area close to the origin of the Z'' versus Z' plot, and can be visualized only by enlarging this area. Therefore, the Nyquist plot of the whole spectrum can be conveniently employed if one is interested in pointing out the resistance R_{max} of the dielectric slab of highest resistance. This is apparent in Fig. 6.5, which shows the Nyquist plot for the tBLM incorporating valinomycin, whose Bode plot is reported in Fig. 6.4. The whole Nyquist plot displays a single semicircle. However, the enlargement of the initial portion of the plot in the inset of Fig. 6.5 reveals the presence of two additional partially fused semicircles.

To better visualize all semicircles, it is convenient to represent impedance spectra on a $\omega Z'$ versus $-\omega Z''$ plot, that is, an M plot (Becucci et al., 2005; Guidelli and Becucci, 2012*b*). A single *RC* mesh yields a semicircle even on this plot. Thus, if we multiply both members of Eq. 6.3 by ω^2 and we combine the resulting equation with Eq. 6.2b, after simple passages, we obtain

$$\omega^2 Z''^2 + \omega^2 Z'^2 - \omega Z''/C = 0 \rightarrow \left(\omega Z'' - \frac{1}{2C}\right)^2 + \left(\omega Z'\right)^2 = \left(\frac{1}{2C}\right)^2. \tag{6.4}$$

This is the equation of a semicircle of diameter C^{-1} and center of coordinates $(1/(2C), 0)$ on a $\omega Z'$ versus $-\omega Z''$ plot. Moreover, ω at the maximum of the semicircle is again equal to the reciprocal of the time constant *RC* of the mesh. While ω decreases along the positive direction of the abscissa on a Z'' versus Z' plot, it increases on an M plot. Therefore, for a series of *RC* meshes, the last semicircle on the M plot is characterized by the lowest time constant. This is, unavoidably, the semicircle simulating the solution that baths the self-assembled film, due to its very low capacitance. Figure 6.6 shows the M plot relative to the same impedance spectrum that yields the Bode plot in Fig. 6.4 and the Nyquist plot in Fig. 6.5.

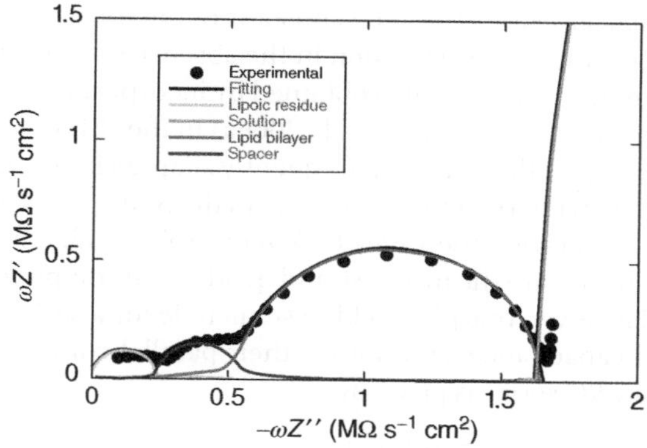

Figure 6.6 Plot of $\omega Z'$ (solid circles) against $-\omega Z''$ for the same tBLM as in Fig. 6.4. The solid black curve is the best fit of the impedance spectrum by the equivalent circuit shown in Fig. 6.4. From left to right, the distorted semicircles of increasing radius refer to $\omega Z'$ contributions from the lipoic acid residue, the spacer, the lipid bilayer, and the solution. Source: Becucci et al. (2005). Reproduced with permission of American Chemical Society.

The solid curve is the best fit of the impedance spectrum by the equivalent circuit in Fig. 6.4, consisting of four *RC* meshes in series.

Proceeding along the positive direction of the abscissa, the distorted semicircles of progressively increasing radius express the contributions to $\omega Z'$ from each of the four different *RC* meshes, namely the quantity $\omega g_i/(g_i^2 + \omega^2 C_i^2)$, where $g_i = 1/R_i$ and C_i are the conductance and capacitance of the *i*th mesh. The deviations of these curves from a pure semicircle measure the extent of their overlapping with the neighboring semicircles. The four semicircles overlap only to a moderate extent, thus allowing their straightforward deconvolution. This is due to an appreciable difference between the time constants of the four *RC* meshes, which are evenly distributed over a frequency range covering seven orders of magnitude. The capacitance of the solution interposed between the working and the counter electrode is of a few tens of nF cm^{-2}. If it is disregarded by simulating the aqueous phase by a pure resistance, the contribution of the solution to the M plot is represented by a vertical straight line. As a matter of fact, the radius of the semicircle simulating the solution is not infinitely large, and the curvature is often clearly visible. The *RC* mesh of the aqueous solution does not depend on the architecture of the tBLM. Hence, the corresponding semicircle can be excluded, at least partially, from the M plot in order to better visualize the contribution from the other meshes. The M plot permits the agreement between an experimental impedance spectrum and the corresponding fit by a series of *RC* meshes to be verified in detail. In this respect, it differs from the Bode plot, which is often almost featureless.

The largest effect of an ion channel or carrier is clearly produced on the lipid bilayer moiety, where these exogenous species penetrate moving ions across it. The different extent of this effect on the three different dielectric slabs decreases the overlapping between the corresponding semicircles, as shown in the M plot of Fig. 6.6 at a tBLM incorporating valinomycin, where the three semicircles are clearly distinguishable (Becucci et al., 2005, Becucci and Guidelli, 2014a; 2014b). Note that the decrease in resistance being higher in the lipid bilayer than in the spacer causes the time constant of the former to be lower than that of the latter, contrary to the situation in the absence of exogenous species. Similar semicircles in the M plot are also obtained by incorporating gramicidin (Becucci et al., 2007a) or melittin (Becucci et al., 2006a) in the tBLM.

A plot that has been frequently adapted in the literature to display an impedance spectrum richer in features than the Bode plot is the Y'/ω versus Y''/ω plot, sometimes called the *Cole–Cole plot* (Toby et al., 1998; Jeuken et al., 2006). Here, Y' and Y'' are the in-phase and quadrature components of the electrode admittance. However, this plot yields a semicircle for a series combination of a resistance and a capacitance, and not for their parallel combination. Thus, the impedance Z of an RC series is given by

$$Z = R + \mathrm{i}/(\omega C). \tag{6.5}$$

Noting that $Y \equiv Y' - \mathrm{i}Y'' = 1/Z$, rearrangement of Eq. 6.5 yields

$$Y''/\omega = C/\left(\omega^2 R^2 C^2 + 1\right), \tag{6.6a}$$

$$Y'/\omega = \omega RC\left(Y''/\omega\right). \tag{6.6b}$$

Eliminating ωRC from Eqs. 6.6a and 6.6b, we get

$$\left(\frac{Y''}{\omega}\right)^2 + \left(\frac{Y'}{\omega}\right)^2 - C\left(\frac{Y''}{\omega}\right) = 0 \rightarrow \left(\frac{Y''}{\omega} - \frac{C}{2}\right)^2 + \left(\frac{Y'}{\omega}\right)^2 = \left(\frac{C}{2}\right)^2. \tag{6.7}$$

On a Cole–Cole plot, this equation yields a semicircle of diameter C and center of coordinates $(C/2, 0)$. Here too, from Eq. 6.6b it follows that ω at the maximum of the semicircle equals $1/RC$. In particular, a Cole–Cole plot is not suitable for verifying the fitting of an experimental impedance spectrum by a series of RC meshes. Thus, if the impedance spectrum of an equivalent circuit consisting of a series of two RC meshes is displayed on a Cole–Cole plot, it yields a semicircle of diameter equal to the capacitance of the RC mesh of higher capacitance, but the corresponding resistance is affected by that of the neighboring RC mesh.

A series of RC meshes with very close time constants yields a single "depressed" semicircular arc on Nyquist, M or Cole–Cole plots, resulting from overlapping of the various RC meshes. In other words, such an overlapping yields a figure resembling an arc whose center lies below the horizontal axis. This arc is often fitted by an equivalent circuit consisting of a parallel combination of a resistance R and of a *constant phase element* (CPE), whose empirical impedance function has the form $Z_{\mathrm{CPE}} = -(\mathrm{i}\omega)^{-\alpha}/A$ (Raistrick et al., 1987). The hybrid CPE finds its justification in a continuous distribution of time constants around a mean. For $\alpha = 1$, it reduces to a pure capacitive impedance. The impedance of this parallel combination, called the ZARC, is given by

$$Z_{\mathrm{ZARC}}^{-1} = R^{-1} - (\mathrm{i}\omega)^{\alpha} A \rightarrow Z_{\mathrm{ZARC}} = \frac{R}{1 - (\mathrm{i}\omega)^{\alpha} RA}. \tag{6.8}$$

For $\alpha = 1$, this expression reduces to the impedance of an RC mesh, and A coincides with the capacitance. The use of ZARC elements is often required for a satisfactory fitting of impedance spectra at biomimetic membranes formed on an insufficiently smooth solid support.

6.2.2 Potential-Step Chronoamperometry (Current–Time Curves)

α-Helical peptides capable of permeating membranes are normally amphiphilic, that is they have one side of the helix relatively hydrophilic by the presence of residues with hydrophilic and/or charged side chains, whereas the other side is hydrophobic. They may permeate a membrane by two main alternative mechanisms, the *barrel-stave* and the *carpet* mechanism (Shai, 1999). According to the first mechanism, the peptide monomers associate to form a bundle inside the membrane, turning their hydrophilic side toward the interior of the bundle and their hydrophobic side toward its exterior, where the latter side interacts attractively with the hydrocarbon tails of the surrounding lipid matrix. The central lumen of the bundle, lined with the hydrophilic side chains, constitutes an ion channel, like a barrel made of helical peptides as staves (Fig. 6.7*a*). A variant of the barrel-stave model is the *toroidal model* (Yang et al., 2001), in which the peptides are associated with the lipid polar heads even when they are perpendicularly inserted into the lipid bilayer. In forming such a pore, the lipid molecules around the pore bend continuously along its side up to assuming a flat orientation at half height of the pore, thus intercalating their polar heads between the embedded peptide molecules (Fig.6.7*b*). In this way, the pore is lined with both peptides and lipid polar heads.

According to the carpet model, the peptide molecules remain in contact with the polar heads of the lipid bilayer during the whole process of membrane permeabilization and do not insert into the hydrophobic core of the membrane. Thus, they do not need necessarily to adopt an amphipathic α-helical structure. Membrane permeabilization takes place only if there is a high local concentration of membrane-bound peptides, with their hydrophilic side facing the lipid polar heads or the water molecules. Ultimately, the peptides disintegrate the membrane by disrupting the bilayer curvature. It has been proposed that this disruption involves a detergent-like micellization, whereby lipid micelles are enveloped by a peptide layer and stripped from the lipid bilayer, forming transient holes in the membrane.

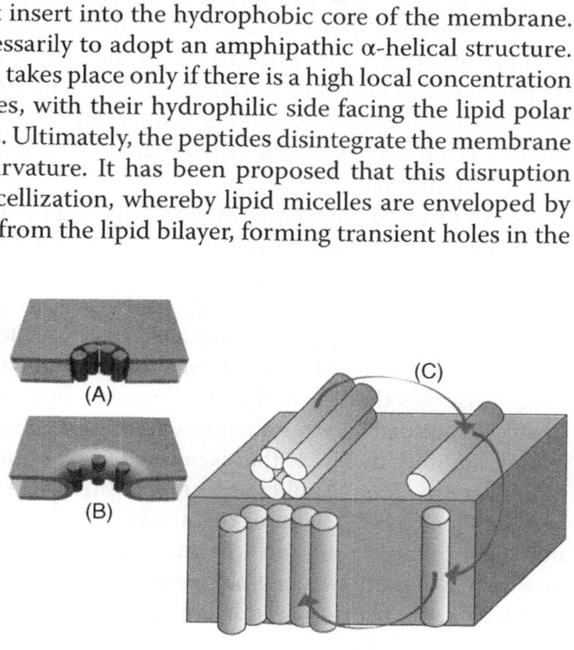

Figure 6.7 Schematic pictures of the barrel-stave model (A) and toroidal model (B) of an ion channel, and of the gradual passage from a "flat cluster" to an "embedded cluster" (i.e., an ion channel) via a "flat monomer" and an "embedded monomer" (C). Source: Reprinted from Becucci and Guidelli (2014b)

Many ion channels resulting from the aggregation of α-helical peptides are voltage-gated, in that they open or close in response to changes in the transmembrane potential (Sansom, 1991). Several molecular models for voltage-gated channels have been proposed, with particular regard to the alamethicin channel. The simplest model assumes that channel formation starts from a small aggregate of peptide monomers (the nucleus) that grows in diameter through the uptake of further monomers. The voltage-dependent step would be the rotation of the monomers from the membrane surface into its interior by the electric field (Bauman and Mueller, 1974). A similar model was proposed by Bruner by assuming that a peptide molecule moves, under the influence of an applied voltage, first from a nonconducting surface state to a nonconducting precursor state and then to a conducting state (Bruner, 1985). This three-state model predicts a sigmoidal shape of current transients in response to voltage steps, as reported for alamethicin (Mauro et al., 1972; Baumann and Mueller, 1974; Boheim and Kolb, 1978a; Bruner and Hall, 1983), melittin (Tosteson and Tosteson, 1984), and monazomycin (Muller and Finkelstein, 1972). Current transients of sigmoidal shape for monazomycin (Becucci and Guidelli, 2007) and potassium channels (Becucci et al., 2003) are also predicted on the basis a model accounting for monomer aggregation by a mechanism of nucleation and growth.

A different group of molecular models assumes the aggregation of peptide monomers at the membrane surface. According to Boheim and Kolb (1978a), the voltage pushes into the membrane two or three monomers simultaneously from a hexamer already preformed at the membrane–solution interface. Salt concentration effects led Hall (1975) to hypothesize two-dimensional alamethicin micelles on the membrane surface, which are pushed by the applied voltage into the hydrocarbon tail region, where they fluctuate among various conformations. As a matter of fact, direct transfer of an aggregate from the membrane surface to its interior is improbable. Surface aggregates have the hydrophobic sides of the helices turned toward each other. Conversely, they must turn them outward to interact attractively with the surrounding lipid chains, if they have to penetrate into the membrane.

A third group of molecular models referring to alamethicin assumes the presence of aggregates of antiparallel dipolar molecules oriented perpendicularly to the membrane plane and bridging the hydrocarbon tail region at zero voltage, with some probability for "flipping" (Boheim et al., 1983). Application of an electric field forces one or more molecules into a parallel orientation, leading to electrostatic repulsion and to the formation of water-filled pores. A variant of this model assumes that the antiparallel peptide dipoles composing an aggregate are hydrogen bonded between themselves through their side chains and contain at its center a dipolar molecule that is not hydrogen bonded to its neighbors (Mathew and Balaram, 1983). Application of an electric field would expel the latter dipole, generating a channel without the need for molecule "flip-flop." These models do not seem to clearly explain why the alamethicin ion channel induces a current flow only when the transmembrane potential is made positive on the membrane side where the peptide is added, at least at BLMs formed with saturated lipids (Eisenberg et al., 1973).

Overall, the majority of evidence seems to be in favor of models in which the voltage-dependent step includes a reorientation of peptide monomers, inducing them to span the whole hydrophobic region of the membrane and to aggregate with ion-channel formation. There are quite different views about the peptide orientation at zero transmembrane potential, based on different experimental data and supported by different calculations. An NMR investigation showed that alamethicin interacts primarily at the water/lipid interphase without significant insertion into the hydrocarbon tail region in the absence of voltage (Banerjee et al., 1985). On the other hand, circular dichroism data in unilamellar vesicles (Schwarz et al., 1986) and infrared attenuated total reflection spectroscopy in multibilayer membranes (Fringeli and Fringeli, 1979) pointed to alamethicin incorporation into the lipid. Site-directed spin-labeling studies showed that alamethicin is in a linear form and normal to the membrane plane, with its C-terminus in the aqueous region and the N-terminus in the membrane hydrocarbon tails, even in the absence of voltage (Barranger-Mathys and Cafiso, 1996). The uncertainty in alamethicin orientation at zero voltage is not clarified by computational approaches. In fact, the difference in energy between superficial and transmembrane orientation is small, and even similar approaches treating both hydrophobic and electrostatic effects can easily produce different results (Kessel et al., 2000; Mottamal and Lazaridis, 2006).

The response of a conventional BLM incorporating a channel-forming peptide to a potential step is usually monitored by recording *current–time* (I–t) *curves*. This technique is referred to as *chronoamperometry* in electrochemical jargon. A remarkable and peculiar example of chronoamperometry is represented by the application of the voltage clamp technique to a squid giant axon (cf. Fig. 4.15). Peptides are often added only on one side of BLMs; in what follows, we refer to this side as the *cis* side, defining the transmembrane potential as the electric potential on the *trans* side with respect to the *cis* side, taken conventionally equal to zero. Figure 6.10 shows two series of I–t curves following transmembrane potential steps of increasing height at a BLM incorporating alamethicin. The current induced by a potential step tends to a limiting value that increases with an increase in step height. The electric field elicited by the potential step grows with its height, causing an increase in the flow rate of permeant ions, which in turn increases the steady-state level of the current.

At ambient pressure, the I–t curves in Fig. 6.8A tend to their steady-state value by constantly maintaining the concavity of the curve turned toward the time axis. Conversely, at an elevated pressure of 100 MPa, the I–t curves of alamethicin have a clear sigmoidal shape, as shown in Fig. 6.8B. Even the I–t curve of alamethicin at ambient pressure, if carefully examined in the range of a few tens of milliseconds from the start of the potential jump, exhibits an initial S-shaped time course, followed by an asymptotic increase toward a steady-state value (Mauro et al., 1972).

With peptides that form ion channels by aggregation of helices turning the hydrophilic side toward the lumen of the channel, the current transients following a potential jump are indeed often characterized by a sigmoidal shape. This shape can be explained on the basis of a general kinetic model that accounts for (i) the disruption of possible clusters of monomeric molecules adsorbed on top of the

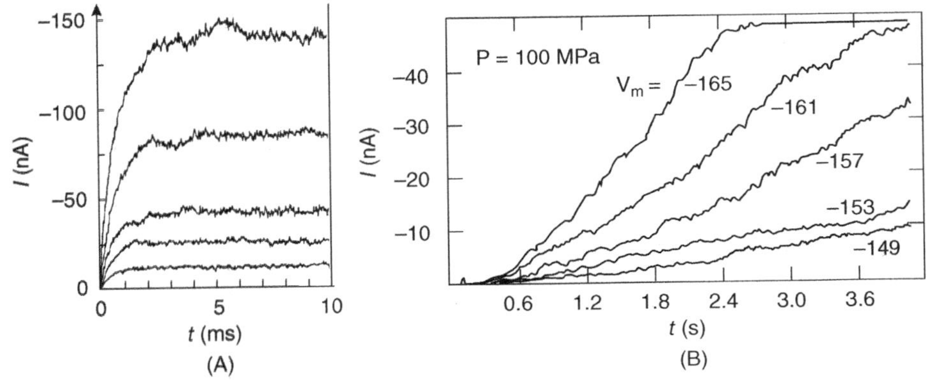

Figure 6.8 (A) Proceeding upward, current versus time curves at a BLM in aqueous solution of 1.0 M NaCl and 5×10^{-7} g mL^{-1} alamethicin, following transmembrane potential steps from 0 to −40, −42, −50, −55, and −58 mV. Source: Eisenberg (1973). Reproduced with permission of Springer. (B) Current versus time curves at a BLM in aqueous solution of 1.0 M NaCl and 4×10^{-8} g mL^{-1} alamethicin, at a pressure of 100 MPa, following transmembrane potential steps of the height (in mV) reported in the figure. In both cases, alamethicin was added only on the *cis* side of the membrane. Source: Bruner (1983). Reproduced with permission of Elsevier.

BLM (henceforth referred to as *flat clusters*); (ii) the potential-dependent penetration of the resulting *flat monomers* into the lipid bilayer; (iii) the aggregation of the monomers incorporated in the lipid bilayer (henceforth referred to as *embedded monomers*) into channel-forming *embedded clusters* (Becucci and Guidelli, 2007). These three consecutive events are strictly correlated and are schematically depicted in Fig. 6.7C.

The kinetic process of disruption of flat clusters is treated as a *nucleation of holes* and *growth of hole aggregates* within the clusters. By nucleation of holes, we mean the quasi-reversible detachment of an initial number of flat monomers from a flat cluster and their random intercalation with the water molecules on top of the BLM. In other words, a hole is just a flat monomer that, upon detachment from a flat cluster, leaves behind a hole in the cluster. These flat monomers are considered to detach from a flat cluster and to reaggregate into it in a quasi-reversible manner, until the number of nearest-neighboring holes in the cluster attains a critical value (n) beyond which this number increases irreversibly up to complete disruption of the flat cluster. This critical number of nearest-neighboring holes will be loosely called *nucleus*. Nucleation is followed by the irreversible disruption of the flat clusters, which can be viewed as an irreversible "*growth* of hole aggregates," according to the nucleation-and-growth terminology.

Let θ_0 denote the fraction of the whole surface of the BLM initially covered by flat clusters. The disruption of the flat clusters starts at the instant at which the transmembrane potential on the *trans* side of the BLM is stepped to a sufficiently negative value; this instant is taken as the origin of the time axis. Let S_h denote the time-dependent ratio of the area covered by holes (i.e., the flat monomers being released from the flat clusters) to that, θ_0, initially covered by flat clusters. Since the nucleation of holes takes place within the surface area covered by clusters, the nucleation rate $\nu_{h,N}$ will be set proportional to this fractional surface area, $\theta_0(1 - S_h)$, according to a proportionality constant $k_{h,N}$. Upon ascribing, for simplicity, a circular shape to the growing hole aggregates, the rate, $\nu_{h,R}$, of their radial growth will be regarded as roughly independent of the surface coverage, $\theta_0(1 - S_h)$, by clusters. This assumption is particularly justified during the initial growth of the hole aggregates within a cluster, when the holes are almost completely surrounded by the flat monomeric units forming the given flat cluster. Such an assumption becomes rougher at later stages of cluster disruption, when, however, this assumption has a minor effect on the rate of the whole process.

In fact, as the flat monomers are gradually formed, they are assumed to penetrate into the lipid bilayer with a potential-dependent probability p that increases as the transmembrane potential is stepped to progressively more negative values. The rate constant of nucleation of holes, $k_{h,N}$, and the rate of radial growth of hole aggregates, $\nu_{h,R}$, are not directly affected by the electric field. However, they depend indirectly upon potential via the penetration probability p of the flat monomers. The ratio, S_h, of the area covered by flat monomers to that initially covered by flat clusters is estimated by first ignoring the overlapping of the growing hole aggregates, and by then correcting for this neglect by using Avrami's approach (Avrami, 1940). This approach is outlined later, in connection with the further stage of the whole process, which involves the nucleation of the embedded monomers and the growth of the resulting embedded clusters with ion-channel formation.

As the embedded monomers are formed, they start nucleating, giving rise to channel-forming embedded clusters. Upon denoting by S the ratio of the area covered by embedded clusters to that covered by both embedded monomers and embedded clusters, the fraction of the whole electrode surface covered by embedded monomers is given by $\Theta = S_h\theta_0 p(1 - S)$, where $S_h\theta_0 p$ is clearly that covered by both the embedded monomers and the embedded clusters. Clusters of embedded monomers resulting from a series of consecutive collisions are characterized by a critical size (the nucleus), below which they have a higher tendency to shrink by releasing one unit than to grow by aggregation of a further unit, and above which they have a practically irreversible tendency to increase. Similarly to the case of nucleation of holes and growth of hole aggregates, the formation of this critical cluster size from embedded monomeric units is referred to as *nucleation*, and the irreversible increase beyond the critical size as *growth*. If we denote by n the number of embedded monomers composing a nucleus, the elementary step yielding the nucleus consists of the incorporation of a monomer into an $(n - 1)$-meric *subcritical nucleus*. This results from $(n - 1)$ elementary steps involving the incorporation of each monomeric unit into the immediately preceding subcritical nucleus, starting from an initial monomer acting as a *nucleation center*. If we assume that all steps preceding the step yielding the nucleus are in quasi-equilibrium, then the nucleation rate, ν_N, will be proportional to the

nth power of the fractional surface coverage, $\Theta = S_h \theta_0 p(1-S)$, by the embedded monomers randomly distributed in the monolayer, according to a nucleation rate constant k_N:

$$dN/dt \equiv v_N = k_N \Theta^n. \tag{6.9}$$

Here, N is the number of nuclei per unit surface area. The irreversible aggregation of monomeric units to a nucleus, just after its formation, gives rise to a *supercritical nucleus*, whose continuous growth ultimately yields a channel-forming embedded cluster. Assuming for simplicity that the cross-sectional area A of a growing supercritical nucleus is a circle of radius R, the rate of growth of A is given by the time derivative of πR^2. Moreover, it is proportional to the frequency of the successful impacts of the embedded monomers, of surface coverage Θ, with the circumference $2\pi R$ of the supercritical nucleus, according to a proportionality constant k_R:

$$dA/dt = d(\pi R^2)/dt = 2\pi R\, dR/dt = k_R 2\pi R\Theta \quad \rightarrow \quad dR/dt \equiv v_R = k_R \Theta. \tag{6.10}$$

It follows that the rate v_R of radial growth of a supercritical nucleus is proportional to Θ according to the rate constant k_R. It should be stressed that the frequency of the successful impacts is not necessarily controlled by the two-dimensional diffusion of the embedded monomers within the lipid bilayer, but rather by the occurrence of a favorable mutual orientation between the supercritical nucleus and the aggregating embedded monomer.

Let us now consider a general approach to the kinetics of nucleation and growth that can allow any of the two quantities S_h and S to be calculated as a function of time. To simplify notations, we use the symbol S, bearing in mind that the approach holds for both S_h and S. Let us denote by $t = 0$ the starting time of the nucleation and growth process under consideration, and let us imagine observing it at a later time $t = t$. Let dN denote the infinitesimal number of nuclei that are forming in the infinitesimal time interval between y and $y + dy$ before the observation time t. Having assumed that the resulting supercritical nuclei have a circular shape, the nuclei formed in this infinitesimal time interval make the following contribution to the area covered by the supercritical nuclei at time t:

$$dN \int_0^{A(t)} dA = \pi \left(\int_0^R dR \right)^2 dN = \pi \left[\int_y^t (dR/dt)_z dz \right]^2 dN. \tag{6.11}$$

R is the radius of the growing supercritical nuclei and z is an auxiliary variable having the dimensions of time, which varies between the time y at which the nuclei form and the time t at which the resulting supercritical nuclei are observed. If the supercritical nuclei do not have a circular shape, the multiplying factor π may be replaced by a different geometrical factor. If we now sum all the aforementioned infinitesimal contributions by integrating Eq. 6.11 over time between the limits of integration $y = 0$ and $y = t$, we obtain the ratio of the area covered by the supercritical nuclei (ultimately yielding the channel-forming clusters) to the area covered by both monomers and supercritical nuclei (Bosco and Rangarajan, 1981):

$$S_x = \pi \int_0^t dy \left[\int_y^t (dR/dt)_z dz \right]^2 (dN/dt)_y = \pi \int_0^t dy \left[\int_y^t \nu_R(z)\, dz \right]^2 \nu_N(y)\,. \tag{6.12}$$

This ratio is denoted by S_x, rather than S, to point out that it ignores the possible overlapping of the progressively growing supercritical nuclei. It is commonly called the *extended area*. This expression, which includes both the rate of nucleation, $(dN/dt) = \nu_N$, and the rate of radial growth, $(dR/dt) = \nu_R$, is entirely general, and can be applied to the kinetics of any nucleation-and-growth process.

The unrealistic overlapping of the progressively growing supercritical nuclei can be avoided by having recourse to Avrami's formalism (Avrami, 1940). In general, Avrami's approach is based on the consideration that the area covered by the growing supercritical nuclei is an extensive property of the system and, as such, must be directly proportional to the *available area*. If we apply this concept *ab absurdo* to the case in which the single growing supercritical nuclei are allowed to grow without being limited by the neighboring growing supercritical nuclei, then the available area is the total surface area, S_T, where the nucleation-and-growth process occurs, and we have

$$dS_x/dt = \text{const.} \times S_T. \tag{6.13}$$

S_x is just the hypothetical extended area, namely the area that would be covered by all the growing supercritical nuclei if they were free to grow without limit. Naturally, this is not possible in reality, and the actual surface area available to the growing supercritical nuclei is that still uncovered, which is given by the difference, $S_T - S$, between the total area, S_T, and that, S, already covered. In this case, Eq. 6.13 becomes

$$dS/dt = \text{const.} \times (S_T - S)\,. \tag{6.14}$$

Upon eliminating the common proportionality constant between Eqs. 6.13 and 6.14, we obtain

$$dS/dt = (1 - S/S_T)\, dS_x/dt \quad \text{or, for } S_T = 1\text{:}\;\; dS/dt = (1 - S)\, dS_x/dt. \tag{6.15}$$

The present model of nucleation and growth is entirely general. Thus, it only assumes that the elementary steps preceding the step yielding the nucleus are in quasi-equilibrium, and that the growth of the supercritical nucleus proceeds irreversibly by activated aggregation of monomers.

Equation 6.12 can be applied to the nucleation of holes and growth of the resulting hole aggregates by inserting the expressions for the corresponding rates, $\nu_{h,N}$ and $\nu_{h,R}$; it can also be applied to the subsequent nucleation of embedded monomers and growth of the resulting supercritical nuclei by inserting the expressions of Eqs. 6.9 and 6.10 for the ν_N and ν_R rates. Repeated differentiation of the two equations so obtained with respect to time via the generalized Leibnitz formula yields a set of differential equations, which are combined with the two differential equations obtained from Eq. 6.15 and relating the extended areas, $S_{h,x}$ and S_x, to the corresponding actual areas, S_h and S. This set of differential equations can be readily solved numerically by the fourth-order Runge–Kutta method, thus obtaining S_h and S as a function of time. It should be noted that the kinetics of the two nucleation-and-growth stages depends exclusively on the products, $k_{h,N}v^2_{h,R}$ and $k_N k^2_R$, of each rate constant of nucleation by the square of the corresponding rate constant of radial growth, thus reducing the number of adjustable parameters to only four, that is, θ_0, n, $k_{h,N}v^2_{h,R}$, and $k_N k^2_R$.

Finally, the current density j across the BLM induced by the transmembrane-potential jump is obtained by setting it proportional to the fractional surface coverage by the channel-forming embedded clusters, $S_h\theta_0 pS$, since each newly formed channel makes a contribution to this current:

$$j \propto S_h\theta_0 pS. \tag{6.16}$$

It should be noted that the fractional area $S_h\theta_0 pS$ should by no means be regarded as homogeneous. Rather, it should be regarded as consisting of separate channel-forming clusters of possibly different size dispersed in the BLM. The limiting value attained by j is clearly given by $\theta_0 p$, since both S_h and S will ultimately tend to unity.

The polyene-like antibiotic monazomycin incorporated in a BLM yields sigmoidal $I-t$ curves with a long foot at ambient temperature and pressure (Muller and Finkelstein, 1972), as shown in Fig. 6.9. Its shape can be quantitatively predicted on the basis of the aforementioned general treatment. The initial long foot is ascribed to the disruption of the flat clusters originally adsorbed on top of the BLM and only indirectly affected by the applied potential; its length is longer the lower the $k_{h,N}v^2_{h,R}$ product is. The subsequent concave upward section of the $I-t$ curve is due to the induction period stemming from the nucleation of embedded monomers and growth of the resulting supercritical clusters; its is modeled by ascribing an appropriate value to the "overall" kinetic parameter $k_N k^2_R$. Finally, the last concave downward section of the curve is determined by the spontaneous tendency of the current to attain its time-independent steady-state value, which increases with an increase in the value ascribed to the probability p of penetration of flat monomers into the hydrocarbon tail region of the BLM. The dashed curves in Fig. 6.9 were calculated from this model by increasing progressively the penetration probability p, while leaving the $k_{h,N}v^2_{h,R}$ adjustable parameter constant.

The same approach was adopted by Becucci et al. (2003) to simulate the experimental conductance versus time curves obtained by Hodgkin and Huxley (1952) with a squid axon (cf. Fig. 4.17), as shown in Fig. 6.10. Note the difference between the procedure adopted by Hodgkin and Huxley, who calculated the current as the sum of the three contributions in Eq. 4.90, where the specific conductances of sodium and potassium channels were determined experimentally, and the present approach. Here, no mention is made of the presence of the four potassium subunits that must properly aggregate to form an ion channel. This approach merely shows that the potassium channels form spontaneously by nucleation and growth

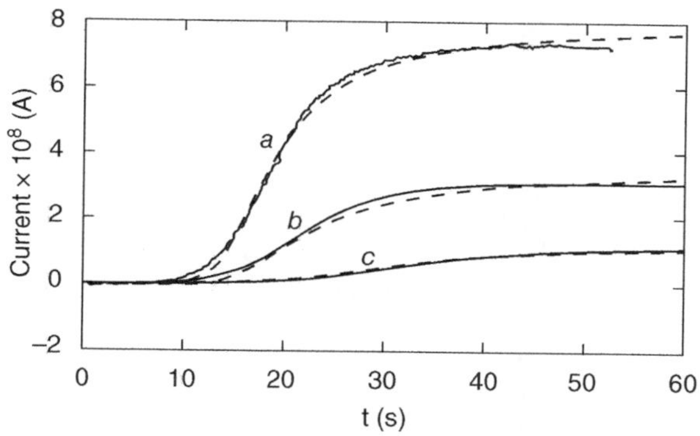

Figure 6.9 The solid curves are three successive current–time curves on the same BLM following transmembrane potential steps from 0 to −60 mV (*a*), −55 mV (*b*), and −50 mV (*c*), in aqueous 0.1 M KCl containing 0.625 μM monazomycin (Muller et al., 1981). The corresponding dashed curves were calculated by the nucleation-and-growth model outlined in the text using the parameters $\theta_0 = 0.1$, $n = 2$, and $k_{h,N}v^2_{h,R} = 0.016\ s^{-3}$ for all three curves; *p* and $k_N k_R^2$ were given the values: (*a*) 1 and 15 s^{-3}; (*b*) 0.438 and 130 s^{-3}; (*c*) 0.163 and 1250 s^{-3}. The three calculated currents were matched to the experimental ones by multiplying them by the same factor 80. Source: Becucci and Guidelli (2007). Reproduced with permission of American Chemical Society.

Figure 6.10 Markers are conductance values in squid giant axon at 6–7 °C brought about by the following depolarizations: 32 (*a*), 38 (*b*), 51 (*c*), 63 (*d*), 76 (e), 88 (*f*), and 100 mV (*g*). The solid curves were calculated for $\theta_0 = 0.03$, $n = 2$, $k_N^{1/2} k_R = 1.95 \times 10^7\ s^{-3/2}$, $p = 0.47$ (*a*), 0.55 (*b*), 0.67 (*c*), 0.74 (*d*), 0.83 (*e*), 0.946 (*f*), 1 (*g*), and for a normalizing factor of 697 mS cm^{-2}. Source: Becucci et al. (2003). Reproduced with permission of American Chemical Society.

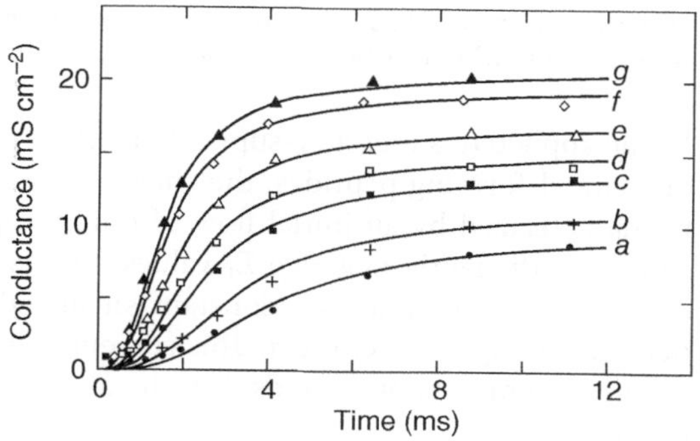

of an unspecified number of monomeric units (the potassium subunits), and that the steady-state value of the current increases with an increase in the penetration probability *p*.

6.2.3 Potential-Step Chronocoulometry

Potential-step chronocoulometry is the technique of choice for the investigation of the mechanism of formation of ion channels at mercury-supported tBLMs. It consists in subjecting the electrochemical system under investigation to a potential jump from an initial value E_i to a final value E_f and in recording the charge $Q(t)$ that flows as a consequence of this jump as a function of time.

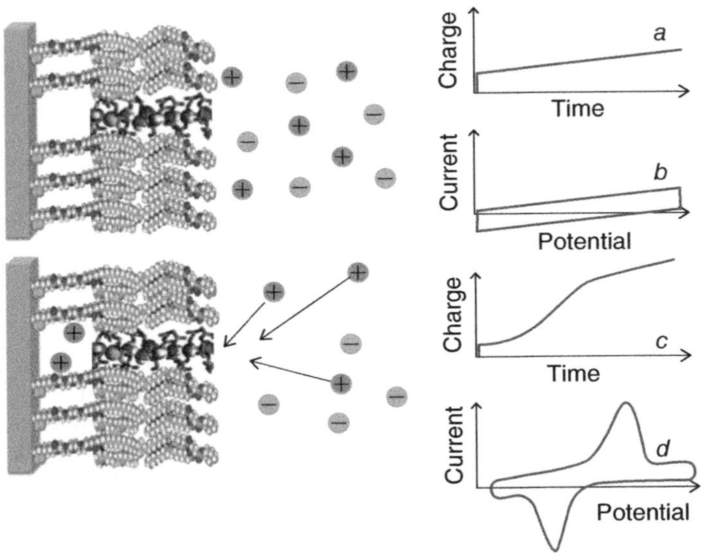

Figure 6.11 Schematic pictures of a DPTL/DOPC tBLM incorporating a gramicidin channel in the closed state (top left) and in the open state (bottom left). The typical shape of charge versus time and current versus potential curves at a closed ion channel (or in the absence of ion channels) and at an open ion channel is depicted on the right-hand side of the corresponding schematic pictures. Source: Becucci and Guidelli (2014b).

When applied to a mercury-supported DPTL/phospholipid tBLM in the absence of channel-forming peptides, the charge transient triggered by a potential jump is characterized by an initial flow of purely capacitive charge, lasting less than 1 ms; at sufficiently negative E_f values, this is followed by a linear increase of charge with time, due to a constant small reduction current, as schematically depicted in Fig. 6.11, curve *a*. This current, which is clearly detectable only over a timescale of tens of seconds, is to be ascribed to the electroreduction of trace impurities.

Upon incorporating cation-selective or nonselective ion channels, either ohmic or voltage-gated, a potential jump from an initial potential E_i, at which the ion channel is closed or not yet formed, to a sufficiently negative final potential E_f, at which it is open, causes the channel to induce an ion flow into the tetraethyleneoxy (TEO) hydrophilic spacer. In particular, ohmic channels start moving ions into the TEO spacer in the proximity of the zero value of the potential difference across the lipid bilayer moiety of the tBLM (i.e., the transmembrane potential $\Delta\phi$), while voltage-gated channels show this behavior only at sufficiently negative $\Delta\phi$ values. The ion flow is revealed by a negative charge that adds to the background charge recorded in the absence of channels. This negative charge is due to electrons that flow along the external circuit and accumulate on the metal surface, to maintain the electroneutrality of the whole electrified interface interposed between the bulk metal and the bulk aqueous solution.

The electron charge density σ_M accumulating on the metal surface is practically equal in magnitude and opposite in sign to the cation charge density accumulating in the hydrophilic spacer. In fact, the only other charge density located within the electrified interface is that present in the diffuse layer adjacent to the tBLM, which is much smaller than the other two (Becucci and Guidelli, 2009). As the final potential E_f is made progressively more negative, an E_f value is ultimately attained at which the hydrophilic spacer is completely saturated by the cations of the electrolyte. This gives rise to a charge step with a well-defined plateau, which adds to the background charge, as shown schematically in Fig. 6.11, curve *c*. In the presence of potassium ions, the height of this charge step ranges from -45 to $-50\,\mu C\,cm^{-2}$ and corresponds to the charge associated with the maximum amount of potassium ions that can be accommodated in the TEO spacer, as verified for different ion channels, such as gramicidin (Becucci and Guidelli, 2009), melittin, monazomycin (Becucci and Guidelli, 2007), distinctin, alamethicin (Becucci et al., 2011a), trichogin GA IV (Becucci et al., 2012a), dermcidin (Becucci et al., 2014), and syringomycin E (Becucci et al., 2015).

The constant maximum value of $45-50\,\mu C\,cm^{-2}$, corresponding to spacer saturation by K^+ ions, is explained by the aforementioned ion channels being either closed or not yet formed at the initial rest potential, E_i, which precedes the potential jump (Becucci and Guidelli, 2014a; 2014b). However, it has been recently observed that the lipodepsipeptide syringopeptin 25 is open at the initial potential E_i (Becucci et al., 2016) and is permeable to both positive and negative monovalent inorganic ions, such as K^+ and Cl^-. In this case, the potential jump yields a charge transient appreciably higher than $50\,\mu C\,cm^{-2}$. This is due to the fact that the electrons that flow along the external circuit to ensure the electroneutrality of the whole electrified interface must compensate not only the positive charge of the K^+ ions flowing into the hydrophilic spacer, but also the negative charge due to the concomitant outflow of the Cl^- ions that are present in the spacer at the initial potential E_i.

The current flowing across a tBLM, as a consequence of a potential jump inducing ion-channel formation, behaves differently from that across a BLM. In fact, the ions moving along the lipid bilayer moiety of a tBLM are not free to diffuse in a semi-infinite aqueous medium on the *trans* side of the bilayer, as in the case of a BLM. Rather, they accumulate within the hydrophilic spacer moiety of the tBLM, spreading radially from the mouth of each newly formed channel, until they completely saturate it. Thus, the current attains a maximum and then decays to zero, as soon as saturation is reached. Consequently, it is the charge versus time curve that exhibits a sigmoidal shape, rather than the $I-t$ curve, attaining a plateau that depends exclusively on the spaciousness of the hydrophilic spacer, irrespective of the number density of the ion channels. The model outlined in connection with potential step chronoamperometry must, therefore, be modified by associating the nucleation and growth of ion channels to the concomitant radial diffusion of the translocating ions into the hydrophilic spacer (Becucci and Guidelli, 2007).

The charge density $Q(t)$ is calculated by assuming that the formation of each newly formed channel starts a radial diffusion of ions from its mouth. To this end, the fractional surface area covered by the ion channels formed in any infinitesimal time interval from y to $(y + dy)$ is multiplied by the area of a circle whose radius, R_d, equals zero at time y and increases in such a way that the time derivative of its square, dR_d^2/dt, is constant. The product of these two areas is integrated over time for y varying from 0 to the observation time t, yielding a quantity that, upon correction for the overlapping of the growing diffusion circles by Avrami's approach, yields the ratio, $S_d(t)$, of the surface area covered by radial diffusion of ions from the channel mouths to the available surface area. The charge $Q(t)$, normalized to unity, is then obtained by summing the fractional surface area, $S_h\theta_0 pS$, covered by the channel mouths (cf. Eq. 6.16) to the product of S_d by the remaining fractional surface area, $(1 - S_h\theta_0 pS)$. The assumption of a constant time derivative dR_d^2/dt is justified by the expression for the concentration c of a species diffusing on an infinite plane surface from a source point, in which c maintains the same value for a constant R_d^2/t ratio, upon disregarding the time-dependent pre-exponential factor as a first approximation (Crank, 1956).

An example of application of the aforementioned approach is provided by the Q versus t curves of the antimicrobial peptide dermcidin 1L (DCD-1L) (Becucci et al., 2014). Curve a in Fig. 6.12 shows a charge transient obtained at a mercury-supported DPTL/DOPC tBLM in an unbuffered solution of 0.1 M KCl and 4 μg mL^{-1} DCD-1L upon carrying out a potential jump from $E_i = -0.15$ to $E_f = -1.00$ V(SCE) and by then subtracting the corresponding background charge. The rest time at E_i before the potential jump was fixed at 30 s. The charge involved in the sigmoidal step amounts to about −45 μC cm^{-2} and corresponds to complete saturation of the hydrophilic spacer by potassium ions. If the potential jump yielding the sigmoidal charge step is repeated after a rest time of 30 s at E_i, a much steeper charge step of the same height and lacking the sigmoidal shape is obtained. This indicates that a rest time of 30 s at E_i is not sufficient to dismantle the ion channels formed at E_f as a consequence of the potential jump. If the rest time at E_i is of only 3 s, the charge step is as steep as in the previous case, albeit smaller. This proves that a time period of 3 s at E_i is not sufficient to expel all the positive ions from the hydrophilic spacer. Figure 6.12 also shows a charge transient (curve b) recorded at a DPTL/DOPS tBLM in a pH 7 buffer solution, upon subtracting the background charge, where DOPS stands for dioleoylphosphatidylserine (DOPS). The dashed curves are the best fits of the two charge transients by the model. The adjustable parameters used for the fitting are the same for the two calculated curves, except for the rate of disaggregation of the flat clusters, which is three orders of magnitude lower for curve b. Quite probably, this rate is much lower at the DPTL/DOPS tBLM in a pH 7 buffer because of the much lower rate at which DCD-1L is found to undergo incorporation at this pH; this allows the peptide molecules initially adsorbed on the surface of the lipid bilayer to aggregate in a more organized and compact way.

If only flat monomers are adsorbed on top of a tBLM, in the absence of flat clusters, then the first stage of ion-channel formation, which models flat

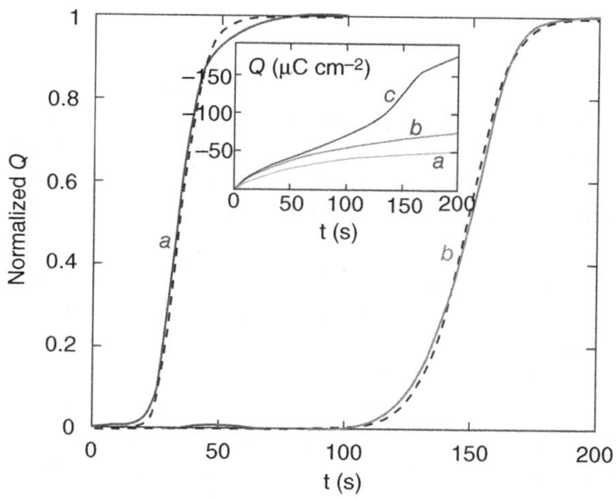

Figure 6.12 Charge transients recorded in 0.1 M KCl and 4 μg mL^{-1} DCD-1L at a DPTL/DOPC tBLM in an unbuffered solution (curve *a*) and at a DPTL/DOPS tBLM in a pH 7 buffer solution (curve *b*), by stepping the applied potential from −0.15 to −1.00 V(SCE); the curves are corrected for the background current and normalized to unity. The corresponding dashed curves are the best fits by the model outlined in the text. The inset shows charge transients at a DPTL/DOPS bilayer in the pH 7 buffer solution recorded by stepping the potential from −0.15 to −0.75 V (*a*), −0.85 V (*b*), and −0.95 V (*c*). Source: Becucci et al.(2014). Reproduced with permission of Royal Society of Chemistry.

cluster disruption via nucleation of holes, must be skipped from the treatment. In this case, the calculated charge transient lacks the long initial foot, while still retaining the typical sigmoidal shape. This is the case encountered with melittin at a DPTL/DPhPC tBLM, where DPhPC stands for diphytanoylphosphatidylcholine (Becucci and Guidelli, 2007). If, however, the tBLM is kept in the presence of a melittin solution for a long period of time, for example, 15 min, at an initial potential E_i at which the melittin molecules are adsorbed flat on top of the lipid bilayer, the charge transients following potential jumps negative enough to allow ion-channel formation exhibit a long foot.

These examples demonstrate that potential-step chronocoulometry provides useful information on the mechanism of membrane permeabilization by amphiphilic α-helical peptides. Thus, it may monitor the formation of ion channels by the barrel-stave or toroidal mechanism via the cooperativity in the peptide aggregation process. Moreover, it allows an approximate estimate of the time required for ion-channel formation and for its disruption. It permits one to verify whether the peptide is initially adsorbed on the membrane surface as a monomer or an oligomer, and whether monomers adsorbed flat on the membrane surface have a tendency to form flat clusters over time. Finally, the finite and well-defined spaciousness of the hydrophilic spacer allows the charge accommodated in it to be "dosed" as a function of time and of applied potential.

6.2.4 Cyclic Voltammetry

Valuable hints about the conformation of ion channels can also be gained by using cyclic voltammetry. This technique consists in scanning the applied potential at a constant scan rate from an initial to a final value and in then inverting the potential scan; this potential cycling can be repeated several times during a single experiment. The current is plotted against the applied potential to yield the cyclic voltammetry trace.

When applied to both BLMs and tBLMs in the absence of channel-forming peptides, the negative-going scan gives rise to a roughly constant negative capacitive current. In fact, the capacitive current density j is given by $\mathrm{d}Q/\mathrm{d}t = (\mathrm{d}Q/\mathrm{d}E)(\mathrm{d}E/\mathrm{d}t) = Cv$, where Q is the capacitive charge density, C is the differential capacitance, and v is the constant potential scan rate. In the case of BLMs, the capacitive current is due to a gradual accumulation of negative ions in the diffuse layer adjacent to the *trans* side and of positive ions in the diffuse layer adjacent to the *cis* side; in the case of tBLMs, it is due to a gradual accumulation of electrons on the metal support and of positive ions in the diffuse layer adjacent to the tBLM. The current also includes a resistive contribution due to the noninfinite resistance of the lipid bilayer moiety, which imparts a slight tilt to the current trace. The reverse positive-going potential scan yields a current with the same resistive contribution but an opposite capacitive contribution with respect to the zero-current axis. This results in a trace whose shape is similar to that of a parallelogram (see curve b in Fig. 6.11). Under these conditions, the vertical distance between the negative and positive voltage scans is clearly equal to $2Cv$.

Cyclic voltammograms (CV) of conventional BLMs incorporating channel-forming peptides are simply referred to as *current–voltage* ($I-V$) curves in biophysical jargon. In BLMs, the absolute value of the negative current increases exponentially with the absolute value of the negative transmembrane potential $\Delta\phi$, as shown in Fig. 6.13 for a BLM incorporating alamethicin. Thus its slope, which measures the conductance G, is such that its logarithm increases linearly with increasing $|\Delta\phi|$. When the peptide is added on only one side of the BLM, the transmembrane potential $\Delta\phi$ is usually referred to that side (the *cis* side) with respect to the opposite one (the *trans* side) in the biophysical literature; moreover, the current is regarded as positive when cations flow from the *cis* to the *trans* side. In this section, we use the term *voltage* (V) to denote the transmembrane potential $\Delta\phi$, to comply with the biophysical usage. However, V denotes the electric potential on the *trans* side of the BLM with respect to its *cis* side, still regarded as the side where the peptide is added. Moreover, the current is taken as negative when cations move from the *cis* to *the* trans side, at variance with the biophysical usage. This will facilitate a direct comparison with the transmembrane potential at metal-supported tBLMs, which is always referred to the metal side, just as the corresponding applied potential E.

An increase in the concentration of a channel-forming peptide in the bathing solution shifts the current–voltage curve toward less negative V values. For a fixed average conductance G, such as that marked by the dashed line in Fig. 6.13,

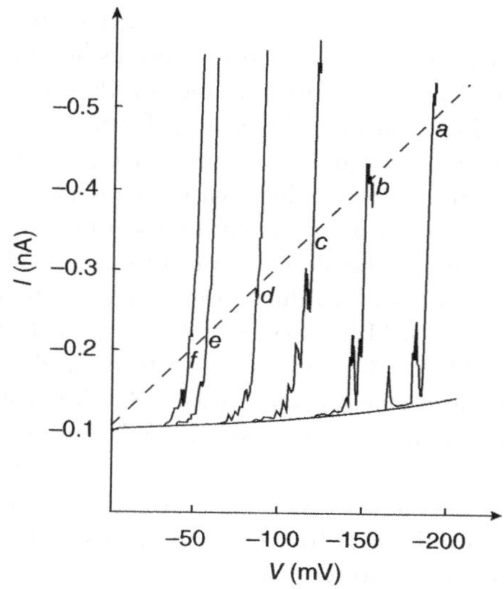

Figure 6.13 Current–voltage curves at a phosphatidylethanolamine BLM in aqueous 0.1 M NaCl, with the *cis* side solution containing different alamethicin concentrations: (*a*) 0.05, (*b*) 0.1, (*c*) 0.2, (*d*) 0.5, (*e*) 1, and (*f*) 1.5 μg mL^{-1}. Scan rate = 12 mV/s. The baseline current is the current through the bare membrane. The dashed line intersects the curves at a current whose ratio to the corresponding voltage is constant and measures the average conductance *G*. Source: Eisenberg et al. (1973). Reproduced with permission of Springer.

the transmembrane potential decreases linearly with the logarithm of the peptide concentration.

Alamethicin and melittin are the two most thoroughly investigated voltage-gated channel-forming peptides in conventional BLMs. In spite of their structural differences, they have several functional features in common. Thus, under suitable experimental conditions, they both exhibit two distinct relaxation processes, a fast relaxation process with weakly voltage-dependent conductance, and a slow relaxation one with strongly voltage-dependent conductance. In particular, this situation is encountered when the peptide is added to only the *cis* side of the membrane, and the current flows only in the direction from the *cis* to the *trans* side, provided the latter is made sufficiently negative with respect to the former (i.e., under strict *voltage-gating* behavior). The conductance *G* associated with any of the two different relaxation processes depends upon the voltage *V* and the peptide concentration c_{pep} according to the following equation (Latorre and Alvarez, 1981; Sansom, 1991):

$$G \propto c_{pep}^{\delta} \exp(-\gamma FV/RT). \tag{6.17}$$

The fast relaxation process has a time constant two orders of magnitude smaller than that of the slow relaxation one and occurs in the millisecond range, which is of the same order of magnitude as the mean lifetime of single pore states (Boheim and Kolb, 1978a); it appears to arise from a shift in the probability distribution of the different conductance levels of a pore as voltage is changed. Conversely, the slow relaxation process is attributed to the fluctuation in the number of channels present in the membrane (Boheim and Kolb, 1978a; 1978*b*) and correlates well with the lifetime of single current bursts. The conductance of the fast relaxation process is from one to two orders of magnitude smaller than that of the slow relaxation process. The passage from a weakly to a strongly voltage-dependent conductance with a progressive negative shift in voltage is shown by a number of synthetic Aib-based polypeptides of different length incorporated into palmitoyloleoylphosphatidylcholine (POPC) BLMs (Menestrina et al., 1986). The critical voltage marking the passage from a weakly to a strongly voltage-dependent conductance regime is accompanied by a dramatic rise of current, and shifts toward more negative values with a decrease in the polypeptide chain length and in its concentration in the bathing solution.

During the slow relaxation process, alamethicin yields γ values ranging from 4.4 to 6 and δ values ranging from 6 to 11, depending on the lipid composition of the BLM (Latorre and Alvarez, 1981, Boheim and Kolb, 1978a), while melittin yields $\gamma = 4.3$ and $\delta \approx 4$ (Tosteson and Tosteson, 1984) and synthetic Aib-based peptides yield $\gamma = 4.7$ and $\delta = 8.3$ (Menestrina et al., 1986). Conversely, during the fast relaxation process, alamethicin yields $\gamma = 0.96$ and $\delta \approx 2$ (Boheim and Kolb, 1978a), synthetic Aib-based peptides yield γ values ranging from 2.8 to 3.3 and $\delta \approx 2$ (Menestrina et al., 1986), whereas melittin yields γ values ranging from 1.3 to 1.6 and $\delta \approx 4$ (Tosteson and Tosteson, 1984), although a δ value of 2.5 at zero voltage was also reported (Stankowski et al., 1991). The δ parameter for melittin was found to decrease from ~2 to ~0.5 with a decrease in the chain length of the lipid molecules forming monoglyceride/squalene BLMs (Hall et al., 1984). Summarizing, the dependence of the current upon the peptide concentration increases by several orders of magnitude in passing from the fast to the slow relaxation process.

When melittin is added on the *cis* side of the BLM, the steady-state $I-V$ curve shows both a positive and a negative branch, albeit not perfectly symmetrical with respect to $V = 0$ (Pawlak et al., 1991). Adding alamethicin on the *cis* side yields a single negative branch of the $I-V$ curve if the BLM is formed with a saturated lipid (Eisenberg et al., 1973; Vodyanoy et al., 1983). On the other hand, with unsaturated or halogenated membrane lipids, alamethicin yields both branches, with the positive branch more shifted with respect to zero transmembrane potential than the negative one, as shown in Fig. 6.14 (Vodyanoy et al., 1983). This suggests that the presence of saturated lipids in the BLM hinders the diffusion of alamethicin across the membrane.

The salient features of the experimental behavior of voltage-gated channels at BLMs can be accounted for by a general approach (Guidelli and Becucci, 2016) that treats the time-dependent formation of ion channels as a nucleation of embedded monomers spanning the lipid bilayer and growth of the resulting supercritical nuclei, as already examined in connection with potential-step $I-t$ curves in Section 6.2.2. Prior to the voltage scan inducing the $I-V$ exponential curve, the monomers are regarded as located on the *cis* side of the membrane, bathed by the solution where they were initially added. For the peptide dipolar molecule to span the hydrocarbon tail region of the BLM during the negative voltage scan, the positive pole of the dipole will have to move across this region up to the attainment of the polar head region on the *trans* side, with the negative pole remaining on the *cis* side.

The probability p of this charge movement can be expressed by the one-sided Boltzmann equation (Eq. 4.66). To this end, let us consider the BLM in Fig. 4.11,

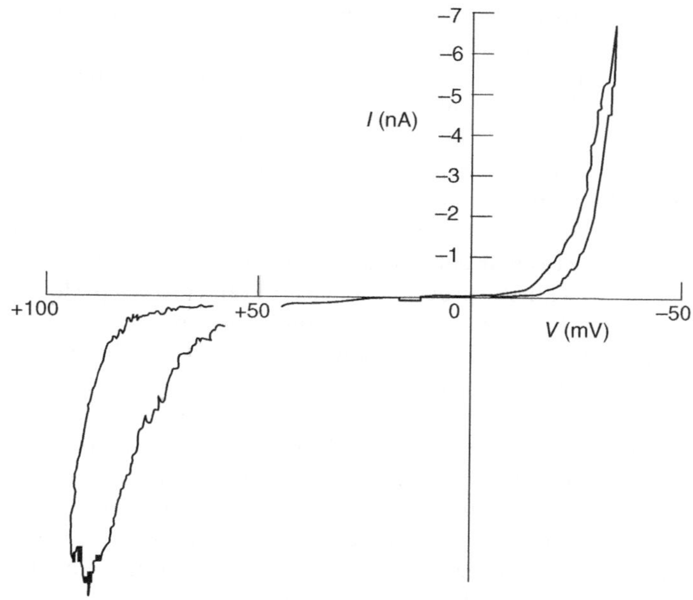

Figure 6.14 Current–voltage curve at a dibromostearylphosphatidylcholine BLM in a pH 5.5 unbuffered solution of 1 M KCl and 0.3 μg mL^{-1} alamethicin, added on the *cis* side of the membrane (right upper quadrant). Source: Vodyanoy et al. (1983). Reproduced with permission of Elsevier.

in which the external solution is now identified with the *cis* side of the BLM, and the $x = x_c$ and $x = x_o$ planes are considered to be located in the *cis* and *trans* polar head regions, respectively. The alignment of the peptide dipole along the direction of the electric field following a negative voltage scan will move its positive pole q' from the x_c to the x_o position. The charge q in Eq. 4.63 will now be given by $q = -q'(x_o - x_c)/d = -\Delta\mu/d$, where $\Delta\mu$ is the change in the dipole moment of the peptide as a consequence of its alignment along the direction of the electric field. Replacing q into the expression of Eq. 4.66 for the probability p yields

$$p = 1/\left\{1 + \exp\left[\Delta\mu V/(dkT)\right]/a\right\} \quad \text{with } a \equiv \exp\left[-\left(\mu_o^{\emptyset} - \mu_c^{\emptyset}\right)/kT\right]. \tag{6.18}$$

Here, $(\mu_o^{\emptyset} - \mu_c^{\emptyset})$ is the difference between the standard chemical potentials of the peptide in the embedded and flat orientations, that is, the difference in Gibbs energy between these two conformational states in the absence of the electric field (Honig et al., 1986; Sansom, 1991). Since the parameter a is always much less than unity, it measures the probability p at zero voltage, in view of Eq. 6.18.

The contribution to an α-helix dipole from the peptide backbone is roughly equivalent to a partial charge of $+e/2$ at the N-terminal of the helix and of $-e/2$ at its C-terminal, where e is the proton charge. Hence, the dipole moment of an α-helix spanning a lipid bilayer of thickness $d = 30$ Å is about equal to 70 D, if any dipole contributions arising from an asymmetric distribution of charged side chains along the length of the axis can be disregarded. As an example, the dipole moment of alamethicin was reported to vary from 40 to 75 D, depending on the polarity of the organic solvent (Yantorno et al., 1982). Naturally, if a peptide molecule is not completely embedded in the *cis* polar head region of the membrane at zero voltage, the $\Delta\mu$ value in Eq. 6.18 is less than its maximum estimated value, causing a decrease in the probability p.

Irrespective of the actual $\Delta\mu$ value, the passage of the peptide molecule from its initial stand at zero voltage to an orientation spanning the whole hydrocarbon tail region is accompanied by a kinetic process of nucleation and growth perfectly analogous to that outlined in Eqs. 6.9–6.15 for the potential-step I–t curves. In the present case, even assuming the presence of any flat aggregates on the *cis* side of a newly formed BLM prior to the pristine negative voltage scan, their continuous restoration after repeated voltage cycles is highly improbable. Consequently, the nucleation of holes within flat aggregates will be removed from the model, and the fractional surface coverage by the embedded monomers will take the simpler form $\Theta = \theta_0 p(1 - S)$. Moreover, differently from the potential-step chronoamperometry case, p varies with V. This point must be correctly accounted for when expressing the extended area S_x from Eq. 6.12, after replacing the rates of nucleation and of radial growth from Eqs. 6.9 and 6.10, respectively:

$$S_x = \pi \int_0^t dy \left[\int_y^t k_R \Theta(z)\, dz \right]^2 k_N \Theta(y)^n$$
$$= \pi k_N k_R^2 (\theta_0 p)^{2+n} \int_0^t \left[\int_y^t (1 - S(z))\, dz \right]^2 (1 - S(y))^n dy. \qquad (6.19)$$

In particular, the last member of this equation is obtained by removing the probability p from under the integral signs. This removal is necessary, since p is an explicit function of time t and is operative from the instant $t = 0$, thanks to the constant voltage scan rate $v \equiv dV/dt$. Conversely, the quantity S_x is a function of the integration variables z and y and refers to the events occurring throughout the various nucleation processes starting at different times y and followed by the growth of the resulting supercritical nuclei, up to the final observation time $t = t$. Three consecutive differentiations of Eq. 6.19 with respect to t via the generalized Leibnitz formula, in combination with Avrami's differential equation 6.15 relating S_x to S, yield four differential equations that can be readily solved by the first-order Runge–Kutta method, thus providing the time dependence of S. The fractional surface coverage by the embedded channel-forming clusters is then given by $\theta_0 pS$.

The last step consists in calculating the current density j from the rate theory of ion transport across membranes (Schultz, 1980), as applied to ion translocation across the potential energy barrier located in the hydrocarbon tail region of the BLM and expressed by Eq. 4.86. In applying this equation, the potential energy barrier will be regarded as symmetrical, for simplicity, by ascribing a common value, k_t, to the forward and backward rate constants at zero voltage, k_f and k_b. Moreover, the ion channel will be regarded as ion unselective, by setting the current density equal to the sum of the contributions from the flow of a monovalent cation in one direction and from that of a monovalent anion in the opposite direction. Finally, the ionic current density j across the membrane, which would be equal to zero in the absence of ion channels, is set proportional to the fractional surface coverage by the embedded channel-forming clusters, $\theta_0 pS$, since each newly formed channel makes a contribution to this current. With these assumptions, Eq. 4.86 takes the form

$$j = (\theta_0 pS)\, Fk_t \left[\left(-c^+_{\mathrm{cis}}\, e^{-\frac{\alpha FV}{RT}} + c^+_{\mathrm{trans}}\, e^{\frac{(1-\alpha)FV}{RT}} \right) + \left(c^-_{\mathrm{cis}} e^{\frac{\alpha FV}{RT}} - c^-_{\mathrm{trans}}\, e^{-\frac{(1-\alpha)FV}{RT}} \right) \right]. \tag{6.20}$$

Here, α is the transfer coefficient, c^+_{cis}, c^+_{trans} are the concentrations of a monovalent cation on the *cis* and *trans* sides of the membrane, and c^-_{cis}, c^-_{trans} are those of a monovalent anion. The first and second expressions between parentheses in Eq. 6.20 measure the cation and anion currents, respectively.

The present model of nucleation and growth is entirely general. Thus, it only assumes that the elementary steps preceding the step yielding the nucleus are in quasi-equilibrium, and that the growth of supercritical nuclei proceeds irreversibly by activated aggregation of monomers. Upon setting for simplicity $c^+_{\mathrm{cis}} = c^+_{\mathrm{trans}} = c^-_{\mathrm{cis}} = c^-_{\mathrm{trans}} \equiv c$, j turns out to be proportional to c, besides being proportional to k_t. The behavior of the dimensionless quantity $j/(Fk_t c)$ will serve to show the way in which the quantities $\Delta\mu$, θ_0, a and the kinetic parameters of nucleation and growth affect the shape of $I-V$ curves and the dependence of $\ln G$ upon V and the concentration, c_{pep}, of the peptide in the bathing solution.

In particular, setting the current density proportional to the number density of the embedded clusters that have undergone a kinetic process of nucleation and growth with ion-channel formation amounts to assuming that only a single type of aggregate develops during the whole voltage scan. However, this assumption also holds if the single-channel state distributions are independent of voltage and the macroscopic conductance induced by the peptide exclusively arises from the increase in the number of channels as the membrane voltage increases. These requirements are approximately fulfilled by the alamethicin channel in a phosphatidylethanolamine (PE) BLM (Eisenberg et al., 1973) and by melittin in a DOPC BLM (Pawlak et al., 1991).

Figure 6.15 shows plots of ln G against V for different values of the probability, $a \equiv \exp[-(\mu^{\circ}_{\mathrm{o}} - \mu^{\circ}_{\mathrm{c}})/kT]$, of the peptide dipoles being perpendicular to the membrane plane at zero voltage, and for $\Delta\mu = 70$ D, $\theta_0 = 0.1$, $k_N k_R^2 = 1 \times 10^6$ s^{-3} and $n = 1$. All plots refer to a temperature of 298 K, a voltage scan rate of 10 mV/s and a membrane thickness $d = 30$ Å. The plots show $\ln G$ both along the negative-going voltage scan and along the subsequent positive-going one. The sharp dip marks the maximum negative current attained at the beginning of the reverse voltage scan.

During the negative-going voltage scan, the plots exhibit a region of weakly voltage-dependent conductance, with $\gamma = 0.95$, at lower negative voltages, as well as a region of strongly voltage-dependent conductance, with $\gamma = 3.85$,

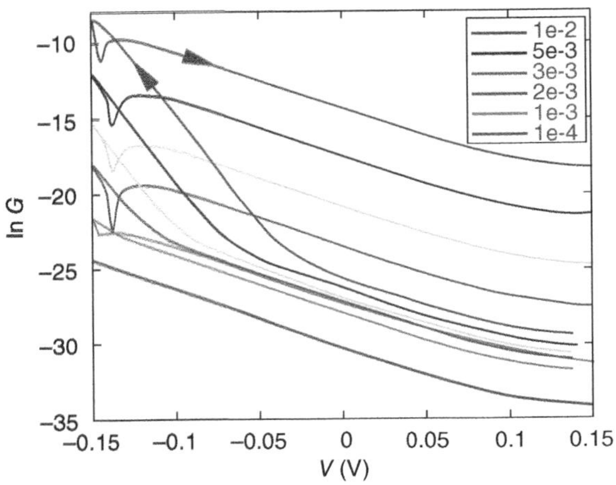

Figure 6.15 Plots of ln G versus V for different values of a and for $\Delta\mu = 70$ D, $\theta_0 = 0.1$, $k_N k_R^2 = 1 \times 10^6$ s^{-3}, and $n = 1$. The curves shift toward progressively lower ln G values with decreasing a. Source: Guidelli and Becucci (2016). Reproduced with permission of Elsevier.

at higher negative voltages. At extreme positive voltages, G tends to attain a voltage-independent minimum value. During the positive-going voltage scan, the voltage dependence of the conductance is characterized by a γ value of 0.95, other than in the proximity of the voltage reversal. A progressive decrease of a lowers the ln G versus V curve and gradually shifts its "elbow" toward more negative voltages. This causes the region of weakly voltage-dependent conductance to cover the whole negative voltage range usually spanned by experimental $I-V$ curves, which rarely exceeds 200 mV. As a result of this trend, the $I-V$ curves falling over this voltage range and calculated for a values varying from 1×10^{-2} to 5×10^{-3} exhibit a strongly voltage-dependent conductance and an appreciable hysteresis, whereas those with $a < 2\times10^{-3}$ exhibit a weakly voltage-dependent conductance and a small or negligible hysteresis. This behavior is exemplified by the calculated $I-V$ curves in Figs. 6.16 and 6.17, which refer to the two different situations. Figure 6.16 shows an experimental $I-V$ curve for 0.2 μg mL^{-1} alamethicin in a PE BLM (Vodyanoy et al. 1983) and a corresponding curve calculated for $a = 1\times10^{-2}$, $\Delta\mu = 70$ D, $\theta_0 = 0.1$, $k_N k_R^2 = 1 \times 10^5$ s^{-3} and $n = 1$. The shape of the calculated curve closely simulates those reported for alamethicin (Menestrina et al., 1986; Hall et al., 1984; Vodyanoy et al., 1983; Vodyanoy et al. 1988) and for synthetic Aib-based polypeptides (Menestrina et al., 1986); the γ value of 3.85 is also close to those for the synthetic polypeptides, although that for alamethicin is slightly higher (Boheim and Kolb, 1978a; Latorre and Alvarez, 1981; Menestrina et al., 1986).

On the other hand, Fig. 6.17 shows an experimental $I-V$ curve for 0.4 μg mL^{-1} melittin in a DOPC BLM (Pawlak et al., 1991) and a corresponding curve calculated for $a = 1\times10^{-4}$, $\Delta\mu = 70$ D, $\theta_0 = 0.1$, $k_N k_R^2 = 1 \times 10^6$ s^{-3} and $n = 1$, in which the forward and backward voltage scans practically coincide.

The fact that the experimental $I-V$ curves for alamethicin and melittin in Figs. 6.16 and 6.17 are simulated with $a = 1\times10^{-2}$ and 1×10^{-4}, respectively, is consistent with their different structures. Thus, upon assuming that both

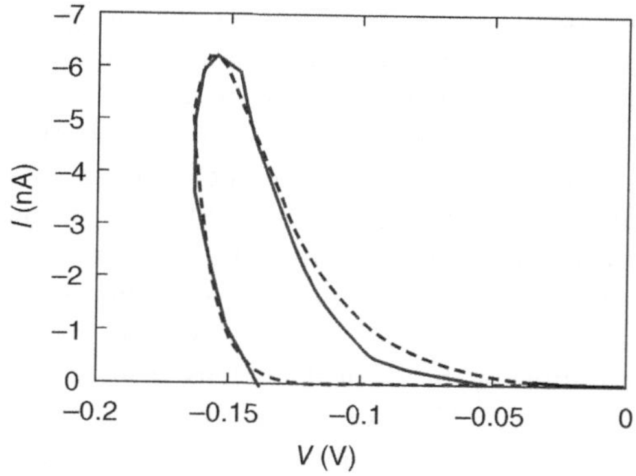

Figure 6.16 Experimental *I–V* curve for 0.2 μg mL^{-1} alamethicin in bacterial PE, taken from curve 2 in Fig. 5 of Vodyanoy et al. (1983) (solid curve) and curve calculated for $a = 1 \times 10^{-2}$, $\Delta\mu = 70$ D, $\theta_0 = 0.1$, $k_N k_R^2 = 1 \times 10^5\ s^{-3}$, and $n = 1$ (dashed curve). The height of the calculated curve was normalized to that of the experimental one. Source: Guidelli and Becucci (2016). Reproduced with permission of Elsevier.

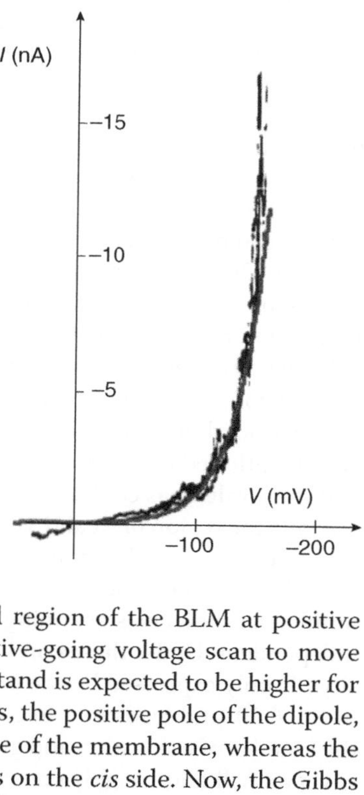

Figure 6.17 Experimental *I–V* curve for 0.4 μg mL^{-1} melittin in DOPC, taken directly from Fig. 7 of Pawlak et al. (1991) (curve with small fluctuations), and curve calculated for $a = 1 \times 10^{-4}$, $\Delta\mu = 70$ D, $\theta_0 = 0.1$, $k_N k_R^2 = 1 \times 10^6\ s^{-3}$, and $n = 1$ (smooth curve). The height of the calculated curve was normalized to that of the experimental one. Source: Guidelli and Becucci (2016). Reproduced with permission of Elsevier.

peptides are incorporated in the *cis* polar head region of the BLM at positive voltages, the Gibbs energy required for a negative-going voltage scan to move the peptide dipoles from a flat to an embedded stand is expected to be higher for melittin than for alamethicin. With both peptides, the positive pole of the dipole, located at the N-terminal, moves to the *trans* side of the membrane, whereas the negative pole, located at the C-terminal, remains on the *cis* side. Now, the Gibbs energy required to force the positively charged N-terminal amino group and the Lys-7 residue of the N-terminal sequence of melittin into the hydrocarbon tail region is much greater than that required to force the neutral N-terminal of the alamethicin molecule, whose only charge (the negative Glu17 residue) resides close to the C-terminal. The regions of weakly and strongly voltage-dependent conductance in the ln *G* versus *V* curves of Fig. 6.15 are also in semiquantitative agreement with the corresponding fast and slow relaxation processes exhibited by the current–time curves in voltage jump experiments, characterized by $\gamma \approx 0.96$ and 6.4 for alamethicin (Boheim and Kolb, 1978a), and by $\gamma \approx 1.3$ and 4.3 for melittin (Tosteson and Tosteson, 1984).

The γ values for the regions of weakly and strongly voltage-dependent conductance are only slightly affected by a change in the combined rate, $k_N k_R^2$, of nucleation and growth. The main effect of an increase in $k_N k_R^2$ from 10 to $1 \times 10^7\ s^{-3}$, for $a = 10^{-2}$ and the other parameters as in Fig. 6.15, consists in a shift of the elbow of the bent ln G versus V curves toward less negative voltages (Guidelli and Becucci, 2016). The effect of a decrease in the number of peptide monomers composing the critical nucleus from $n = 4$ to $n = 1$, with the other parameters as in Fig. 6.15, is qualitatively similar to that produced by an increase in $k_N k_R^2$. Thus, apart for the obvious changes in the absolute value of ln G with varying the adjustable parameters of the model, the slope of ln G against V yields γ values in fairly good agreement with the experimental ones over both regions of strongly and weakly voltage-dependent conductance, irrespective of the values ascribed to the adjustable parameters.

In the regime of strongly voltage-dependent conductance, the γ value predicted by the model increases linearly with the magnitude of the dipole moment and is expressed by the equation: $\gamma = 0.66 + 0.044\,\mu/\mathrm{D}$, for $\theta_0 = 0.1$, $a = 1 \times 10^{-2}$, $k_N k_R^2 = 1 \times 10^7\ s^{-3}$, and $n = 1$. Hence, the model predicts a γ value of 4, close to that reported for several channel-forming α-helical peptides, when $\Delta\mu$ is close to 70 D, which is the value estimated for a α-helical peptide spanning a hydrocarbon tail region 30 Å thick. This strongly suggests that the change, $\Delta\mu$, in the dipole moment normal component undergone by these peptides during a negative-going voltage scan is determined by the movement of their α-helix from the *cis* polar head region, with an orientation parallel to the membrane plane, to one spanning the hydrocarbon tail region (about 30 Å in thickness for a typical BLM). The aforementioned connection between the effective dipole moment of the peptide and the γ value has contributed to regarding γ as an "apparent gating charge."

Along the region of weakly voltage-dependent conductance, the model predicts a γ value about equal to unity. This is due to the fact that over this region the ratio, S, of the number of peptide molecules aggregated into ion channels to the total number of molecules spanning the lipid bilayer is very low and almost voltage independent. Under these conditions, the voltage dependence of the current density j is only expressed by the product of p by the term between square brackets in Eq. 6.20, and leads to a ln G versus V plot of slope very close to unity. Since the kinetics of nucleation and growth exclusively affects the parameter S, and this is very small and roughly constant, the current is clearly insensitive to the $k_N k_R^2$ and n values. Moreover, for S practically constant, the current turns out to be proportional to the fraction, θ_0, of the membrane unit area covered by the peptide. Within the limits in which θ_0 can be regarded as proportional to the peptide concentration, c_{pep}, in the aqueous solution, the current along the low conductance region is proportional to c_{pep}, and the parameter δ in Eq. 6.17 equals unity. This prediction is consistent with the experimental observation that the low γ values characterizing the regime of weakly voltage-dependent conductance are often associated with δ values ranging from 1 to 2.5 (Boheim and Kolb, 1978a; Menestrina et al., 1986; Stankowski et al., 1991), with the exception of melittin in asolectin BLMs (Tosteson and Tosteson, 1984, Tosteson et al., 1987). It must be pointed out that the prediction of a γ value of about 0.95 along the region of weakly voltage-dependent conductance results from a passage of the peptide dipoles from a parallel to a vertical orientation with respect to the membrane plane as voltage becomes progressively more negative, similar to what happens along the region of strongly voltage-dependent conductance. However, in this case no aggregation of vertical monomers occurs. For a typical dipole moment of 70 D, the mere dipole reorientation without aggregation yields a γ value of about 0.95. It is the aggregation of embedded dipoles with nucleation-and-growth kinetics that causes a significant γ increase from about 1 to about 4, when passing from the regime of weakly to that of strongly voltage-dependent conductance. This suggests that, in the regime of weakly voltage-dependent conductance, the ion movement elicited by the electric field occurs within a bunch of vertical monomers, whose side chains are randomly distributed with respect to each other (i.e., without aggregation into a proper ion channel).

The dependence of the conductance upon the peptide concentration predicted by the model in the regime of strongly voltage-dependent conductance is higher than that in the regime of weakly voltage-dependent conductance, as expressed by $\delta = 1$. Thus, the plot of $\ln G$ against $\ln \theta_0$ obtained from $I-V$ curves calculated for $\Delta\mu = 70$ D, $a = 1 \times 10^{-2}$, $k_N k_R^2 = 1 \times 10^6$ s^{-3}, and $n = 1$, that is, under conditions of strongly voltage-dependent conductance, is linear and exhibits a slope of 3.85 over the whole θ_0 range from 0.1 to 1 (Guidelli and Becucci, 2016). This value can be identified with δ, provided we can assume that the peptide is incorporated from the bathing solution into the *cis* polar head region of the membrane according to a Henry's adsorption isotherm. This δ value is close to those reported in the regime of strongly voltage-dependent conductance for certain Aib-based polypeptides (Menestrina et al., 1986) and for melittin (Tosteson and Tosteson, 1984), whereas alamethicin exhibits appreciably higher values ranging from 6 to 11 (Latorre and Alvarez, 1981). The fact that the sole introduction of a mechanism of nucleation and growth for monomer aggregation into ion channels yields a δ value close to 4 as a natural consequence demonstrates the inconsistency of the frequent assumption (Tosteson et al., 1990; Pawlak et al., 1991) that such a value is indicative of the formation a tetrameric ion channel. The δ value for alamethicin being appreciably higher than predicted by the model, especially in certain BLMs, may possibly be ascribed to the failure of the assumption of voltage independence of the single-channel state distributions, on which Eq. 6.20 relies. A higher δ value is expected if an increase in negative voltage tends to favor higher aggregates at the expense of lower ones.

The cyclic voltammograms at mercury-supported tBLMs are quite different from the $I-V$ curves recorded at conventional BLMs incorporating the same peptides. In the presence of an ion channel, the negative-going scan of a cyclic voltammetry curve yields a negative current peak that adds to the background current, as soon as a sufficiently negative transmembrane potential value is attained (curve *d* in Fig. 6.11). This peak is due to cation inflow into the hydrophilic spacer and/or anion outflow. In the case of K^+ inflow into the TEO spacer of a DPTL/phospholipid tBLM, integration of the current peak over time yields a maximum charge of −45 to −50 μC cm^{-2}, corresponding to spacer saturation by this univalent cation (Becucci and Guidelli, 2009). Under steady-state conditions, the positive-going scan yields a positive current whose integration must necessarily provide a charge equal in magnitude and opposite in sign to that obtained by integration of the negative current peak.

As a rule, cyclic voltammetry curves are not controlled by diffusion of the permeating ions in the aqueous solution toward or away from the tBLM surface. Rather, they are controlled by the rate at which the ions overcome the potential energy barrier represented by the lipid bilayer interposed between the aqueous solution and the hydrophilic spacer (Becucci and Guidelli, 2014a; 2014*b*). In other words, the concentration of a permeating ion in direct contact with the external surface of the tBLM does not change to a detectable extent with respect to its bulk value during the functional activity of an ion channel. Hence, cyclic voltammetry provides useful information on the kinetics of inflow and outflow of permeating ions, by also varying the scan rate. In particular, it allows a distinction to be made between ohmic and voltage-gated ion channels.

An example of ohmic ion channel is provided by gramicidin. Figure 6.18 shows stabilized cyclic voltammograms at a DPTL/DOPC tBLM in aqueous solution of 0.1 M KCl and 0.1 μM gramicidin, at a scan rate of 50 mV/s and at pH 3, 5.4, and 6.8 (Becucci and Guidelli, 2015). The solution at pH 3 was obtained by adding 1 mM HCl, that at pH 5.4 was unbuffered, and that at pH 6.8 was buffered with a $HCl-K_2HPO_4$ mixture.

The cyclic voltammograms exhibit a negative peak, due to the flow of K^+ ions into the hydrophilic spacer, and a corresponding positive peak, due to their ejection from the spacer into the aqueous solution bathing the tBLM. The two peaks are roughly centrosymmetric with respect to the midpoint potential, $E_{1/2}$, between them, namely the potential at which K^+ inflow matches its outflow. $E_{1/2}$ shifts toward less negative potentials with decreasing pH, passing from −0.66 V versus Ag/AgCl/0.1 M KCl at pH 6.8 to −0.55 V at pH 5.4 and to −0.44 V at pH 3. The separation between the two peaks decreases with decreasing pH. Almost the same positive $E_{1/2}$ shift with decreasing pH from 6.8 to 3 is observed with mercury-supported DPTL/DOPC tBLMs incorporating the ohmic ion channel formed by the lipodepsipeptide syringopeptin 25A (Becucci et al., 2016). Moreover, another lipodepsipeptide, syringomycin E, which yields a voltage-gated channel in DPTL/DOPC tBLMs, shows an identical positive shift of the zero current potential with decreasing pH (Becucci et al., 2015). Since syringopeptin 25A and syringomycin E are structurally quite different from gramicidin and both positively charged, their common pH-dependent behavior seems to be ascribable more to the nature of the phospholipid than to that of the ion channel.

The positive shift of $E_{1/2}$ with decreasing pH can be explained by a dragging of the lipid molecules immediately adjacent to the mouth of the gramicidin channel toward the hydrocarbon tail region of the lipid bilayer.

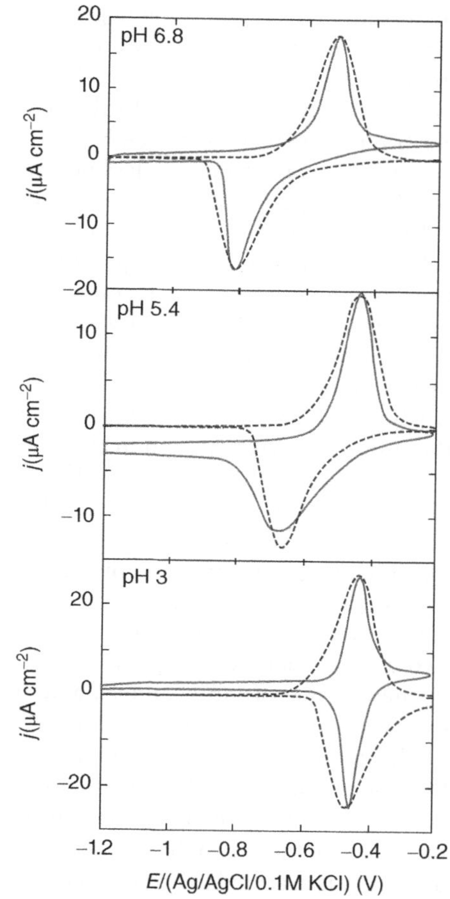

Figure 6.18 The solid curves are experimental cyclic voltammograms at a DPTL/DOPC tBLM in aqueous solution of 0.1 M KCl and 0.1 μM gramicidin A at pH 6.8, 5.4, and 3. Scan rate = 50 mV s^{-1}. The corresponding dashed curves were calculated as outlined in the text. Source: Becucci and Guidelli (2015). Reproduced with permission of Elsevier.

The DOPC distal monolayer is characterized by an arrangement of the $(H_3C)_3N^+{-}CH_2{-}CH_2{-}OPO_3^-$ zwitterion coplanar to the monolayer, due to electrostatic interactions between trimethylammonium and phosphate groups of adjacent lipid molecules, with a resulting stabilization of the unprotonated form of the latter group (Moncelli et al., 1994). Due to the exposure of the phosphate group to the aqueous solution, its intrinsic pK_a is about equal to 0.8, and hence its protonation starts to become appreciable around pH 2. However, if the $(H_3C)_3N^+{-}CH_2{-}CH_2{-}OPO_3^-$ dipoles of the DOPC molecules adjacent to the mouth of a gramicidin channel are dragged closer to the hydrocarbon tail region, they will tilt with respect to the bilayer plane, with the phosphate group (linked to the glycerol backbone) being in a more embedded position with respect to the trimethylammonium group. The decreased polarizability of the local environment surrounding the phosphate group will tend to favor its protonation more than if it were exposed to the aqueous phase, causing a progressive decrease in its negative charge with decreasing pH from 6.8 to 3. At constant applied potential E, the resulting decrease in the negative potential difference across the polar head region is compensated for by an increasing negative potential difference across the hydrocarbon tail region (i.e., the transmembrane potential ϕ_m), causing an increase in the K^+ inflow into the spacer relative to its outflow. Therefore, the electric potential at which the K^+ outflow matches its inflow (i.e., the midpoint potential $E_{1/2}$) is attained at less negative potentials with decreasing pH. Similar conclusions apply to the polar heads of DOPS (Becucci and Guidelli, 2015).

The experimental cyclic voltammetry behavior of the gramicidin ion channel can be interpreted semiquantitatively on the basis of the aforementioned considerations by having recourse to the following approximate model. The mercury-supported DPTL/DOPC tBLM can be regarded as consisting of four consecutive dielectric slabs: (i) the region of the lipoic acid residues; (ii) the TEO hydrophilic spacer; (iii) the lipid bilayer moiety; (iv) the polar head region, which will be considered together with the diffuse layer region, for simplicity. Following the approach used in Section 2.6, the potential difference across each slab is expressed by Eq. 2.18 for a parallel plate capacitor, with the charge density on the plate turned toward the metal equal to the sum of all charge densities located between this plate and the bulk metal. The negative charge density due to the disulfidated groups of the lipoic acid residues attached to the mercury surface, even if partially or totally transferred to the metal, is counterbalanced to a large extent by a defect of conduction electrons on the metal surface; the sum of these two charges, denoted by q_M, is that experienced by the diffuse layer ions when in the absence of ion channels, and amounts to no more than a few $\mu C\,cm^{-2}$ over the potential range of interest (Becucci et al., 2009*a*). The charge density of the monovalent cations entering the hydrophilic spacer by moving along the gramicidin channel is denoted by $F\Gamma$, where Γ denotes the cation surface concentration: as a first approximation, it is regarded as mainly located on the metal side of the hydrophilic spacer (Becucci and Guidelli, 2009). An ionic charge density σ_i is ascribed to any ionizable polar-head groups deeply embedded in the polar head region and located in the immediate vicinity of the hydrocarbon tails. With these assumptions, the absolute potential difference $\Delta\phi$ across the whole mercury/water interface is given by

$$\Delta\phi = \frac{q_M}{C_{il}} + \left(\frac{q_M + F\Gamma}{C_s} + \chi_s\right) + \frac{q_M + F\Gamma}{C_m} + (q_M + F\Gamma + \sigma_i)\left(\frac{1}{C_{ph}} + \frac{l}{\varepsilon_0\varepsilon_w}\right). \tag{6.21}$$

Here, χ_s is the surface dipole potential of the hydrophilic spacer, which contributes to the potential difference across this dielectric slab; C_{il}, C_s, C_m, and C_{ph} are the differential capacitances of the inner-layer region (formed by the lipoic acid residues), the spacer, the lipid bilayer moiety, and the polar heads, respectively; $\varepsilon_0\varepsilon_w/l$ is the diffuse layer capacitance, where l equals $\sqrt{(\varepsilon_0\varepsilon_w RT/2F^2c)}$ (cf. Eq. 2.40) and c is the concentration of a 1,1-valent electrolyte. The ether linkage between the TEO spacer and the phytanyl chains and the ester linkages between the glycerol backbone and the fatty acids of the distal lipid monolayer have dipole potentials that may partially compensate each other; in any case, these linkages have a fixed conformation that is not affected by pH changes in the aqueous solution. Hence, their surface dipole potentials are not included in Eq. 6.21.

Upon extracting q_M from Eq. 6.21 as a function of $\Delta\phi$ and substituting the resulting expression into the equation, $\phi_m = (q_M + F\Gamma)/C_m$, for the transmembrane potential, after straightforward algebraic calculations, we obtain

$$\phi_m = \frac{q_M + F\Gamma}{C_m} = \frac{\Delta\phi + F\Gamma C_{il}^{-1} - \chi_s - \sigma_i \left(C_{ph}^{-1} + l/\varepsilon_0\varepsilon_w\right)}{C_m \left(C_{il}^{-1} + C_s^{-1} + C_m^{-1} + C_{ht}^{-1} + l/\varepsilon_0\varepsilon_w\right)}$$
$$\cong \Delta\phi + F\Gamma C_{il}^{-1} - \chi_s - \sigma_i \left(C_{ph}^{-1} + l/\varepsilon_0\varepsilon_w\right). \quad (6.22)$$

The last member of this equation is justified by the fact that the differential capacitance C_m of the lipid bilayer moiety is always appreciably smaller than the other capacitances, such that the reciprocals of these latter can be disregarded with respect to C_m^{-1} as a good approximation.

The surface dipole potential χ_s of the spacer and the inner-layer capacitance C_{il} for a DPTL/phospholipid tBLM are estimated by independent measurements. Thus, several pieces of experimental evidence interpreted on the basis of a modelistic approach concur in estimating χ_s at about −0.250 V (Guidelli and Becucci, 2011a). Moreover, the analysis of the electrochemical impedance spectra of a DPTL/DOPC tBLM incorporating the gramicidin ohmic channel from aqueous 0.1 M KCl reveals that the capacitance, C_{il}, of the inner layer interposed between the K^+ charge density $F\Gamma$ and the opposite charge density on mercury increases almost linearly with $F\Gamma$, attaining a value of about 200 μF cm^{-2} when $F\Gamma$ attains its maximum saturation value, ranging from +45 to +50 μC cm^{-2} (Becucci and Guidelli, 2009). On the other hand, in the absence of ion channels, and hence of ions in the spacer, the capacitance C_{il} can be approximately identified with that estimated for the region of the lipoic acid residues, which amounts to about 5 μF cm^{-2} (Becucci et al., 2005; Guidelli and Becucci, 2011a). Hence, the experimental dependence of C_{il} upon the charge density $F\Gamma$ can be approximately expressed by the following equation:

$$C_{il} = \left[5 + (F\Gamma/45)\,195\right]\ \mu\text{F cm}^{-2} \quad (6.23)$$

according to which C_{il} varies linearly with $F\Gamma$ from 5 to 200 μF cm^{-2}.

The current I is again obtained from the rate theory of ion transport across membranes (Schultz, 1980), as applied to cation translocation across the potential energy barrier located in the lipid bilayer moiety of the tBLM (cf. Eq. 4.86):

$$I \propto Fk_t \left[c_s \exp\left(\frac{(1-\alpha)F\phi_m}{RT}\right) - c_0 \exp\left(-\frac{\alpha F\phi_m}{RT}\right)\left(1 - \frac{F\Gamma}{45}\right)\right]. \quad (6.24)$$

Here, k_t is the translocation rate constant for $\phi_m = 0$, α is the charge transfer coefficient, c_s is the volume concentration of the cations in the spacer, and c_0 is that just outside the tBLM surface. The $(1 - F\Gamma/45)$ factor is introduced to account approximately for the probability of finding free sites available for ion accommodation at the inner mouth of the ion channel; it vanishes when the K^+ charge density in the spacer attains its saturation value of 45 μC cm^{-2}. The cyclic voltammetry current I is exclusively controlled by ion translocation, and no ion depletion is assumed in the aqueous solution adjacent to the outer mouth of ion channels. Hence, c_0 is set equal to the K^+ bulk concentration. The c_s value, in moles per unit volume, is obtained by dividing the moles of K^+ ions in the spacer by its volume; thus, we have $c_s = \Gamma/d$, where d is the length of the hydrophilic spacer, estimated at 2.25 nm (Becucci and Guidelli, 2009). For simplicity, the potential energy barrier in the lipid bilayer is assumed to by symmetrical by setting $\alpha = 0.5$.

Cyclic voltammetry curves are determined numerically by subdividing the potential range of interest into 1×10^6 $\delta\phi_m$ steps. The current I is calculated at each step from Eq. 6.24 and is integrated by adding the corresponding charge contribution, $\delta(F\Gamma) = I\,\delta\phi_m/v$, to the sum of all preceding contributions, where v is the potential scan rate. The resulting $F\Gamma$ values are used to continuously update $c_s = \Gamma/d$, C_{il} via Eq. 6.23 and ϕ_m via Eq. 6.22, and all these values are feedbacked into Eq. 6.24 for the current.

The $\sigma_i(C_{ph}{}^{-1} + l/\varepsilon_0\varepsilon_w)$ surface dipole term is ascribed to the negative charge density σ_i created by the phosphate group of the DOPC polar heads surrounding the mouth of the gramicidin channel and dragged close to the hydrocarbon tail region. This charge density is expected to decrease with a decrease in pH from 6.8 to 3, due to progressive protonation of the phosphate group, but its actual value cannot be directly determined. The contrivance consisting in provisionally setting the surface dipole term equal to zero at pH 6.8 is, therefore, adopted, and the focus of attention is shifted on the changes of this term with decreasing pH. With $\sigma_i(C_{ph}^{-1} + l/\varepsilon_0\varepsilon_w) = 0$, the best fit of the experimental cyclic voltammogram (CV) at pH 6.8 by the model is obtained for $k_t = 5 \times 10^3$ cm/s and is represented by the dashed curve in Fig. 6.18, top panel. The experimental changes, $\Delta E_p^{n(6.8\rightarrow pH)}$, in the negative peak potentials and those, $\Delta E_p^{p(6.8\rightarrow pH)}$, in the positive peak potentials, when passing from pH 6.8 to pH 5.4 and to pH 3, are exclusively ascribed to corresponding changes, $\Delta\sigma_i(C_{ph}^{-1} + l/\varepsilon_0\varepsilon_w)$, in the surface dipole potential, and are identified with the latter, accordingly.

The dashed curves at pH 5.4 and 3 in Fig. 6.18 are CVs calculated from the model using the surface dipole potentials estimated in this way, while maintaining the same k_t and χ_{sp} values used for the fit of the CV at pH 6.8 and calculating the C_{il} dependence upon Γ via Eq. 6.23. The shape of the calculated curves in Fig. 6.18 is in fairly good agreement with that of the experimental ones. Most importantly, the three calculated curves, which are obtained from the model on the scale of the absolute potential difference $\Delta\phi$, retain the same relative positions along this scale as the corresponding experimental CVs along their E scale, relative to the Ag/AgCl/0.1 M KCl reference electrode. It is, therefore, possible to superimpose the three calculated CVs over the corresponding experimental ones by simply shifting them along their original $\Delta\phi$ axis by the same quantity, which amounts to −0.33 V.

In this connection, we must consider that the absolute potential difference $\Delta\phi$ across the mercury/water interface is more positive than the corresponding electric potential E, as measured against the Ag/AgCl/(0.1 M KCl) reference electrode, by about 0.19 V, as estimated on the basis of mild extrathermodynamic assumptions (Guidelli and Becucci, 2011a). Hence, the actual shift along the $\Delta\phi$ scale required to superimpose the CVs calculated at the three different pH values over the corresponding experimental CVs on the E scale amounts to $(-0.33+0.19)\text{V}=-0.14\text{ V}$. Even though this quantity cannot be directly ascribed to the surface dipole potential $\sigma_i(C_{ph}^{-1}+l/\varepsilon_0\varepsilon_w)$ at pH 6.8, which was provisionally set equal to zero, nonetheless it points to a negative value of the charge density σ_i due to the embedded phosphate groups of the DOPC polar heads at this pH value. In principle, this approach might be extended to DPTL/DOPC tBLMs supported by other noble metals such as Au and Ag, where gramicidin channels have indeed been incorporated (He et al. 2005, Becucci et al., 2008a). If these tBLMs could yield CVs with a well-defined midpoint potential, $E_{1/2}$, between the positive and the negative peaks, then its relative position with respect to the $E_{1/2}$ value for the mercury-supported tBLM at the same pH would provide an approximate estimate of the difference between the corresponding absolute potential scales. In fact, the characteristic parameters, χ_s and C_{il}, of the DPTL/DOPC tBLM are not expected to be appreciably affected by the nature of the supporting metal.

The aforementioned modelistic procedure for calculating CVs at a tBLM can be readily extended to nonselective ohmic channels by simply adding to Eq. 6.24 for the cation current the further contribution from a monovalent anion, in much the same way as it was done in Eq. 6.20 for a BLM. The resulting CV is, however, altogether different from that predicted for a conventional BLM interposed between two bulk aqueous phases, which provide practically semi-infinite ion reservoirs on both sides of the membrane. In fact, the anion flow takes place separately from the cation flow, at less negative potentials. More precisely, the positive peak due to anion inflow into the spacer and the negative peak due to its outflow fall at more positive potentials than the corresponding peaks for the in- and outflow of the monovalent cation. This behavior is shown by the calculated CV in the inset of Fig. 6.19. The solid curve in the inset is calculated by assuming the presence of a potential energy barrier in the lipid bilayer moiety of the tBLM, as expressed by the rate theory of ion transport (Schultz, 1980), and by ascribing the same value, $1\times10^4\text{ cm s}^{-1}$, to the translocation rate constant, k_t, of monovalent anions and cations. The inset also shows the cation and anion charge densities in the hydrophilic spacer as a function of the applied potential E, during both the forward and backward potential scans. The maximum slope of these charge versus E curves falls at the same potential as the corresponding current peak.

This predicted behavior is due to the limited spaciousness of the spacer, which causes cations to accumulate at more negative potentials than anions. A modest overlapping of anion and cation flow is only predicted at the junction between the negative peak due to anion outflow and the negative peak due to cation inflow, and between the reverse positive peak due to cation outflow and the positive peak due to anion inflow. Unfortunately, the positive potential range required to monitor the predicted massive anion in- and outflow is experimentally inaccessible

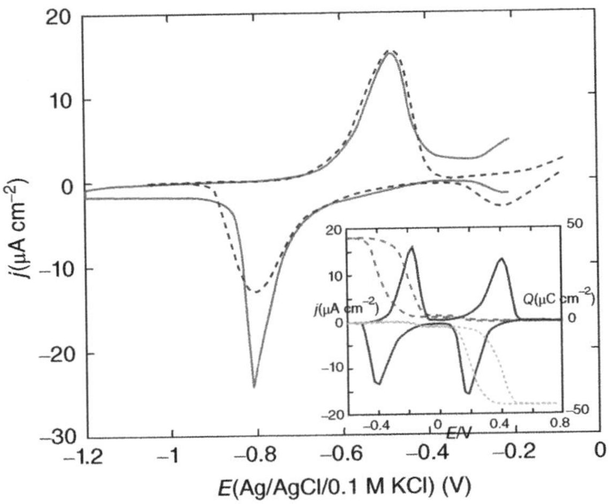

Figure 6.19 CV at a DPTL/DOPC tBLM in a pH 6.8 buffer solution of 0.1 M KCl and 1 μg mL^{-1} SP25A, recorded between −0.20 and −1.20 V at a scan rate of 50 mV s^{-1}. The solid curve is the experimental CV, the dashed curve is the best fit by the model. The inset shows the CV calculated as outlined in the text (solid curve), as well as the corresponding charge densities of the monovalent cation (dashed curve) and anion (dotted curve), against potential. Source: Becucci et al. (2016). Reproduced with permission of Elsevier.

on a Hg-supported tBLM, due to mercury surface oxidation. Nonetheless, the incipient small anion inflow (during the positive-going potential scan) and its subsequent outflow (during the negative-going scan) are revealed by the trumpet-shaped enlargement exhibited at the most positive potentials by the CVs of nonselective ohmic channels. An example is provided by the lipodepsipeptide syringopeptin 25A at a mercury-supported DPTL/DOPC tBLM, whose CV at pH 6.8 is shown by the solid curve in Fig. 6.19 (Becucci et al., 2016). The corresponding dashed curve is the best fit of the experimental CV by the model, obtained by ascribing the same value of 4×10^3 cm s^{-1} to the translocation rate constant k_t for both K^+ and Cl^- ions and by shifting the calculated curve by −0.35 V in passing from the $\Delta\phi$ potential scale to the E scale, in order to achieve the best overlapping. Note that this shift is close to that, −0.33 V, required to superimpose the calculated CV for the gramicidin channel over the experimental one under otherwise identical conditions. Moreover, the midpoint potential $E_{1/2}$ for the CV of syringopeptin 25A shifts gradually toward more positive values with decreasing pH from 6.8 to 5.4 and to 3, assuming values almost identical with those for the CV of gramicidin at the same pH values (Becucci et al., 2016). This confirms that the magnitude of the potential shift from the $\Delta\phi$ scale to the E scale at constant pH, just as the $E_{1/2}$ shift along the E scale with varying pH, are to be ascribed to the DOPC molecules surrounding the mouth of ion channels, more than on the nature of the latter.

A different behavior is exhibited by the CVs of mercury-supported tBLMs incorporating voltage-gated channels, such as that formed by the peptaibol alamethicin. As distinct from the negative-going current, the positive-going one is relatively flat, increasing slightly at the most positive applied potentials, as shown in Fig. 6.20. Integration of the area under the negative peak yields a charge density of about −35 μC cm^{-2} at a scan rate of 10 mV s^{-1}. This charge density decreases with an increase in scan rate. This implies that the positive ionic charge

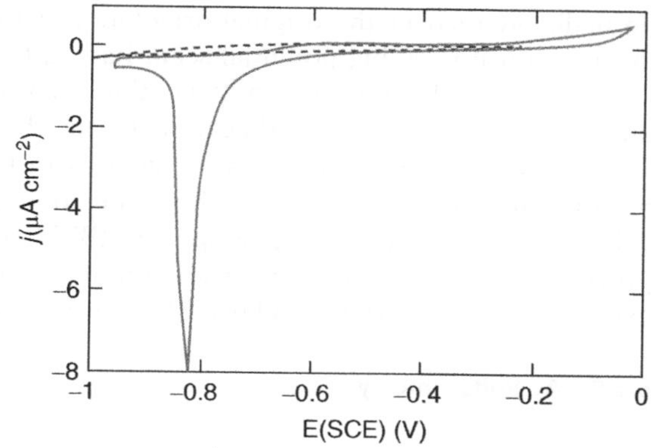

Figure 6.20 Cyclic voltammograms at a DPTL/DOPC tBLM in aqueous 0.1 M KCl at a scan rate of $10\,mV\,s^{-1}$, in the absence (dashed curve) and presence of 0.4 μM alamethicin (solid curve).

stored in the hydrophilic spacer during the negative-going voltage scan is not completely expelled during the positive-going one. Hence, under steady-state conditions, the amount of charge that moves into and out of the spacer during a voltage cycle is less than its saturation value of $45-50\,\mu C\,cm^{-2}$. The charge involved in the positive-going scan cannot be accurately measured because of the flatness of the cyclic voltammogram at potentials positive of −0.60 V.

It should be noted that no negative peak is observed at a freshly prepared tBLM incorporating alamethicin if the voltage scan is confined to potentials positive of about −0.90 V. In view of the correspondence between the applied potential E and the transmembrane potential ϕ_m (Guidelli and Becucci, 2011a), the latter E value corresponds to a ϕ_m value of about −300 mV, which is outside the range of physiological transmembrane potentials. Conversely, after its incorporation, alamethicin permeabilizes the lipid bilayer stably at transmembrane potentials of −80 to −90 mV. These are physiological transmembrane potentials, although their relatively high values are indicative of a voltage-gated behavior, analogous to that exhibited by trichogin GA IV, another peptaibol (Becucci et al., 2012a).

When compared with the cyclic voltammogram of the ohmic ion channel gramicidin, the highly asymmetric cyclic voltammogram of alamethicin can be regarded as typical of a voltage-gated channel, triggered by a negative transmembrane potential. The fact that the first negative-going scan at a freshly prepared tBLM in the presence of alamethicin does not give rise to any ion inflow into the hydrophilic spacer at E values less negative than −0.90 V indicates that alamethicin is inactive over this potential range. This inactive state is usually ascribed to alamethicin molecules being adsorbed flat on the polar heads of the distal lipid leaflet. After a single cyclic voltammogram between −0.05 and −1.00 V, in addition to a sharp negative current peak at about −0.80 V, a flat positive current is recorded, with the charge under this current necessarily equal in magnitude and opposite in sign to that under the negative current peak. This positive current is recorded at potentials much more positive than those at which alamethicin starts to be active during the pristine negative-going voltage scan, although the positive ion outflow is slower than the corresponding ion inflow along the negative current peak. This indicates that the alamethicin helices do not immediately recover the original orientation parallel to the membrane surface during a positive-going potential scan. Rather, they assume some intermediate conformation allowing an ion flow, albeit slower than that in the opposite direction. It is also possible that they do not change conformation but exhibit "sidedness" within the lipid bilayer, such as to favor a higher ion flow in the negative direction than in the positive one. The cyclic voltammetry behavior of voltage-gated channels at Hg-supported tBLMs can be simulated by the same procedure adopted for the gramicidin ohmic channel, by simply multiplying the expression for the current I in Eq. 6.24 by the one-sided Boltzmann equation 4.66.

6.2.5 AC Voltammetry

This technique consists in applying an AC voltage of small amplitude and given frequency to the electrochemical cell, and in recording the quadrature component of the current that flows through the cell with the same frequency, as a consequence of this perturbation. To record the current as a function of the applied potential E, the AC voltage is superimposed on a bias voltage that is varied linearly with time. The current is converted into the differential capacitance C by calibrating the instrument with a high-precision capacitor in place of the electrochemical cell. AC voltammetry can be used to gain insight into the mode of interaction of a biomolecule with the distal lipid leaflet of a biomembrane. This type of interaction can be conveniently investigated at a mercury-supported lipid *self-assembled monolayer* (SAM). As long as the lipid molecules of the SAM do not undergo reorientation and no ion movement across the SAM takes place, the AC voltammogram is flat. Conversely, if the lipid molecules undergo a cooperative reorientation that causes a change in their dipole moment normal component, the potential difference so generated must be compensated for by a flow of electrons along the external electric circuit, to maintain the potential difference across the whole electrified interface constant. This causes a more or less intense pseudocapacitance peak in the AC voltammogram. Similar considerations apply to an ion movement across the monolayer.

AC voltammetry is conveniently employed for investigating the interaction of peptides and small proteins with the hydrocarbon tail region and, most importantly, with the polar heads of mercury-supported phospholipid SAMs. These SAMs can be regarded as the first barrier that a peptide meets in its attempt to penetrate a membrane. The interaction of the peptide with the hydrocarbon tail region of a lipid SAM is generally different from that with a lipid bilayer, especially if its length is appreciably higher than the SAM thickness, as is often the case. In fact, the peptide can only intercalate partially between the hydrocarbon tails, without being capable of assuming the same conformation as in a biomembrane. Conversely, its interaction with the polar heads of a SAM can realistically reproduce that with a biomembrane.

A mercury-supported dioleoylphosphatidylcholine (DOPC) monolayer in a 0.1 M KCl aqueous solution yields the AC voltammogram of the differential capacitance C against the applied potential E shown by curve *a* in Fig. 6.21

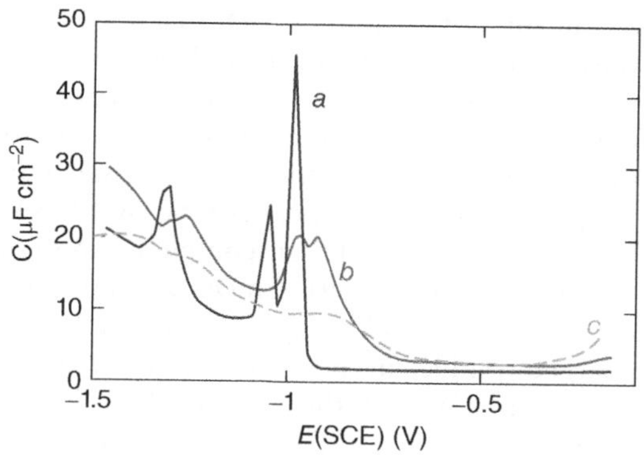

Figure 6.21 AC voltammograms at 75 Hz of a DOPC monolayer in aqueous 0.1 M KCl (*a*), and in aqueous solution of 0.1 M KCl and 2 μg mL^{-1} DCD either unbuffered (*c*) or buffered at pH 7 (*b*). Source: Becucci et al. (2014). Reproduced with permission of Royal Society of Chemistry.

(Nelson and Benton, 1986; Becucci et al., 2014). This phospholipid forms well-defined and tightly packed monolayers on mercury, which are free from defects inducing nonspecific ion leakage. Over the potential region of minimum capacitance, which ranges from −0.15 to −0.75 V/SCE, the DOPC monolayer is impermeable to inorganic ions, whereas it becomes permeable outside this region. The C value over this region amounts to 1.8 μF cm^{-2}, namely twice as high as that for a solvent-free BLM. At positive potentials, the region of minimum capacitance is delimited by a capacitance increase that precedes mercury oxidation; at negative potentials it is delimited by a sharp pseudocapacitance peak that lies at about −0.98 V/SCE, followed by two further peaks at about −1.04 and −1.31 V. The first two peaks are ascribed to a cooperative reorientation of the lipid molecules, whereas the third one is due to their partial desorption (Bizzotto and Nelson, 1998). The first peak generates surface defects that allow a practically uninhibited access of inorganic ions to the mercury surface, while the second peak results from nucleation and growth of the defects formed during the first peak, causing their coalescence.

As a rule, species capable of penetrating the hydrocarbon tail region of the phospholipid monolayer increase its capacitance over the potential range of the flat capacitance minimum with respect to its value in the absence of foreign species, if their polarizability is appreciably higher than that of the lipid molecules. Conversely, they affect the monolayer capacitance only slightly if they have a low polarizability, or they may even decrease it if they contribute to thickening or stiffening the monolayer (Nelson et al., 1990). In both cases, if their concentration in the lipid monolayer is sufficiently high, their intercalation between the lipid molecules prevents the latter molecules from undergoing a sufficiently sharp cooperative reorientation, thus broadening and depressing the pseudocapacitance peaks of curve *a* in Fig. 6.21. Molecules adsorbed on top of the lipid monolayer, but unable to penetrate it, alter and depress the pseudocapacitance peaks by interacting with the polar heads, but exert a negligible effect on the capacitance over the potential range from −0.2 to −0.8 V (Lecompte et al., 1998).

The AC voltammogram of a DOPC monolayer in aqueous 0.1 M KCl is unaltered in passing from an unbuffered solution to a pH 7 buffer solution. Addition of 2 µg mL^{-1} dermcidin (DCD) to a pH 7 buffer solution of 0.1 M KCl depresses the pseudocapacitance peaks and increases the capacitance along the flat region of minimum capacitance, as shown by curve *b* in Fig. 6.21 (Becucci et al., 2014). This effect is enhanced in an unbuffered solution of 0.1 M KCl, where the pseudocapacitance peaks are practically suppressed (curve *c* in Fig. 6.21). The depression of the pseudocapacitance peaks of the DOPC SAM by DCD in the pH 7 buffer solution of 0.1 M KCl and their complete suppression in unbuffered solution denote a strong interaction between DCD and the polar heads of this phospholipid. This can possibly be explained by the presence of three pairs of contiguous oppositely charged residues in the DCD peptide chain, namely Glu^{5}-Lys^{6}, Lys^{23}-Asp^{24}, and Lys^{41}-Asp^{42}. We can, therefore, envisage a strong dual electrostatic interaction of a negative glutamate (Glu) or aspartate (Asp) residue with the trimethylammonium group of a DOPC polar head and of the contiguous positive lysine (Lys) residue with the negative phosphate group of the same polar head. The complete suppression of the DOPC pseudocapacitance peaks by DCD in unbuffered solution denotes a stronger interaction of the peptide in this solution than in a pH 7 buffer solution. It is possible that the phosphate ions of the pH 7 buffer interact with the Lys residues of DCD, thus competing with the phosphate groups of the DOPC polar heads.

6.3 Lipid Bilayers Interposed Between Two Aqueous Phases

A BLM consists of a lipid bilayer that occludes a small hole, about 1 mm in diameter, in a septum of Teflon or another plastic material, such as polycarbonate and polystyrene, which separates two chambers (Fig. 6.22). Prior to bilayer formation, the aperture on which the bilayer is to be formed is "primed" with a small quantity of a phospholipid solution in a liquid hydrocarbon, such as decane or squalene. The aperture is allowed to dry, and both *cis* and *trans* chambers are then filled with the desired aqueous solution. Additional phospholipid solution in the hydrocarbon is then drawn across the hole using a "stick." This implement varies considerably from laboratory to laboratory and may be a small brush or a plastic rod. This procedure for forming the lipid bilayer is called the *painting method* and was devised by Mueller et al. (1962).

The spontaneous assembly of the amphiphilic lipid molecules into a bilayer is driven by the strong self-association of water molecules, which inhibits the mixing of water and amphiphile. Strong lateral intermolecular forces between the long hydrocarbon tails of the lipid molecules also contribute to this self-assembly. The layer of the lipid solution in the nonpolar solvent becomes gradually thinner, with rainbow interference colors appearing on it, followed by black spots. Thinning of the film is monitored under reflected light. Finally, the whole layer becomes completely black. The blackening marks the transition of the lipid layer from a multimolecular to a bimolecular film, 25–50 Å thick, called *bilayer* (or *black*) *lipid membrane* (BLM), which is a nonreflecting optically black film.

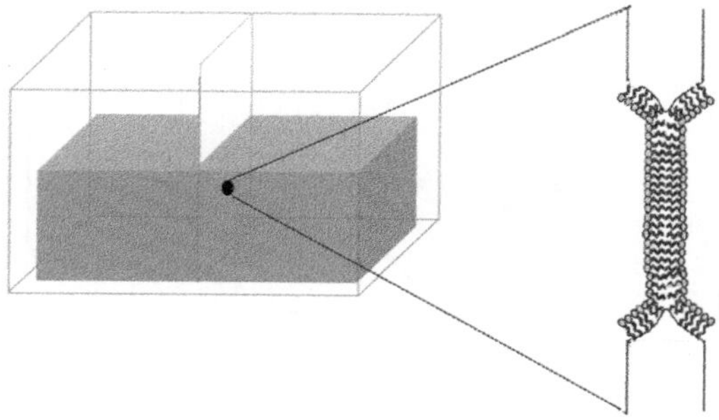

Figure 6.22 Schematic picture of a cell consisting of two compartments separated by a septum, with an enlarged picture of a lipid bilayer spanning a small hole in the septum, according to the method of BLM formation by Mueller et al. (1962).

The resulting system consists of a lipid bilayer in quasi-equilibrium with an annulus of lipid solution in the nonpolar solvent, called *torus* or *Plateau–Gibbs border*, which forms a transition zone between the bilayer and the septum. Thinning of the film my also be monitored by measuring the lipid capacitance. The final capacitance of a painted bilayer should be in the region of 0.4 $\mu F\,cm^{-2}$. It has been shown that the amount of solvent contained in these lipid bilayers depends strongly on the length of the alkyl chains of the nonpolar solvent, varying from about 47% for *n*-decane to about 17% for *n*-hexadecane. This system has been extensively studied as a model for the lipid bilayer of cell membranes. Nonetheless, its usefulness has been questioned, because the thin lipid film contains solvent molecules dispersed within the bilayer and in the form of microlenses floating in the bilayer. To what extent the presence of the organic solvent limits the usefulness of these BLMs has not been established, but it is clearly important to be able to form solvent-free planar bilayers.

Two main procedures are used to form solvent-free lipid bilayers: the Montal–Mueller (1972) method and the patch-micropipette method. Both methods present some advantages over the conventional Mueller–Rudin method (1962). A first important advantage consists in the possibility of forming asymmetrical lipid bilayers, by opposing two lipid leaflets with different compositions. In the second place, these two methods allow the simultaneous deposition of lipids and proteins in the form of a bilayer, starting from a monolayer of these two components at the air/water interface. This process of bilayer formation resembles the mechanism by which natural membranes are self-assembled. It is also noteworthy that it is possible to insert proteins of various types and size in solvent-free bilayers, whereas the Mueller–Rudin method does not allow the incorporation of bulky proteins, probably due to the presence of solvent in the lipid bilayer.

According to the method devised by Montal and Mueller (1972), the cell consists of two communicating compartments separated by a septum that can be lowered vertically and has a hole 1 cm in diameter at its center, as shown in Fig. 6.23. The hole is covered by a Teflon film, whose thickness amounts to a few tens of micrometers. The film contains a small aperture, obtained by piercing it with a very thin, electrically heated platinum wire or a syringe needle. The two compartments are then filled with an electrolytic solution, in such a way that the aperture

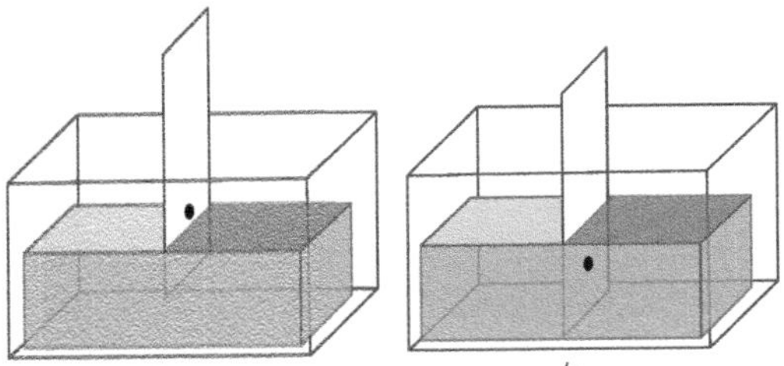

Figure 6.23 Schematic picture of a cell containing a solution whose surface is separated by a movable septum, with two different lipids spread on the two portions of the solution surface. The solvent-free BLM is formed by lowering the septum so as to bring the hole in the septum below the level of the solution surface.

remains above the aqueous surface. Two lipid monolayers are subsequently prepared by spreading two lipid solutions in a volatile solvent, such as pentane or hexane, on the two separate aqueous surfaces. The septum electrically isolates the two lipid monolayers and may be lowered at a preestablished rate. Once the solvent is evaporated, the aperture is lowered below the solution level, to allow the two monolayers to come into contact and to form a bilayer by interaction between their alkyl chains. The rate at which the lipid bilayer is formed is usually higher than the immersion rate of the aperture, due to an abrupt increase in the aqueous solution level driven by capillarity. This often occurs as soon as the aqueous surface reaches the lower boundary of the aperture. If the surface area of the aperture is small enough to incorporate a limited number of ion channels, it may allow the recording of the currents due to single-channel opening, which are of the order of few picoamperes (1 pA = 10^{-12} amperes).

In a simplified version of the Montal–Mueller method, the septum interposed between the two compartments, hermetically separated from each other except for the very small aperture, is kept fixed. Aqueous solutions are then introduced into the two compartments, maintaining their levels below the aperture in the Teflon film. The solution level in one of the two compartments is then raised above the aperture by adding further solution, and the same procedure is subsequently followed with the solution in the other compartment. In this way, the lipid bilayer is formed by a mechanism similar to that previously described. The instability of the ultrathin lipid bilayer spanning the aperture in the Teflon film increases with an increase in its surface area.

The electrical resistance of lipid bilayers formed by the Montal–Mueller method is similar to that of the bilayers formed by the Mueller–Rudin method, and ranges from 10^6 to $10^8\ \Omega\,\mathrm{cm}^2$, but its differential capacitance is about twice as high, assuming a value of $0.9 \pm 0.1\ \mu\mathrm{F\,cm}^{-2}$. This value is comparable with that of plasma membranes and can be justified by simulating the hydrocarbon tail region and the polar head region by two parallel-plate capacitors in series, C_{ht} and C_{ph}, respectively. Since the polar head region is thinner than the hydrocarbon tail region and its dielectric constant is one order of magnitude higher, its contribution to the overall capacitance, $C = 1/(C_{\mathrm{ph}}^{-1} + C_{\mathrm{ht}}^{-1})$, is practically negligible. Setting the thickness l of the hydrocarbon tail region equal to 2 nm and its dielectric constant ε_{m} equal to 2, the differential capacitance per unit surface is given by the equation for a parallel plate capacitor: $C \approx C_{\mathrm{ht}} = \varepsilon_0\varepsilon_{\mathrm{m}}/l = 0.885\varepsilon_{\mathrm{m}}/d$

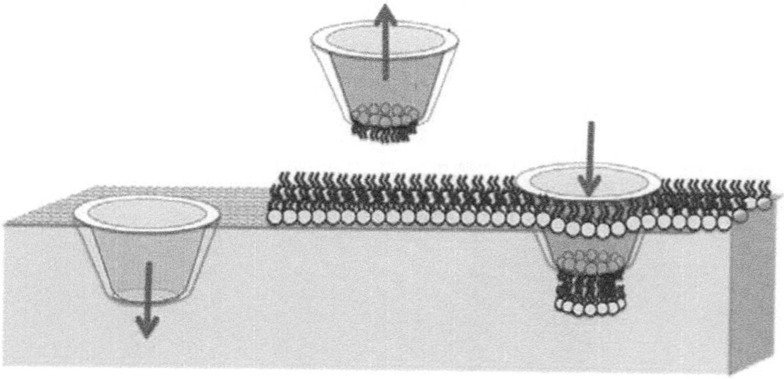

Figure 6.24 The three stages of BLM formation at the tip of a micropipette: (i) immersion before spreading the lipid at the solution/air interface; (ii) emersion after spreading the lipid; (iii) reimmersion.

(in nm) = 0.885 μF cm^{-2}. The presence of an organic solvent in Mueller–Rudin lipid bilayers increases their thickness and hence decreases their capacitance.

With the patch-micropipette method, bilayers are formed at the end of a micropipette with a tip diameter in the range from 0.5 to 5 μm, either with or without fire polishing. The tip of the micropipette is immersed in the working aqueous solution and a phospholipid monolayer is then formed at the air/solution interface by spreading a lipid solution in a volatile organic solvent. A portion of the monolayer is transferred to the micropipette tip by raising the micropipette into the air (Fig. 6.24). The phospholipid molecules turn their polar heads toward the aqueous solution filling the micropipette and the glass wall of the micropipette, while their hydrocarbon tails face the air. A bilayer is constructed by reimmersion of the micropipette into the bath solution. As the tip of the micropipette crosses the monolayer at the air/solution interface, a second portion of monolayer interacts with the monolayer in the micropipette to form a bilayer. Bilayers can be formed from several phospholipids; however, seal formation with negatively charged phospholipids may require the presence of divalent cations both in the micropipette and in the bath solution. The use of this method appears to have declined in recent years.

The most significant applications of a micropipette are directly carried out on the plasma membrane of a cell, rather than on a lipid bilayer self-assembled at its tip. This electrophysiological technique, called *patch-clamp technique*, was devised by Neher and Sakmann (1976) and allows the recording of the currents of single ion channels. This technique can be applied to a wide variety of cells, but is especially useful in the study of excitable cells, such as neurons and muscle fibers. Use is made of a glass micropipette with a tip inner diameter of about 1 μm, which contains an aqueous solution and an AgCl-coated silver wire. The composition of this solution can be changed, and drugs can also be added to study ion channels under different conditions. The micropipette is placed next to a cell, and a gentle suction is applied through it to draw a piece of the cell

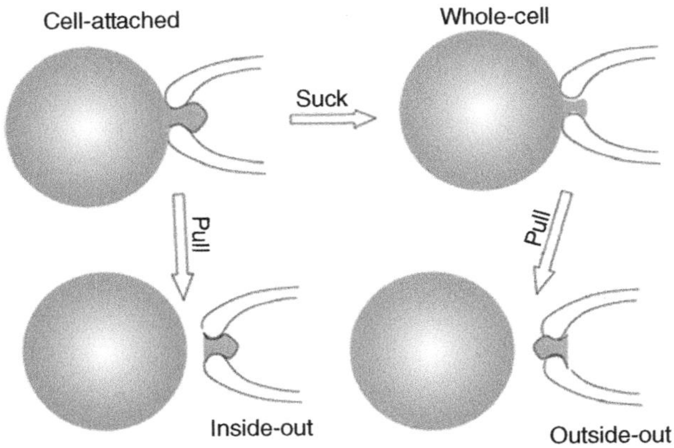

Figure 6.25 Different configurations of the patch-clamp technique, showing the micropipette tip and a cell. First, a seal in the cell-attached configuration is formed. From this, whole-cell, perforated-patch, and inside-out configurations can be generated. The outside-out configuration requires starting from the whole-cell configuration. Source: Guidelli and Becucci (2011b). Reproduced with permission of Springer.

membrane (the *patch*) into the micropipette tip, as shown in Fig. 6.25; the patch seals the micropipette tip imparting to it a resistance of the order of gigaohms ($1\,G\Omega = 10^9\,\Omega$), which reduces the background electrical noise to one or few picoamperes. The resulting configuration, with the intact cell clamped to the micropipette tip, is called *cell attached*. If the pipette is pulled away from the cell, the patch may be detached from the cell but remains at its place, continuing to seal the tip. This new configuration is called *inside-out*, because the intracellular side of the patch is exposed to the external solution, while the extracellular side is in contact with the solution inside the pipette.

Even under the most favorable experimental conditions, the current noise is unavoidably determined by the thermal agitation of electrical charges, and is inversely proportional to the resistance R of the film. Ohm's second law states that R is given by $R = \rho d/A$, where d is the film thickness, A is its surface area and ρ is a proportionality constant called *resistivity*. Since the membrane thickness is not accurately known, its resistance is usually expressed by $RA = \rho d$, in $\Omega\,cm^2$. In practice, to record single-channel currents, a membrane resistance of at least $2\,G\Omega$ is required. Since the resistance of a typical cell containing intrinsic proteins amounts to about $10^3\,\Omega\,cm^2$, the membrane area must be less than about $50\,\mu m^2$. Hence, the inner diameter of the pipette tip must be less than $8\,\mu m$.

Starting from the cell-attached configuration, a strong suction or a short voltage pulse of 1 V perforates the patch without altering the *gigaseal*. In this way, the Ag/AgCl electrode inside the micropipette comes in direct contact with the cytosol. With this configuration, called *whole-cell* configuration, the content of the cell equilibrates rapidly with the solution inside the pipette, allowing a control of the ionic concentrations inside the cell. With this configuration, the area of the membrane is much greater than that of the patch; consequently, the level of noise is too high to allow the recording of single-channel currents, and only the overall current due to all the ion channels in the cell membrane can be recorded. An alternative method for recording the current that flows across the whole cell membrane consists in starting again from the cell-attached configuration and inserting *amphotericin B*, a channel-forming antibiotic, in the solution inside the pipette. The antibiotic makes the patch permeable to monovalent ions. This configuration, called *perforated patch*, lowers the resistance of the patch to such an extent as to allow the recording of the current that flows across the whole cell membrane. As distinct from the whole-cell configuration, however, many constituents of the cytosol cannot diffuse into the solution inside the pipette through the patch, thus preserving several properties of the cytosol.

Starting from the whole-cell configuration, it is possible to pass to a configuration, called *outside-out* configuration, which exposes the extracellular side of a membrane patch to the external solution and the intracellular one to the internal solution. It is obtained by pulling the pipette away from the cell. In this way, the "neck" of the membrane remains attached to the pipette tip and this membrane fragment forms spontaneously a patch that seals the tip, with the extracellular side exposed to the external solution. Outside-out patches must be larger than the inside-out ones, otherwise it is impossible to detach the neck from the rest of the membrane by pulling away the pipette.

In both inside-out and outside-out configurations, the patch area is sufficiently small to contain only a few ion channels, thus permitting the recording of the opening and closing of single ion channels by application of a suitable potential difference between the Ag/AgCl electrode inside the pipette and an identical electrode in the external solution. If, in addition to the protein under study, other proteins capable of interfering with the recording are present in the patch or in the clamped cell, specific inhibitors are used to block their function. The protein of interest is often *expressed* in the oocytes of the *Xenopus laevis* frog, which transcribe and translate the injected genetic information. Due to their great availability, large size and lack of interfering proteins, these cells are particularly convenient for patch-clamp investigations.

Other widely employed biomembrane models are the *liposomes* (also called *unilamellar lipid vesicles*). They are spherical lipid bilayers, a few tens of nanometer in diameter, which enclose an aqueous solution (see Fig. 1.5). Lipid vesicles may be obtained by treating a lipid suspension in water with ultrasounds, a procedure called *sonication*. Due to their small size, they cannot be normally studied by electrochemical techniques. Rather, they can be investigated by spectroscopic techniques using fluorescent- or spin-labeled molecular probes. Remarkable exceptions are *giant unilamellar vesicles* (GUVs), clamped to the tip of an ultramicropipette, whose transmembrane potential can be varied by the patch-clamp technique, or aqueous suspensions of vesicles subjected to strong external electric fields for the investigation of pore formation in biomembranes induced by electric fields (cf. Section 2.8).

6.4 Biomimetic Membranes Noncovalently Supported by Metals

6.4.1 Lipid Monolayers Self-Assembled on Mercury

A simple and easily prepared biomembrane model obtained by noncovalent self-assembly consists of a phospholipid monolayer supported by a hanging mercury drop electrode. The use of a lipid-coated mercury electrode as a biomembrane model was introduced by Miller (1981) and subsequently adopted in a modified version by Nelson and Benton (1986). The lipid coating is obtained by spreading a solution of the lipid in pentane on the surface of an aqueous electrolyte, allowing the pentane to evaporate and immersing a hanging mercury drop electrode in the electrolyte. This procedure gives rise to a lipid monolayer, with the hydrocarbon tails directed toward the hydrophobic mercury surface and the polar heads directed toward the solution. The defect-free support of the lipid film provided by liquid mercury and the complete absence of pentane in the film impart to the film high mechanical stability, resistance to electric fields, and reproducibility, which are not shared by BLMs. Over the potential region of minimum capacitance, which ranges from −0.15 to −0.75 V/SCE, the film is impermeable to inorganic metal ions, whereas it becomes permeable outside this region. The differential capacitance C of a lipid monolayer on mercury over this region is about equal to 1.7 – 1.8 $\mu F\ cm^{-2}$, namely twice the value for a solvent-free BLM.

Phospholipid-coated mercury electrodes have been used to measure the intrinsic protonation constants of DOPC, dioleoylphosphatidylethanolamine (DOPE), and DOPS (Moncelli et al., 1994), the surface dipole potential (χ) of DOPC, DOPS, and dioleoylphosphatidic acid (DOPA) (Becucci et al., 2000), as well as the change in the dipole potential of DOPS and DOPA with a change in the charge density of the polar heads of these phospholipids (Moncelli et al., 1998) and upon adsorption of certain antitumor drugs (Herrero et al., 2000). These investigations have been conducted by measuring the differential capacitance C and the charge density σ_M of lipid-coated mercury at a special homemade hanging mercury drop electrode (Moncelli and Becucci, 1997). In particular, σ_M was measured by contracting a lipid-coated mercury drop while keeping its neck in contact with the lipid reservoir spread at the water/argon interface (Fig. 6.26), so as to allow a free exchange of lipid material between the lipid monolayer that coats the mercury drop and the lipid film spread on the aqueous solution (Becucci et al., 1996). This procedure ensures that the thickness and all other properties of the lipid monolayer remain unaltered during the contraction. The charge following the contraction divided by the decrease in drop area yields directly the charge density σ_M on the mercury surface.

The experimental data were analyzed on the basis of a model of the membrane/solution interphase that accounts for the presence of charged ionizable groups, either exposed to the aqueous phase or buried well inside the polar head region. The behavior of DOPS was found to be particularly involved, with its phosphate

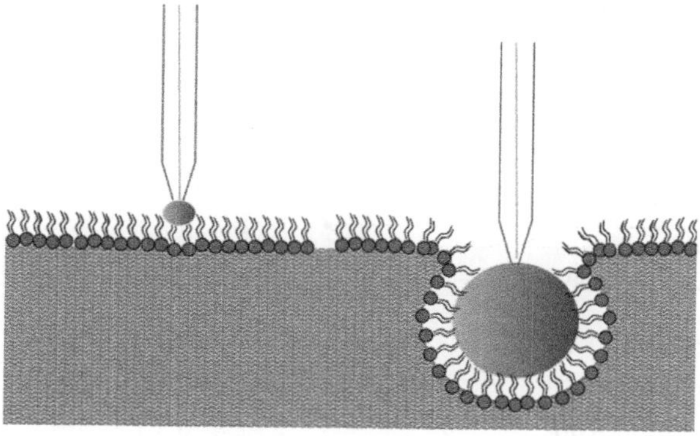

Figure 6.26 Schematic picture of a hanging mercury drop before its gradual expansion in contact with a lipid film spread on the surface of an aqueous solution (left); hanging mercury drop almost completely immersed in the solution, in such a way as to prevent the solution from wetting the capillary tip (right). The thickness of the lipid monolayer is enormously enlarged with respect to the mercury drop radius, for clarity.

group buried deep inside the polar head region, and the carboxyl and ammonium groups exposed to the aqueous phase. It was concluded that the overall charge density of a DOPS monolayer varies from slightly negative to slightly positive values as the bulk pH of the bathing solution is varied from 7 to 3 (Moncelli et al., 1994). Generalizing this model to account for the expansion of a lipid-coated Hg drop totally immersed in the aqueous solution and for the resulting tilt in the self-assembled phospholipid molecules, the dipole potentials χ of DOPS and DOPC monolayers were estimated at +140 to +150 mV, positive toward the interior of the film (Becucci et al., 2000). The fact that these monolayers have very similar dipole potentials indicates that this potential is not to be ascribed to the serine or choline group of their polar heads, but rather to a group common to these two lipids and buried deeper inside the polar head region. This can be reasonably identified with the ester linkages between the carboxyl groups of the two fatty acids and the glycerol backbone.

Ubiquinone-10 (UQ) is an important, ubiquitous biomolecule that is present in many membranes and acts as a proton and electron carrier in the respiratory chain of the mitochondrial membrane (cf. Section 3.4.2). It consists of a quinone ring, which can be reduced to the corresponding quinol, and is provided with a long, rigid isoprenoid chain imparting a high affinity for lipids. The mechanism of its reduction inside a phospholipid monolayer supported by mercury was investigated by carrying out a series of consecutive potential steps from a fixed initial value E_i, where UQ is still electroinactive, to progressively more negative potentials, E_f, and by measuring the charge, $Q(t, E_f)$, following each step, as a function of both time t and potential E_f (Fig. 6.27) (Moncelli et al., 1996). The faradaic charge, $Q_f(t, E_f)$, was obtained by subtracting the capacitive charge measured in the absence of UQ from $Q(t, E_f)$. From the linear dependence of E_f upon $\log t$, at constant $Q_f(t, E_f)$ and pH, and upon pH, at constant $Q_f(t, E_f)$ and t,

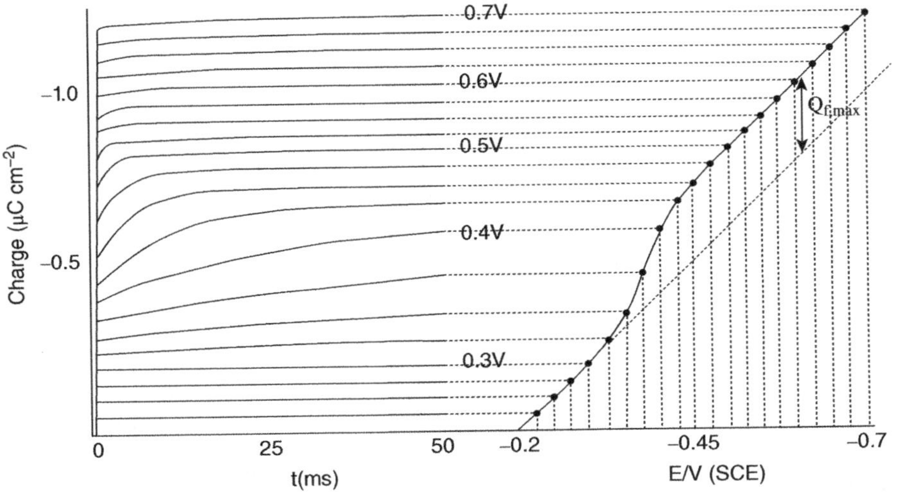

Figure 6.27 The left-hand side of the figure shows *Q*(*t*) versus *t* curves for the reduction of 1 mol% UQ in a mercury-supported DOPC self-assembled monolayer immersed in aqueous 0.075 M borate buffer of pH 9.4, as obtained by stepping the potential from a fixed initial value $E_i = -0.200$ V to final values *E* varying from −0.225 to −0.700 V by 25 mV increments. The right-hand side shows the corresponding $Q(t = 50\ \text{mV})$ versus *E* curve. Source: Moncelli et al. (1996). Reproduced with permission of Elsevier.

it was concluded that UQ electroreduction to ubiquinol, UQH_2, takes place via a one-electron transfer in quasi-equilibrium, followed by the rate-determining uptake of a proton; this rate-determining step is followed by the rapid uptake of a further electron and a further proton, yielding UQH_2. The chronocoulometric procedure employed for UQ was also adopted to study the electrochemical behavior of vitamin K_1 (Herrero et al., 1998), an essential constituent for blood coagulation, upon incorporating this physiological quinone in a self-assembled DOPC monolayer supported by mercury.

Even though a mercury-supported lipid monolayer has no ionic reservoir, under certain conditions, it can be used for investigating the properties of channel-forming peptides such as gramicidin (Nelson, 1991a; Becucci et al., 2002) and amphotericin B (Stoodley et al., 2002; Becucci et al., 2013a), which have a length comparable with the monolayer thickness (~3.2 nm). The ability of short peptides to form pores in the membrane can be tested by verifying whether they allow the electroreduction of inorganic cations such as Tl^+ (Nelson, 1991a; Becucci et al., 2002) and Cd^{2+} (Nelson, 1991b; Becucci et al., 2013a), which are reduced over the potential range of stability of the lipid monolayer. In this case, mercury itself provides an unlimited "ionic" reservoir to the incorporated peptides, by forming an amalgam with the reduced metal ion. The effect of longer peptides or small proteins on a lipid monolayer may also provide useful information on their initial interaction with the outer leaflet of a biological membrane.

A widely investigated ion channel is gramicidin, a linear pentadecapeptide with helical structure that turns its hydrophobic groups toward the exterior of the helix and its hydrophilic carboxyl groups toward the interior. The length of a gramicidin channel is of 1.3 nm, about one half of the thickness of the hydrocarbon tail region of a biomembrane. To span a biomembrane, two helical monomers of gramicidin form a helical dimer, with the N-terminals of the dimer interacting in the center of the membrane. The resulting hydrophilic pore allows the transport of monovalent cations, such as alkali metal ions, across the membrane. The monomer of gramicidin incorporated in a mercury-supported phospholipid monolayer was reported by Nelson (1991a) to act as an ion channel toward Tl^+ ion, thus allowing its penetration across the monolayer and its electroreduction to thallium amalgam. From a cyclic voltammetric and a chronoamperometric investigation of Tl^+ ion electroreduction through the gramicidin channel, Nelson (1996;1997) proposed a chemical–electrochemical (CE) mechanism, in which the rate-determining step is a preceding homogeneous chemical step associated with Tl^+ entry into the channel, prior to Tl^+ electroreduction. This electrode process at a mercury-supported phospholipid monolayer incorporating gramicidin was employed by Nelson as a model system to probe the effect of lipid charge, solution composition and incorporation of biologically active compounds on ion-channel transport. Thus, it was shown that a negatively charged DOPS monolayer increases the rate of Tl^+ transport with respect to a neutral DOPC monolayer, while adsorption of polyvalent cations, such as Mg^{2+} and Dy^{3+}, has an opposite effect (Nelson, 1997). This is ascribed to the electrostatic interaction of the charge of the aforementioned species with that of Tl^+ ions in the immediate vicinity of the lipid film, with a resulting influence on their local concentration. The behavior of the same system by potential-step chronocoulometry was interpreted by Becucci et al. (2002) on the basis of a mechanism that includes the diffusion of thallous ion toward the lipid film and a potential-independent heterogeneous step consisting in the dehydration of the ion and its binding to a site located at the mouth of the ion channel. The further step, involving the surmounting of a potential energy barrier located somewhere in the middle of the channel, is considered to be in quasi-equilibrium, and hence its potential dependence is expressed by the Nernst equation.

6.4.2 Solid-Supported Bilayer Lipid Membranes (sBLMs)

While the mercury surface is hydrophobic, freshly formed surfaces of Pt, Au, Ag, Cu, Ni, or stainless steel are hydrophilic. In this case, the minimum adsorption Gibbs energy is attained when the polar heads of the lipid are turned toward the metal surface, with a resulting noncovalent self-assembly of a lipid monolayer; the repulsion between the hydrocarbon tails of this monolayer and the aqueous phase is circumvented by the self-assembly of a second lipid monolayer, with its hydrocarbon tails in contact with those of the inner monolayer and its polar heads turned toward the aqueous solution. These biomimetic membranes are typically formed by immersing the hydrophilic solid support in an aqueous dispersion of small unilamellar vesicles (SUVs), which slowly rupture and spread on the surface of the support. Alternatively, they can be formed by Langmuir–Blodgett and Langmuir–Schaefer transfers. When appropriately formed, these sBLMs are separated from the support surface through a water layer, with an estimated thickness between 6 and 15 Å (Mossman et al., 2005).

To obtain vesicles, lipids are usually dissolved in an organic solvent. The solvent is then evaporated using a nitrogen stream or vacuum, so that a thin lipid film is produced on the glass surface of a vial. The lipid film is hydrated with an aqueous solution, whose temperature should be above the gel to liquid–crystalline transition temperature of the lipid with the highest transition temperature in the mixture. Multilamellar vesicles (MLVs) are quickly generated by this general protocol. Starting from MLVs, large unilamellar vesicles (LUVs, 100–1000 nm in diameter), with a narrow size distribution around a desired value, are produced by freeze–thaw cycling the vesicles, followed by extrusion; this consists in pressing the vesicle suspension repeatedly through a membrane of defined pore size. Small unilamellar vesicles (SUVs, 20–50 nm in diameter) are prepared by extrusion through membranes with smaller pore size (about 30 nm) or by supplying ultrasound energy to the MLV suspension by using an ultrasonic bath or an ultrasonic probe (*sonication*). The procedure for vesicle fusion consists of adsorbing and fusing SUVs on a suitable substrate from their aqueous dispersion. On a hydrophilic substrate, vesicle fusion gives rise to a lipid bilayer by rupture of the vesicles and their "unrolling" and spreading onto the substrate. This process is schematically depicted in Fig. 6.28b.

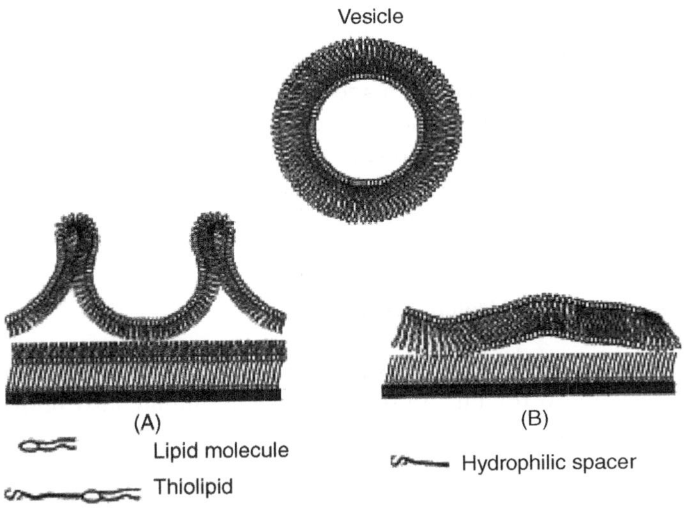

Figure 6.28 Scheme of the splitting and spreading of a vesicle onto a metal-supported self-assembled hydrophobic thiolipid monolayer (A), and of its unrolling and spreading on a self-assembled monolayer of a thiolated hydrophilic spacer tethered to the metal (B). Source: Guidelli et al. (2001). Reproduced with permission of Elsevier.

The kinetics of vesicle fusion is often followed by monitoring the position of the minimum of surface plasmon resonance (SPR) reflectivity curves (cf. Section 7.2). As a rule, bilayer formation by vesicle unrolling onto a hydrophilic surface is faster than monolayer formation by vesicle fusion onto a hydrophobic surface. This is probably due to the fact that the processes involved in forming a planar bilayer starting from a vesicular bilayer are considerably less complex than those involved in forming a planar monolayer (Williams et al., 1997a; 1997b). The pathway of vesicle fusion on hydrophilic surfaces depends on several factors: the nature of the support (its surface charge, chemical composition, and roughness), the nature of the lipid vesicles (their composition, charge, size, and physical state), and the aqueous environment (its composition, pH, and ionic strength). It has been suggested that the initial rapid stage of vesicle adsorption on hydrophilic surfaces is controlled by vesicle adsorption at free sites of the surface, according to a Langmuirian-type behavior. A second, lower stage is ascribed to vesicle unrolling and spreading processes. Rupture occurs when strong adhesive forces cause the tension in the membrane of a partially fused vesicle to exceed the threshold for disruption of the membrane (Lipowsky and Seifert, 1991).

Among hydrophilic substrates, those allowing easier formation of lipid bilayers by vesicle fusion are freshly oxidized surfaces of silica, glass, quartz, and mica (Kalb et al., 1992). However, hydrophilicity is a necessary but not a sufficient condition to promote vesicle fusion. Surfaces of oxidized metals and metal oxides (e.g., TiO_2, Pt and Au) allow adsorption of intact vesicles but resist the formation of bilayers, presumably due to weak surface interactions (Reimhult et al., 2003). Electrostatic, van der Waals, hydration, and steric forces cause the noncovalently supported lipid bilayer to be separated from the solid surface by a nanometer layer of water (Sackmann, 1996). This water layer prevents the support from interfering with the lipid bilayer structure, thus preserving its physical attributes, such as lateral mobility of the lipid molecules.

A different procedure for fabricating sBLMs on noble metals was devised by Tien et al. (Wang et al., 1995). In this case, self-assembly of a lipid bilayer is realized by cutting the end of a Teflon-coated hydrophilic metal wire while keeping it dipped in a decane or squalene solution of the lipid, and by then immersing the freshly cut metal surface in an aqueous solution for 5–10 min, to allow the lipid bilayer to self-assemble. During this period, the lipid solution in excess creeps between the metal wire and its Teflon coating, leaving a self-assembled lipid bilayer on the metal surface. The differential capacitance of this film, $0.3-0.5\,\mu F\,cm^{-2}$, is slightly lower than that, $\sim 0.7-0.8\,\mu F\,cm^{-2}$, of solvent-free BLMs, thus denoting the presence of alkane molecules, and its resistance ($\sim 0.5\,M\Omega\,cm^2$) is also lower. The potential range covered by this solid-supported bilayer before its breakdown is appreciably higher than that at conventional BLMs, attaining values as high as 1.5 V. Since the metal surface can be rough after cutting, it can induce defects in the overlying lipid bilayer.

These freely suspended, noncovalently attached lipid bilayers supported by hydrophilic metals have been used with the main aim of realizing electrodes for biosensor applications. In this connection, nonbiological liposoluble molecules such as ferrocene, tetracyanoquinodimethane, and tetrathiafulvalene have been incorporated in the bilayer and employed as electron carriers between the metal electrode and a hydrophilic redox couple present in the aqueous solution (Ottova-Leitmannova and Tien, 1992). As an example, vinyl-ferrocene incorporated in a platinum-supported lipid bilayer acts as an electron carrier between the electrode and the peripheral, water-soluble protein ferricytochrome *c*. The lipid bilayer provides a natural, biocompatible, surface for cytochrome *c* binding; this results in an enhancement both in the equilibrium constant for cytocrome *c* adsorption on the electrode and in the electron-transfer rate between electrode and protein, with respect to a gold electrode simply coated with the monolayer of a *promoter*, such as bipyridyl.

The aforementioned biomembrane models are not suitable for the study of the function of integral proteins. Quite often, these proteins have hydrophilic sections protruding by over 60 Å outside the lipid bilayer. To avoid their denaturation and to promote their function, the incorporation of integral proteins into biomembrane models must ensure that their protruding hydrophilic sections are accommodated in a hydrophilic medium on both sides of the lipid bilayer. In an attempt to meet this requirement, Tien et al. devised a biomembrane model formed at the end of an agar salt-bridge tubing. Briefly, a freshly cut salt-bridge tubing attached to the end of a Ag/AgCl/KCl reference electrode is first dipped into an alkane solution of the lipid and then transferred into an aqueous solution, so as to form a lipid bilayer (Lu et al., 1996). In this case, the hydrophilic surface required for bilayer formation is provided by the freshly cut agar gel containing KCl. The resistance of these supported BLMs is usually irreproducible from one membrane to another.

A different procedure for the realization of a lipid bilayer noncovalently attached to a silver surface was adopted by Salamon et al. (1994). It consists in spreading a small amount of solution of a lipid in a suitable organic solvent across a small orifice in a Teflon sheet sandwiched between a thin silver film, deposited on the external surface of a prism, and an aqueous solution. The hydrophilic surface of silver attracts the polar heads of the lipid molecules, thus forming a lipid monolayer with the hydrocarbon tails directed toward the bulk lipid phase. This phase becomes progressively thinner through the accumulation of the excess lipid and its solvent at the edge of the Teflon orifice, with formation of an annulus (the *Plateau–Gibbs border*), until the orifice is covered by a lipid bilayer sandwiched between the silver surface and the aqueous solution. The authors claim that a water layer remains in contact with the silver electrode during this process, thus favoring the incorporation of integral proteins into the bilayer. The aforementioned freely suspended lipid bilayers, noncovalently attached to the metal support, are reported to be stable for 25–35 h. However, the capacitance C of these bilayers, measured by EIS, is about one order of magnitude greater than that of a conventional BLM, denoting an appreciable amount of defects (Salamon et al., 2003). By a detailed analysis of the minimum, depth, and half-width of the SPR spectra of these silver-supported lipid bilayers, Salamon and Tollin studied the structural changes following the incorporation of a number of integral proteins into the bilayer and the interaction of these proteins with their associated water-soluble proteins (e.g., incorporation of rhodopsin and its interaction with the G-protein transducin (Salamon et al., 1996), incorporation of cytochrome *c* oxidase (COX) and its interaction with cytochrome c (Salamon and Tollin, 1996)). This allowed the authors to distinguish conditions under which nonspecific electrostatic interactions prevail over those characterized by the predominance of specific hydrophobic interactions.

Salamon et al. (1997) subsequently improved the technique by interposing a thin dielectric layer of SiO_2, acting as a waveguide, between the silver layer and the lipid bilayer, which was now self-assembled on the hydrophilic SiO_2. This system allows an estimate of the refractive index and the extinction coefficient of the composite film not only with p-polarized light as in conventional SPR, but also with *s*-polarized light (cf. Section 7.2). These parameters measure the anisotropic character of the lipid membrane structure. This technique, called *coupled plasmon-waveguide resonance* (CPWR) *spectroscopy*, has provided useful pieces of information on the structural changes accompanying the incorporation of the integral protein cytochrome b_6f in a lipid bilayer and its interaction with the peripheral, water-soluble protein plastocyanin (Salamon et al., 1998). The structural changes involved in the incorporation of different receptors and in their ligand binding have also been investigated. The CPWR spectroscopy was also used to monitor the formation of lipid rafts in mixed sphingomyelin/phosphatidylcholine mixtures and the sorting of receptors into each of these microdomains (Salamon et al., 2005). The microdomain size evaluated from the lateral resolution of the CPWR sensor is surprisingly high (100–300 μm). While providing useful structural information on (lipid bilayer)-(integral protein)-(peripheral protein) systems, this technique cannot provide direct evidence for the functional activity of these systems and a verification that the differential capacitance of the lipid film is that expected for a well-behaved lipid bilayer.

Phospholipid bilayers on Au(111) single crystal faces have been prepared by Lipkowski et al. either by vesicle fusion (Zawisza et al., 2003) or by withdrawing the electrode vertically through a lipid monolayer spread on the surface of a Langmuir trough, and by then bringing the electrode into horizontal contact with the lipid monolayer (Zawisza et al., 2007), according to the Langmuir–Schaefer technique (cf. Section 7.7). The latter technique yields bilayers with a higher packing density. These bilayers have been characterized by charge density measurements, *polarization modulation infrared reflection absorption spectroscopy* (PM IRRAS) (cf. Section 7.3), and *neutron reflectivity* (Burgess et al., 2004) (cf. Section 7.4). The minimum differential capacitance of these bilayers is approximately equal to $2\,\mu F\,cm^{-2}$, and is attained at charge densities σ_M on the metal greater than $-8\,\mu C\,cm^{-2}$. The differential capacitance being greater than that of a solvent-free BLM denotes the presence of a number of defects, while neutron reflectivity reveals the presence of water molecules within the lipid bilayer. As σ_M becomes negative of $-8\,\mu C\,cm^{-2}$, the lipid bilayer starts to detach from the electrode, but remains in close proximity to the electrode surface.

In fact, neutron reflectivity indicates that the lipid bilayer is suspended on a thin cushion of the aqueous electrolyte, which screens the metal charge and depresses the potential difference across the bilayer to a notable extent. The detachment of the lipid bilayer from the electrode surface is accompanied by a decrease in the tilt angle of the hydrocarbon chains with respect to the surface normal from about 55° to 35°, with a resulting increase in bilayer thickness. The PM IRRAS measurements indicate that, as long as the lipid bilayer remains attached to the metal surface, changes in the local electric field by several orders of magnitude have only a small effect on the orientation of the phospholipid molecules.

6.4.3 S-Layer Stabilized Bilayer Lipid Membranes (ssBLMs)

Monomolecular crystalline arrays of protein subunits, called S-layers, are common surface structures of archaea and bacteria. They constitute the outermost component of the cell envelope of these prokaryotic organisms. S-layer subunits can be aligned in lattices with oblique, square, or hexagonal symmetry. Since S-layers are monomolecular assemblies of identical protein subunits, they exhibit pores of identical size and morphology. S-layer subunits of most bacteria interact with each other through noncovalent forces and can be set free with high concentrations of agents that break hydrogen bonds, such as guanidine hydrochloride or urea. Once the S-layer lattice of a bacterial cell is completely disintegrated and the disintegrating agent is removed by dialysis, the S-layer subunits have the unique ability to reassemble spontaneously in suspension, at the liquid/air interface, on solid surfaces, on spread lipid monolayers and on liposomes. Recrystallization starts at several distant nucleation points on the surface and proceeds until neighboring crystalline areas meet. The natural tendency of S-layers to interact with membranes has been exploited to insert them as an intermediate layer between a lipid bilayer and a substrate, giving rise to the so-called *S-layer stabilized* BLMs (ssBLMs; see Fig. 6.1d). These architectures have a stabilizing effect on the associated lipid bilayer, leading to an improvement in its lifetime and robustness. The assembly of S-layer structures from solution to a solid substrate, such as an Au-coated glass slide, can be followed by SPR or by a QCM-D (Knoll et al., 2008b). The self-assembling process is completed after approximately 45 min. The mass increase followed by quartz crystal microbalance with dissipation monitoring (QCM-D) (cf. Section 7.8) corresponds to a thickness of about 8–9 nm, in agreement with the value estimated by SPR.

In the case of bacterial S-layer proteins, it has been demonstrated that protein domains or functional groups of the S-layer lattice interact via electrostatic forces with some head groups of lipid molecules. A well-characterized S-layer protein, SbqA from *Bacillus sphaericus* CCM 2177, was used as an ultrathin crystalline, water containing hydrophilic layer between a gold electrode and a lipid bilayer (Gufter et al., 2004). The SbqA protein recrystallizes in monomolecular square lattices. A morphological unit, about 170 nm in diameter, consists of four protein monomers. The pores are of identical size and morphology, with a diameter of about 3.5 nm.

6.5 Biomimetic Membranes Covalently Supported by Metals

Great interest has been focused on self-assembled films attached to a solid support by formation of a covalent linkage between the self-assembled molecules and the solid support, yielding structures of long-term mechanical stability. Monolayers of alkanethiols on gold are probably the most widely used and best characterized of all self-assembled films to date. Self-assembly involves anchoring of the thiol to the gold surface through the sulfhydryl group, with its possible partial or total deprotonation and electron transfer from the sulfur atom to the metal (Guidelli and Becucci, 2012a; Schmickler and Guidelli, 2014). The hydrocarbon chains are, therefore, directed toward the aqueous solution. The possibility of self-assembling monolayers and bilayers covalently on metals with formation of rugged functionalized electrodes has stimulated a research aiming at exploiting self-assembly for the realization of biomembrane models capable of incorporating integral proteins in a functionally active state. This has potential not only for fundamental research on protein functions, but also for biosensor applications.

To achieve this goal, metal-supported biomimetic membranes consisting of lipid bilayers should meet a number of requirements: (i) they should be robust enough for long-term stability, and be easily and reproducibly prepared; (ii) they should be as fluid as lipid films in the liquid crystalline state; (iii) they should have water on both sides of the lipid bilayer, or at least a hydrophilic medium on one side and an aqueous solution on the other; (iv) they should be sufficiently free from pinholes and other defects that might provide preferential pathways for electron and ion transport across the lipid bilayer. Requirements (ii) and (iii) are necessary for the incorporation of integral proteins into the lipid bilayer in a functionally active state (Guidelli et al., 2001). In fact, integral proteins have a hydrophobic domain buried inside the lipid leaflet of the biomembrane, which must be sufficiently fluid to accommodate this domain. Often, they also have hydrophilic sections protruding outside the lipid bilayer. To avoid their denaturation and to promote their function, the incorporation of integral proteins into biomimetic membranes must ensure that their protruding hydrophilic sections are accommodated in a hydrophilic medium on both sides of the lipid bilayer. Moreover, the transport of hydrophilic ions across the lipid bilayer via ion channels or ion carriers is only possible if an aqueous or hydrophilic layer is interposed between the bilayer and the electrode surface. Requirement (iv) is needed to make the biomimetic membrane sufficiently blocking as to characterize ion-channel activity by electrochemical means, without the disturbing presence of stray currents due to defects.

6.5.1 Alkanethiol/Lipid Hybrid Bilayers

The deposition of a phospholipid monolayer on top of a chemisorbed alkanethiol monolayer can be accomplished by several different preparation techniques, through the exploitation of the attractive hydrophobic interactions between the hydrocarbon tails of the alkanethiol and those of the phospholipid. The architecture of the mixed alkanethiol/lipid bilayer, in which the lipid forms a well-ordered, noninterdigitated monolayer on the alkanethiol monolayer, with the polar heads of the lipid directed toward the aqueous solution, has been confirmed by surface-enhanced Raman spectroscopy, reflection absorption infrared spectroscopy, neutron reflectivity, and EIS. The most common technique for lipid deposition onto the alkanethiol monolayer consists in immersing a thiol-coated gold electrode into a suspension of small lipid vesicles.

Initial adsorption of vesicles on hydrophobic surfaces is energetically disfavored, due to the presence of the hydrophilic polar heads on the outer surface of the vesicular membrane. Therefore, the vesicular membrane must split to allow its inner hydrophobic tails to get in contact with the hydrophobic surface. A possible pathway for vesicle fusion involves vesicle splitting, unrolling, and spreading on the hydrophobic surface, as shown in Fig. 6.28*a*. The kinetics of vesicle fusion on the hydrophobic surface of gold-supported alkanethiol SAMs was followed by SPR (Lingler et al., 1997). In the initial stage, the adsorbed layer thickness d increases linearly with the square root of time t, denoting control by vesicle diffusion to the surface according to Fick's first law. In a second stage, d increases roughly linearly with $\log t$. Finally, the time dependence of d becomes typical of an adsorption process on an almost fully occupied surface. The curve of the surface coverage by vesicles against time was monitored by SPR at different vesicle concentrations (Plant, 1999); it was fitted by an equation practically identical with that derived for an electrode process controlled by diffusion and by a heterogeneous electron transfer step (Delahay and Strassner, 1951). The resulting kinetic constant was ascribed to some surface reorganization of the vesicles. The final increase in thickness following vesicle fusion amounts to 2.0–2.5 nm, as expected for a lipid monolayer, provided the vesicle concentration is not too high (Williams et al., 1997*a*).

The resistance of these bilayers ranges from 1×10^6 to $5 \times 10^7\ \Omega\,\text{cm}^2$. The kinetics of inorganic redox couples exhibiting a Nernstian behavior on bare gold is almost completely suppressed by these alkanethiol/lipid hybrid bilayers for alkanethiol chains with more than seven carbon atoms, indicating an assembly that provides a substantial barrier to electron transfer and to hydrophilic ions (Plant et al., 1994). Electron tunneling across these films takes place only at very high overpotentials. Addition of the pore-forming polypeptide melittin increases the faradaic current for ferricyanide electroreduction, rendering it practically identical with that observed across the sole alkanethiol monolayer (Plant, 1993); this suggests that melittin may only permeate the more fluid lipid monolayer.

Solid-supported alkanethiol/phospholipid bilayers are unsuitable for incorporation of integral proteins, due to the lack of fluidity of the thiol monolayer and to the absence of a hydrophilic spacer interposed between the bilayer and the electrode surface. Nonetheless, a thiol-coated electrode may adsorb membrane fragments whose proteins retain their normal activity. Thus, erythrocyte ghosts, that is, fragments of the plasma membrane obtained by osmotic lysis of erythrocyte cells, are readily adsorbed on alkanethiol-coated gold (Rao et al., 1997). The thickness of the adsorbed erythrocyte ghost membrane, as measured by ellipsometry, SPR, and atomic force microscopy, ranges from 3 to 4 nm. This is not an unreasonable estimate for the thickness of a single leaflet of the erythrocyte membrane in view of the presence of transmembrane proteins, which are expected to contribute to the total thickness of the layer. Acetylcholinesterase, an enzyme anchored to the outer leaflet of erythrocyte cells, retains its activity on this alkanethiol/erythrocyte hybrid for at least 8 days. This suggests that, on interaction with the thiol monolayer, the outer leaflet of ghost membranes opens up and spreads out, incorporating in this process the enzyme molecules exposed at the cell surface.

Alkanethiol-coated gold was also used to adsorb *antigens* for screening antibody production (Boncheva et al., 1996). Incidentally, an antigen is any macromolecule foreign to the living organism that receives it, thus provoking an immune response. This response consists in the production by the organism of special proteins called *antibodies*, often with an elongated Y-shaped structure, which bind in a specific way to the foreign antigens that have induced their production. Usually, peptide sequences representing the major antigenic sites of the antigen (called *epitopes*) are adsorbed onto the alkanethiol monolayer by vesicle fusion or by Langmuir–Blodgett transfer, thus creating a structured antigen-bearing surface. Antibodies raised against the peptide sequence show notable preferential binding to the peptide-covered part of the surface. The resulting increase in the thickness of the gold-supported film is followed by SPR. This procedure can also be used for studying receptor–ligand binding events.

Mixed alkanethiol/phospholipid monolayers are readily prepared on a hanging mercury drop electrode. Thanks to the high affinity of mercury for sulfur, a first alkanethiol monolayer is obtained by immersing the mercury drop for 1 or 2 min in a thiol solution in ethanol (Tadini Buoninsegni et al., 1998). The differential capacitance of the resulting SAM agrees with that calculated from Eq. 2.19 for a parallel plate capacitor, $\varepsilon_0\varepsilon/d$; here d is the length of the fully extended, *all-trans* alkanethiol, ε is its dielectric constant, which is set equal to that, 2.1, of decane, and ε_0 is the electric permittivity of free space. The phospholipid monolayer on top of the alkanethiol monolayer is then obtained by immersing the thiol-coated mercury drop into an aqueous solution on whose surface a phospholipid monolayer has been previously spread. The differential capacitance of the mixed bilayer so obtained agrees with the value calculated from the equation $C = C_t C_m/(C_t + C_m)$, where C_t and C_m are the experimental values of the capacitances of the thiol and phospholipid monolayers on mercury.

Alkanethiol/phospholipid bilayers tethered to gold have been extensively employed to adsorb membrane fragments or liposomes incorporating ion pumps and transporter proteins. These biomimetic membranes are commonly referred to as *solid-supported membranes* (*SSMs*), and their use for the investigation of the function of ion pumps is described in Section 5.6 (see Fig. 5.13).

6.5.2 Tethered Bilayer Lipid Membranes (tBLMs)

In order to incorporate peptides and proteins in metal-supported biomimetic membranes, one must strive to fulfill the four requirements listed at the beginning of Section 6.5. This problem has been tackled by replacing the thiol monolayer in metal-supported thiol/lipid hybrid bilayers by a *thiolipid* monolayer. Thiolipids are synthetic molecules consisting of a hydrophilic spacer terminated with a sulfhydryl or disulfide group for anchoring to the metal surface at one end, and covalently linked to two alkyl chains at the other end (cf. Section 6.1). As a rule, a polyethyleneoxy or oligopeptide chain acts as the spacer. In a thiolipid monolayer tethered to the surface of a gold or mercury electrode, the spacer is interposed between the electrode surface and the linked hydrocarbon tails, which are necessarily turned toward the aqueous solution. By self-assembling a phospholipid monolayer on top of the tethered thiolipid monolayer, two goals are simultaneously achieved, that is, the construction of a whole lipid bilayer and the interposition of a short hydrophilic spacer between the electrode surface and the lipid bilayer. Thiolipids consisting of an oligopeptide chain terminated with a thiol anchor at one end and with a carboxyl group at the other, for *in situ* condensation with the amino group of phosphatidylethanolamine (PE), have also been employed. The second lipid monolayer is usually obtained by immersing a thiolipid-coated electrode in a suspension of small vesicles, which are considered to split and spread on the thiolipid monolayer (Fig. 6.28*a*).

Biomimetic membranes consisting of a lipid bilayer anchored to an electrode through a thiolated hydrophilic spacer are commonly referred to as tBLMs. TBLMs fabricated with polyethyleneoxy hydrophilic spacers exhibit capacitances ranging from 0.5 to 0.7 $\mu F\,cm^{-2}$ and resistances ranging from 5 to 10 $M\Omega\,cm^2$, which are comparable with the values found with conventional BLMs; however, slightly higher capacitance values (1 $\mu F\,cm^{-2}$) have been reported for very long spacers. The monomeric unit of a particularly convenient tBLM of this type, based on the DPTL thiolipid, is shown in Fig. 6.2. On the other hand, gold-supported tBLMs fabricated with oligopeptide hydrophilic spacers, which have been employed for the incorporation of a number of integral proteins, exhibit higher capacitances (from 2 to 10 $\mu F\,cm^{-2}$) and much lower resistances, of the order of $10^4\,\Omega\,cm^2$ (Naumann et al., 1995). This is ascribed to a not perfectly homogeneous coverage of the thiolipid monolayer by the outermost lipid monolayer, which makes the tBLMs insufficiently insulating.

A structural and functional characterization of a DPTL monolayer tethered to gold was reported by Vockenroth et al. (2008) using neutron reflectivity (NR) and EIS (cf. Sec. 7.4 and Fig. 7.14). The TEO moiety was found to be only partly hydrated at the more positive potentials. However, at −0.600 V versus Ag/AgCl(0.1 M KCl), a pronounced increase in the neutron scattering length density of the spacer was observed, denoting an increased amount of water transferred into this region. Leitch et al. (2009) drew similar conclusions using PM-IRRAS. Thus, the fraction of nonhydrated C=O of the lipoic acid ester group was found to be about 50% at the more positive potentials and to attain a value of about 30% at −0.600 V, which denotes an increasing hydration of the spacer at these negative potentials. Analogous conclusions were also drawn by McGillivray et al. (2007) by using NR, EIS, and Fourier-transform IRRAS (FT-IRRAS) to investigate a gold-supported thiolipid monolayer similar to DPTL, with a hydrophilic spacer moiety consisting of a hexaethyleneoxy chain directly bound to a sulfhydryl group. FT-IRRAS revealed a significant disorder in the spacer region and a substantial order in the hydrocarbon tail region. Moreover, NR showed that the spacer region had a thickness smaller than its fully extended length and only 5 vol% exchangeable water, despite its significant disorder. Since the incorporation of proteins with extramembrane domains requires a significant hydration of the spacer, the thiolipid monolayer was then diluted with short β-mercaptoethanol (βME) molecules. This permitted the water molecules to be accommodated in the more spacious thiolipid-βME mixture. By self-assembling a lipid monolayer on top of this mixed monolayer, McGillivray et al. (2007) obtained a tBLM with a differential capacitance comparable with that of conventional BLMs. Moreover, NR data revealed the presence of an appreciable amount of exchangeable water in the spacer moiety of this tBLM.

When fabricating tBLMs, it is important to verify whether the hydrophilic spacer may act as a satisfactory reservoir of inorganic ions. A simple and extremely useful test consists in incorporating in the lipid bilayer the ion carrier valinomycin, a cyclic depsipeptide with a 36-membered ring, and in verifying to what extent it increases the conductance of the biomembrane model (Raguse et al., 1998). Valinomycin selectively transports cations according to the sequence $Ba^{2+} < Na^{+} < Li^{+} < Cs^{+} < K^{+}$; thus the conductance of a bilayer containing valinomycin should increase in the same order. The conductance is conveniently measured by EIS.

Gold-supported tBLMs have been reported to incorporate a few bulky proton pumps. Thus, the proton pump ATPase CF_0F_1 from chloroplasts and that, ATPase EF_0F_1, from *Escherichia coli*, were incorporated in a tBLM fabricated with a hydrophilic oligopeptide spacer (Naumann et al., 1995, 1997). The square-wave voltammogram of the lipid film incorporating the ATPase, as recorded at pH 7.4 in the absence of the activating compound ATP, shows a small reduction peak at −0.7 V versus Ag/AgCl/KCl(sat), due to hydrogen evolution. Progressive additions of ATP to the aqueous solution bathing the biomimetic membrane cause a gradual increase in the reduction peak; this is ascribed to an increase in the proton concentration within the hydrophilic oligopeptide spacer adjacent to the electrode surface, due to proton pumping from the aqueous solution by the activated ATPase. A confirmation comes from the suppression of the reduction peak by tentoxin, an extremely specific inhibitor of the ATPase from chloroplasts. COX was incorporated in an analogous Au-supported, oligopeptide-based tBLM (Naumann et al., 1999). The integral protein COX is the terminal component of the respiratory electron-transport chain and spans the inner mitochondrial membrane (cf. Section 3.4.2). It catalyzes the redox reaction between the small peripheral protein cytochrome *c* in its reduced form, that is, ferrocytochrome *c*, and oxygen, with formation of ferricytochrome *c* and water; it also pumps protons from the matrix to the intermembrane space, where it interacts with ferrocytochrome *c*. In the absence of ferrocytochrome *c*, the square-wave voltammogram of the tBLM incorporating COX shows a reduction peak due to the electroreduction of protons on the gold surface at −0.7 V, as usual. Increasing additions of ferrocytochrome *c* cause a progressive decrease of this peak; this decrease is ascribed to a decrease in the proton concentration within the oligopeptide spacer, due to proton pumping from the spacer to the aqueous solution. This direction of the proton flux is explained by the fact that ferrocytocrome *c* can only interact with the COX protein molecules that turn their intermembrane-space side toward the aqueous phase, which is also the direction of COX proton pumping in biomembranes. The addition of cyanide, which is known to inhibit proton transport by binding to the binuclear center of COX, eliminates the effect of ferrocytochrome *c*.

The lipid monolayer on top of a gold-supported thiolipid monolayer shows no detectable mobility. This is demonstrated by fusing lipid vesicles labeled with a fluorescent probe, photobleaching a spot of the resulting gold-supported tBLM with a laser beam and monitoring the *fluorescence recovery after photobleaching* (FRAP) in this spot (cf. Section 7.5 and Fig. 7.17); no recovery is observed after 10 min (Naumann et al., 2002a). This is due to the fact that solid supports, such as Au or Ag, block the lateral movement of the thiolipid molecules, which are linked to the surface atoms of the metal by covalent bonds. In principle, the lipid molecules on top of the thiolipid monolayer might be free to move laterally. In practice, however, their lateral mobility is hindered by the presence of adsorbed or hemifused vesicles and by the roughness of the metal support (Baumgart et al., 2003). Moreover, the hydration of the polyethyleneoxy moiety of thiolipids anchored to gold is low (Leitch et al., 2009), while the incorporation of proteins with extramembrane domains requires a significant hydration of the spacer. It is, therefore, quite difficult to succeed in incorporating bulky membrane proteins in Au-supported tBLMs. Only small peptides, such as gramicidin (Becucci et al., 2008a), and small ionophores, such as valinomycin (Naumann et al., 2003), can be accommodated in the lipid bilayer moiety of solid-supported polyethyleneoxy-based tBLMs, via incorporation from their aqueous solutions.

It should be noted that vesicles have a low propensity to fuse on the hydrophobic surface exposed to the aqueous solution by a gold-supported thiolipid monolayer, especially if they incorporate an integral protein; rather, they are adsorbed or partially fused (Jeuken et al., 2006). In this respect, even in the absence of a proper incorporation of proton pumps in gold-supported tBLMs from their solutions in detergent, a successful verification of their functional activity may well be ascribed to their incorporation in the membrane of adsorbed or partially fused vesicles (Becucci et al., 2011*b*). In fact, the activation of proton pumps in the membrane of partially fused vesicles is expected to increase the proton concentration on top of the thiolipid monolayer (in the case of F_0F_1 ATPase activated by ATP) or to decrease it (in the case of COX activated by ferrocytocrome *c*). In view of the relative permeability of the leaky oligopeptide-based thiolipid monolayers to protons, this may cause an increase or a decrease, respectively, in the proton electroreduction current on gold, as actually observed.

Incorporation of relatively bulky membrane proteins into gold-supported tBLMs can be realized by diluting the molecules of the thiolipid monolayer with those of a short hydrophilic spacer anchored to gold via a sulfhydryl or disulfide group (e.g., mercaptoethanol or dithiodiglycolic acid). Deposition of a lipid bilayer onto such a mixture of short hydrophilic molecules and thiolipid molecules tethered to the gold electrode is usually carried out by vesicle fusion or by *rapid solution exchange*. The latter procedure consists in placing a small quantity of the bilayer-forming lipid, dissolved in a water-miscible solvent such as ethanol, onto this mixture. The lipid in excess with respect to that required for bilayer formation is then rinsed away vigorously with an aqueous solution. Relatively bulky membrane proteins with extramembrane domains can be accommodated in the tBLM so formed, if the areas covered by the short thiolated spacer are sufficiently large. Differences in chain length and chemistry between thiolipids and short spacers may hopefully favor the formation of phase-demixed domains on the nanoscale, facilitating protein incorporation. This seems to be the case with mixed monolayers of cholesteryl-based thiolipids and short thioalcohol spacers on gold (Jeuken et al., 2007). The resulting lipid bilayer has a capacitance of about 0.5 – 0.6 $\mu F\ cm^{-2}$ and a resistance higher than 5 $M\Omega\ cm^2$. In this case, the areas occupied by the thiolipid are considered to be covered by a lipid monolayer formed by vesicle splitting and spreading; conversely, the areas occupied by the short hydrophilic spacer are assumed to be covered by a lipid bilayer formed by vesicle unrolling and spreading, without splitting into two lipid monolayers, as schematized in Fig. 6.28b. Figure 6.29 shows a picture of a tBLM consisting of a mixture of thiolipid and mercaptoalcohol molecules anchored to gold, with patches of lipid mono- and bilayers on top. The thiolipid represented in the figure, called EO3C, is a triethyleneoxy chain terminated with a sulfhydryl group at one end and covalently linked to a cholesterol molecule at the other end.

This tBLM was used to incorporate cytochrome *bo3* (*cbo3*) (Jeuken et al. 2006). This is the primary membrane-bound terminal oxidase of *E. coli*, which catalyzes the oxidation of ubiquinol-8 and the reduction of oxygen to water in a process coupled to proton translocation across the cell membrane. Upon incorporating *cbo3* and ubiquinone-8 into this tBLM in the presence of oxygen, a reduction cyclic voltammogram was recorded at potentials negative enough to cause ubiquinone-8 electroreduction to ubiquinol-8 on gold. In this way ubiquinol-8 transfers electrons continuously from the gold electrode to *cbo3*, where it is reoxidized by oxygen to ubiquinone-8, which moves back to the electrode surface.

To generate well-defined areas either covered by thiolipids or by short hydrophilic thiols, micropatterned gold-supported tBLMs were also fabricated, using microcontact printing (Toby et al., 1998) or UV lithography (Heyse et al., 1998). The electrode areas initially accessible were filled by a thiolipid monolayer, while those initially masked and made accessible later were filled with short thiols. A micropatterned layer, consisting of regions of thiolipid molecules with a lipid monolayer on top alternated with regions of mercaptoundecanoic acid with a lipid bilayer on top, was used to incorporate rhodopsin (Heyse et al., 1998).

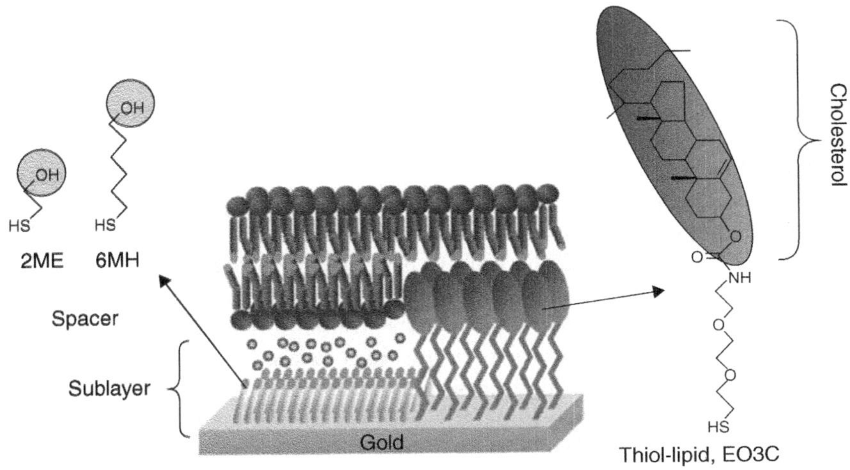

Figure 6.29 Schematic picture of a gold-supported tBLM consisting of a mixture of mercaptoalcohol and thiolipid molecules, with a patch of lipid monolayer on top of the thiolipids and a patch of lipid bilayer on top of the mercaptoalcohols. The space between the mercaptoalcohols and the lipid bilayer is occupied by water molecules and is the site over which proteins are preferentially incorporated. Source: Jeuken et al. (2007). Reproduced with permission of Elsevier.

This integral protein contributes to the closing of Na^+ channels of the plasma membrane of the outer segment of the vertebrate rod cell. In the presence of the peripheral protein transducin, this binds to the rhodopsin molecules incorporated in the regions filled with mercaptoundecanoic acid, with an increase in film thickness. The rapid decrease in film thickness following rhodopsin photoactivation was monitored by SPR. The OmpF porin, a pore-forming protein present in the outer membrane of *E. coli*, was incorporated in a micropatterned tBLM consisting of a mixture of a diethyleneoxy-based thiolipid and of a sulphanylpropionic acid spacer (Stora et al., 1999). This ligand-gated channel acts as a receptor for the antibacterial toxin colicin N, which blocks the channel by binding to its mouth. The gradual decrease in conductance induced by progressive additions of colicin N was monitored by EIS, and the conductance dependence upon colicin N concentration was found to agree with that reported at conventional BLMs.

As distinct from Au-supported thiolipid-based tBLMs, Hg-supported thiolipid-based tBLMs may incorporate bulky integral proteins in a functionally active state, thanks to the fluidity imparted to the thiolipid monolayer by the liquid mercury surface. Incidentally, no hazard is involved in the manipulation of the small amount of mercury used for the preparation of tBLMs. Upon incorporating gramicidin (Becucci et al., 2007a) or valinomycin (Becucci et al., 2005), the TEO moiety of DPTL in aqueous KCl solution undergoes a conformational change ascribable to its elongation, as the applied potential is stepped from a fixed initial value of −0.20 V/SCE to a final value of −0.50 V/SCE. As the final value of this potential step becomes progressively more negative, the charge of K^+ ions accommodated in the TEO spacer increases rapidly, attaining a maximum limiting value of about 45 μC cm^{-2} at −0.8 V/SCE (Becucci and Guidelli, 2009). This corresponds to two potassium ions per DPTL molecule, denoting an appreciable hydration of the spacer.

A drawback in the use of mercury-supported tBLMs is represented by the notable difficulty in using surface sensitive techniques for their structural characterization. Thus, for example, scanning tunneling microscopy (STM) allows a direct visualization of pores of antibiotic peptides such as alamethicin (Pieta et al., 2012) and gramicidin (Sek et al., 2009) in gold-supported planar phospholipid matrixes, SPR measures the optical thickness of tBLMs self-assembled on gold and silver (Knoll et al., 2008a), NR estimates the profile of water content along the hydrophilic spacer and the lipid bilayer moiety of gold-supported tBLMs (Junghans and Köper, 2010) and PM-FTIRRAS is used to investigate the electric field driven transformations of gold-supported phospholipid bilayers (Zawisza et al., 2003).

Nonetheless, mercury-supported tBLMs are definitely superior to gold-supported biomimetic membranes for the investigation of the function of membrane peptides and proteins, thanks to a fluidity and lipid lateral mobility comparable with those of BLMs, but with a much higher robustness and resistance to electric fields. With respect to solid metal supports, mercury has the advantage of providing a defect free, fluid, and readily renewable surface to the self-assembling thiolipid/lipid bilayer. Moreover, the self-assembly of a lipid monolayer on top of a thiolipid monolayer is readily carried out by simply immersing a thiolipid-coated mercury drop in an aqueous electrolyte on whose surface a lipid film has been previously spread. Thanks to the hydrophobic interactions between the alkyl chains of the thiolipid and those of the lipid, this simple procedure gives rise to a lipid bilayer anchored to the mercury surface via the hydrophilic spacer moiety of the thiolipid. By avoiding the use of vesicles, this procedure excludes any artifacts due to partially fused vesicles. These advantageous features make the incorporation of membrane proteins in mercury-supported tBLMs easier and safer than in solid-supported tBLMs. The different features of Hg-supported tBLMs reconstituted with functionally active peptides and membrane proteins of physiological, bacteriological, or pharmacological interest may be disclosed by a judicious choice of the most appropriate electrochemical techniques. A number of applications of mercury-supported tBLMs are described in Section 6.2 to exemplify the use of these techniques.

Hg-supported tBLMs have been used to incorporate and investigate the functional activity of the antimicrobial peptides melittin, monazomycin, distinctin, dermcidin, amphotericin B, alamethicin, trichogin GA IV, and gramicidin (Becucci and Guidelli, 2014a; 2014*b*), as well of small proteins of physiological importance, such as sarcolipin (Becucci et al., 2007*b*; 2009*b*) and phospholamban (Becucci et al., 2012b; 2013*b*). They have also allowed the incorporation of detergent-solubilized bulky channel-forming proteins, such as the OmpF porin from *E. coli* (Becucci et al., 2006*b*) and the HERG potassium channel (Becucci et al., 2008*b*), thanks to the fluidity and smoothness of this liquid metal.

The analysis of the impedance spectra indicates that OmpF porin increases the conductivity of the tBLM over a narrow potential range straddling the zero value of the potential difference across the lipid bilayer moiety. The functional activity of the HERG potassium channel incorporated in the tBLM was monitored by recording the current transient following a suitable potential step.

6.5.3 Polymer-Cushioned Bilayer Lipid Membranes (pBLMs)

The spaciousness of the ionic reservoir of tBLMs may not be sufficient to accommodate bulky extramembrane domains of membrane proteins. The problem is particularly serious with cell-adhesion receptors, whose functional extracellular domains can extend to several tens of nanometers. This problem can be circumvented by separating the lipid bilayer from the solid substrate using soft polymeric materials of typically less than 100 nm thickness, which rest on the substrate and support the bilayer, as shown schematically in Fig. 6.1c. These stratified films are often referred to as *polymer-cushioned* or polymer-supported BLMs. This approach eliminates or appreciably reduces the nonspecific binding of proteins to the solid support and the frictional coupling between proteins and the support, preventing the risk of protein denaturation due to direct contact between protein subunits and the bare support surface. In some cases, the cushion may assist self-healing of local defects in lipid bilayers deposited on macroscopically large supports. To increase the stability of polymer-cushioned membranes, the polymer may be covalently tethered both to the substrate and to the membrane (Naumann et al., 2002*b*). Usually, polymer cushions are anchored to supports such as glass, silica, and mica, by using polymers derivatized with alkyl silanes or triethoxysilane for covalent linkage to silanols at the surface of the silicate substrates. More rarely, they are anchored to gold (Erdelen et al., 1994) or to semiconductors such as indium tin oxide (ITO) (Hillebrandt et al., 1999) via thiolated polymers.

6.5.4 Protein-Tethered Bilayer Lipid Membranes (ptBLMs)

In all the biomimetic membranes previously described and allowing the incorporation of proteins, the protein orientation in the membrane is purely casual. At most, if one of the two extramembrane domains of the protein is much bulkier than the other, incorporation in a tBLM occurs preferentially with the bulkier domain turned toward the aqueous phase, in view of the limited spaciousness of the hydrophilic moiety of the tBLM. Moreover, the packing density of the reconstituted proteins in the lipid bilayer is not well controlled. The need for a well-defined protein orientation with respect to the electrode surface is particularly felt with redox membrane proteins, in which the electrons involved in a chain of redox centers are conveyed across the membrane in a well-defined direction.

To overcome this problem, Knoll, Naumann, et al. have developed a novel methodology based on tethering proteins, rather than thiolipids, to the electrode surface; the lipids are then allowed to self-assemble around the tethered proteins

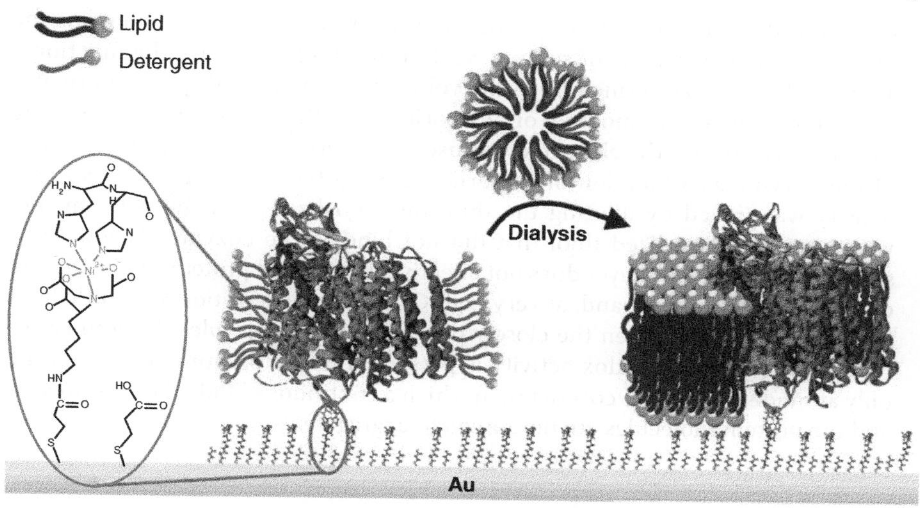

Figure 6.30 Schematic picture of a gold electrode coated with short spacer molecules (dithiobis propionic acid, DTP) and longer molecules with the nitrilotriacetic (NTA) functionality. A Ni^{2+} ion is coordinated by both this functionality and the hystidine tag of a cytochrome *c* oxidase (COX) protein, thus anchoring the latter to the gold surface. A lipid bilayer is formed around the protein, leaving a water layer between the bilayer and the spacer. Source: Friedrich (2008). Reproduced with permission of American Chemical Society.

(Ataka et al., 2004). To this end, a recombinant membrane protein is engineered to bear a stretch of six consecutive histidine residues. A gold surface is functionalized by attaching a molecule terminated with a nitrilotriacetic (NTA) moiety at one end and with a sulfhydryl group for anchoring to gold at the other end. Complexation of Ni^{2+} ions to both the NTA functionality and the histidines of the stretch causes the protein to be anchored to the gold surface from its solution in detergent, as shown schematically in Fig. 6.30. To retain full functional integrity, the membrane protein is incorporated into a lipid bilayer. For this purpose, the protein layer tethered to gold is mixed with detergent-destabilized lipid vesicles. By removing the detergent with microporous biobeads, the tethered proteins are surrounded by lipid molecules that form a lipid bilayer around them, as verified by SPR and EIS; a water layer remains interposed between the lipid bilayer and the NTA moiety, acting as an ionic reservoir. The capacitance and resistance of the bilayer amount to about $6\,\mu F\,cm^{-2}$ and $800\,k\Omega\,cm^2$ (Giess et al., 2004). This high capacitance and low resistance denote a loosely packed lipid bilayer, partly ascribable to the presence of a high protein content.

This approach has been adopted to investigate the function of COX from the proteobacterium *Rhodobacter sphaeroides*, the last enzyme in the respiratory electron-transport chain of bacteria. In this ptBLM, the orientation of the protein with respect to the membrane normal depends on the location of the histidine stretch (his tag) within the protein. Two opposite orientations of the protein were investigated, either with the cytochrome *c* binding side pointing away from the electrode surface or directed toward the electrode, by simply engineering the his tag on the C terminus of subunit SU I or SU II, respectively. The functional activity of COX was verified by cyclic voltammetry with both protein orientations. The cyclic voltammogram of COX oriented with the cytochrome *c* binding side directed toward the electrode was used to determine the functional activity of the enzyme as a function of its surface density (Friedrich et al., 2008). This density was varied by diluting the thiol functionalized with the NTA moiety with a nonfunctionalized thiol that did not bind to the enzyme. At low COX surface densities, the bilayer does not effectively form, and protein aggregates are observed; on the other hand, at very high COX surface densities, very little lipid is able to intrude between the closely packed protein molecules. In both cases, redox activity is low. Redox activity is preserved in the biomimetic membrane only at moderate surface coverages, in which a continuous lipid bilayer is present and the protein molecules are not forced to aggregate.

6.6 Conclusions

The use of electrochemical techniques such as cyclic voltammetry, EIS, and current transient recordings for the investigation of biological systems is becoming increasingly popular, just as the application of the concepts of electrochemical kinetics and of the structure of electrified interfaces to the interpretation of the electrochemical response. Thus, the use of functionalized electrodes for the immobilization of water-soluble redox proteins with the aim of investigating their redox centers is now widely employed and has permitted to monitor complex reactivities and to elucidate some subtle properties of redox enzymes (Rusling et al., 2008).

Several efforts are presently made to realize biomembrane models consisting of a lipid bilayer anchored to a solid electrode through a hydrophilic spacer and satisfying the requirements of ruggedness, fluidity, and high electrical resistance that are necessary for the incorporation of integral proteins in a functionally active state. A unique feature of these biomembrane models is the achievement of the maximum possible vicinity of a functionally active integral protein to an electrode surface (the electrical transducer). The capacitive currents resulting from the activation of ion pumps, transporters, ion channels, and channel-forming peptides incorporated in these biomembrane models can be analyzed over a broad potential range by electrochemical techniques, which are by far less expensive than other techniques presently adopted.

The realization of these biomembrane models allows fundamental studies on the function of integral proteins. These biomimetic membranes are ideally suited to elucidate many problems in molecular membrane biology, by permitting a reliable and rapid functional screening of a large number of mutant receptor proteins. This will open the way to the elucidation of structure–function relationships in ligand–receptor and protein–protein interactions. Moreover, the development of biomimetic systems that incorporate therapeutically or diagnostically important natural proteins will open the door to the realization of sensors targeting biological analytes. Making biomembrane models sufficiently insulating and free from pinholes and other defects that might provide preferential pathways for electron and ionic transport across the lipid bilayer is a particularly challenging goal in the characterization of ion-channel activity. It can be tackled by making the solid support as smooth as possible, by using micropatterned solid-supported lipid bilayers formed via microcontact printing on gold, or by synthesizing hydrophilic and amphiphilic spacers with an architecture that may favor highly compact monolayers.

Microarray technology allows the simultaneous analysis of hundreds of parameters within a single experiment. DNA microarrays, which have revolutionized many areas of biology, have pointed out the generic advantages of microarray technology. The notable importance of quantifying all proteins of a cell and of determining how their translational modifications depend on cell state and environmental influences has raised an enormous interest in transferring microarray technology from DNA to protein molecules. At present, efforts in this direction have been made almost exclusively with water soluble proteins. To the best of our knowledge, no microarrays specifically addressed to membrane proteins by electrochemical means have been reported in the literature so far, even though a large part of proteins are incorporated in biomembranes; moreover, integral membrane proteins represent about 50% of the targets to which drugs are currently addressed. Thus, for example, about one-third of yeast proteins are membrane proteins.

The major difficulty encountered in fabricating microarrays of membrane proteins is that they are normally functionally active only in their natural lipid environment; it is, therefore, necessary to reconstitute them in this environment after purifying them. In order to retain the desired supramolecular structure, a supported lipid bilayer must remain hydrated at all times. This creates a significant challenge for fabricating arrays of supported lipid bilayers. Groves et al. (1997) developed the first procedure for patterning surfaces with solid-supported lipid bilayers. A typical procedure involves the patterning of photoresist on fused quartz wafers by means of standard photolithographic techniques. Several materials that do not readily support continuous lipid bilayers can function as barriers to lateral fluidity, when patterned onto a membrane compatible substrate. Grids of barriers partition a lipid bilayer into an array of isolated "corrals," where the lipid bilayer is continuous and fluid. In addition to simple membrane partitioning, "spatially addressed" arrays of solid-supported lipid bilayers have been fabricated. Spatial addressing allows complete control over the chemical composition of each address in a supported bilayer array. A method for addressing lipid bilayers is the direct pipetting of solutions of SUVs into photolithographically pattern arrays. In this way, each bilayer can contain any desired composition of lipids or proteins, irrespective of the chemical composition of its neighbors (Castellana and Cremer, 2006). These microarrays use nonconducting supports and, hence, can be "individually addressable," for example, by fluorescence microscopy, upon using fluorophore-labeled lipids.

Microarrays individually addressable by electrochemical means are still in their infancy. If micropatterns of supported membranes with different ion channels could be deposited on arrays of field-effect transistors with a comparable sensor area, then such composite material would allow parallel monitoring of channel activity from individual membrane patches (Tanaka and Sackmann, 2006). Combined with microfluidic devices enabling controlled delivery of analytes to each compartment, this would provide a powerful tool for high-throughput screening. A first attempt to realize a microarray for membrane proteins individually addressable by electrochemical means was made by Duran et al. (Andersson et al., 2008). The microarray consists of a number of gold (100 × 100 μm) sensor pads coated with a DPTL/DPhPC tBLM (cf. Fig. 6.2), as shown in Fig. 6.31. The distal DPhPC monolayer is obtained by fusing vesicles incorporating the protein or peptide under study. Each sensor pad is electrically connected to a probe pad. A drop of buffer solution is deposited on the sensor pad, and a conventional patch micropipette containing an Ag/AgCl electrode and filled with the buffer solution is brought into contact with the drop. A tungsten tip is positioned onto the probe pad connected to the sensor pad where single-channel current measurements are to be carried out, with the tungsten tip and the Ag/AgCl electrode connected to a patch-clamp amplifier. Even though this device cannot be easily automated

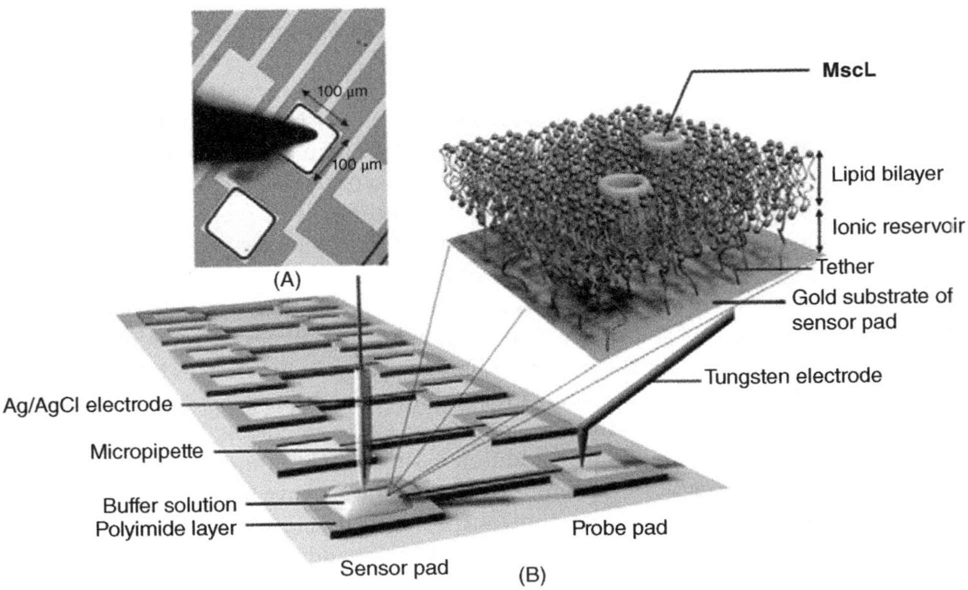

Figure 6.31 Schematic picture of a tethered bilayer lipid membrane (tBLM) array. (A) An optical microscope image of the probe pad and the tungsten electrode tip. (B) Graphical representation of the tBLM array. The lower left corner shows the gold sensor pad covered with a tBLMs that incorporates ion channels. The inset shows a tBLM formed at the gold surface of the sensor pad. Source: Andersson et al. (2008). Reproduced with permission of Elsevier.

for parallel monitoring of channel activity, it may represent a first step in this direction. The ultimate goal is the realization of a membrane-protein microarray platform allowing very small quantities of individual channel proteins to be effectively screened against a large set of drugs or diagnostic targets, without the need of labeling the protein or its target.

Attempts to fabricate microarrays of cell membranes by the patch-clamp technique are presently made by the *planar patch clamping* technique. This technique makes use of glass-based planar patch-clamp chips with a micrometer-sized aperture. In conventional patch clamping, the micropipette tip is placed onto the surface of the membrane of a cell under optical control via a microscope. Then, a tight seal is established by gentle suction, and an "omega"-shaped protrusion of membrane is drawn into the patch pipette. In the case of a planar patch-clamp chip with a single aperture, the procedure is very similar, but still slightly different. Here, a suspension of cells is placed on top of the chip. Then, a single cell is positioned onto the aperture in the chip by application of suction. In contrast to the classic patch-clamp technique, it is the cell that is moved to the aperture and not the pipette that is moved to the cell (Behrends and Fertig, 2007). As in conventional patch clamping, a seal is obtained by application of suction and the membrane can then be ruptured for whole-cell access either with suction or voltage pulses. The result is an electrical connection to the inside of the cell allowing for current recording. Nowadays, microfluidic cartridges containing a glass substrate with several patch-clamp apertures, each of which is individually addressable by microfluidic channels on both sides of the substrate, are employed (Bao et al., 2008). This design allows perfusion of cells and compounds by robotic pipetting means, making the whole approach very suitable for automation. In the case of multiple apertures on a single chip, individual, feedback-controlled, suction lines are required for positioning and sealing of the cells. In addition to scaling up the number of recording channels, the throughput capability is increased by automated application of drugs by a pipetting robot.

Many practical applications are foreseen for these sensors, such as the detection of drug candidates modulating the function of ion channels and pumps or targeting membrane receptors. In this respect, there is strong need to develop novel, rapid, and highly sensitive methods for drug screening, capable of selecting and analyzing a huge number of compounds. At present, screening of pharmacologically active compounds follows traditional procedures, which apply time-consuming ligand-binding studies and receptor-function tests separately. Thus, for instance, the function of ion channels and transporters is traditionally characterized in detail by conventional patch-clamp studies. These assays are tedious to perform, making the investigation of many samples difficult. The lack of knowledge about the different functions of these channels is due to a lack of specific inhibitors, which are unavailable due to the lack of efficient measuring systems. Present ligand-binding experiments identify only ligands to already known binding sites on the protein(s) of interest and neglect other potentially more interesting sites. Moreover, they cannot easily differentiate between agonists and antagonists. Thus, the direct, predominantly electrochemical determination of the function of ion pumps and channels in biomembrane models reconstituted from purified components addresses a strongly felt need for the development of new drug candidates or diagnostic test systems.

References

Andersson, M., G. Okeyo, D. Wilson, H. Keizer, P. Moe, P. Blount, D. Fine, A. Dodabalapur, R. S. Duran, *Biosen. Bioelectron.* 23 (2008) 919–923.

Ashcroft, R. G., H. G. L. Coster, J. R. Smith, *Biochim. Biophys. Acta* 643 (1983) 191–204.

Avrami, M., *J. Phys. Chem.* 8 (1940) 212–224.

Banerjee, U., R. Zidovetzki, R. R. Birge, S. I. Chan, *Biochemistry* 24 (1985) 7621–7627.

Bao, N., J. Wang, C. Lu, *Anal. Bioanal. Chem.* 391 (2008) 933–942.

Barranger-Mathys, M., D. S. Cafiso, *Biochemistry* 35 (1996) 498–505.

Barsoukov, E., J. R. Macdonald, *Impedance Spectroscopy: Theory, Experiments and Applications*, John Wiley & Sons, New York, 2005.

Baumann, G., P. Mueller, *J. Supramol. Struct.* 2 (1974) 538–557.

Baumgart, T., M. Kreiter, H. Lauer, R. Naumann, G. Jung, A. Jonczyk, A. Offenhäusser, W. Knoll, *J. Colloid Interface Sci.* 258 (2003) 298–309.

Becucci, L., M. R. Moncelli, R. Guidelli, *J. Electroanal. Chem.* 413 (1996) 187–193.

Becucci, L., M. R. Moncelli, R. Herrero, R. Guidelli, *Langmuir* 16 (2000) 7694–7700.

Becucci, L., M. R. Moncelli, R. Guidelli, *Biophys. J.* 82 (2002) 852–864.

Becucci, L., M. R. Moncelli, R. Guidelli, *J. Am. Chem. Soc.*, 125 (2003) 3784–3792.

Becucci, L., M. R. Moncelli, R. Naumann, R. Guidelli, *J. Am. Chem. Soc.* 127 (2005) 13316–13323.

Becucci, L., R. Romero León, M. R. Moncelli, P. Rovero, R. Guidelli, *Langmuir* 22 (2006a) 6644–6650.

Becucci, L., M. R. Moncelli, R. Guidelli, *Langmuir* 22 (2006b) 1341–1346.

Becucci, L., A. Santucci, R. Guidelli, *J. Phys. Chem. B* 111 (2007a) 9814–9820.

Becucci, L., R. Guidelli, *Langmuir* 23 (2007) 5601–5608.

Becucci, L., R. Guidelli, C. B. Karim, D. D. Thomas, G. Veglia, *Biophys. J.* 93 (2007b) 2678–2687.

Becucci, L., M. Innocenti, E. Salvietti, A. Rindi, I. Pasquini, M. Vassalli, M. L. Foresti, R. Guidelli, *Electrochim. Acta* 53 (2008a) 6372–6379.

Becucci, L., M. V. Carbone, T. Biagiotti, M. D'Amico, M. Olivotto, R. Guidelli, *J. Phys. Chem. B* 112 (2008b) 1315–1319.

Becucci, L., R. Guidelli, *Soft Matter* 5 (2009) 2294–2301.

Becucci L., A. Schwan, E. E. Sheepwash, R. Guidelli, *Langmuir* 25 (2009a) 1828–1835.

Becucci, L., R. Guidelli, C. B. Karim, D. D. Thomas, G. Veglia, *Biophys. J.* 97 (2009b) 2693–2699.

Becucci, L., M. Papini, D. Muller, A. Scaloni, G. Veglia, R. Guidelli, *Biochim, Biophys. Acta* 1808 (2011a) 2745–2752.

Becucci, L., M. D'Amico, S. Cinotti, S. Daniele, R. Guidelli, *Phys. Chem. Chem. Phys.* 13 (2011b) 13341 – 1334.
Becucci, L., F. Maran, R. Guidelli, *Biochim. Biophys. Acta* 1818 (2012a) 1656 – 1662.
Becucci, L., M. Papini, R. Verardi, G. Veglia, R. Guidelli, *Soft Matter* 8 (2012b) 3881 – 3888.
Becucci, L., M. Innocenti, S. Bellandi, R. Guidelli, *Electrochim. Acta* 112 (2013a) 719 – 726.
Becucci, L., M. L. Foresti, A. Schwan, R. Guidelli, *Biochim. Biophys. Acta* 1828 (2013b) 2682 – 2690.
Becucci, L., R. Guidelli, *An electrochemical platform for studying biomebrane processes. Mercury-supported biomimetic membranes*, Lambert Academic Publishing, Saarbrücken, Germany, 2014a.
Becucci, L., R. Guidelli, *Pharmaceuticals* 7 (2014b) 136 – 168.
Becucci, L., D. Valensin, M. Innocenti, R. Guidelli, *Soft Matter* 10 (2014) 616 – 626.
Becucci, L., V. Tramonti, A. Fiore, V. Fogliano, A. Scaloni, R. Guidelli, *Biochim. Biophys. Acta* 1848 (2015) 932 – 941.
Becucci, L., R. Guidelli, *Bioelectrochemistry* 106 (2015) 343 – 352.
Becucci, L., M. Rossi, A. Fiore, A. Scaloni, R. Guidelli, *Bioelectrochemistry* 108 (2016) 28 – 35.
Behrends, J. C., N. Fertig, in: *Patch Clamp Analysis. Advanced Techniques*, W. Walz (Ed.); 2nd Ed., Chap. 14, Humana Press, New York, 2007.
Bizzotto, D., A. Nelson, *Langmuir* 14 (1998) 6269 – 6273.
Boheim, G., H.-A. Kolb, *J. Membr. Biol.* 38 (1978a) 99 – 150.
Boheim, G., H.-A. Kolb, *J. Membr. Biol.* 38 (1978b) 151 – 191.
Boheim, G., W. Hanke, G. Jung, *Biophys. Struct. Mech.* 9 (1983) 181 – 191.
Boncheva, M., C. Duschl, W. Beck, G. Jung, H. Vogel, *Langmuir* 12 (1996) 5636 – 5642.
Bosco, I., S. K. Rangarajan, *J. Electroanal. Chem.* 129 (1981) 25 – 51.
Bruner, L. J., J. E. Hall, *Biophys. J.* 44 (1983) 39 – 47.
Bruner, L. J., *J. Theor. Biol.*, 117 (1985) 265 – 276.
Burgess, L., M. Li, S. L. Horswell, G. Szymanski, J. Lipkowski, J. Majewski, S. Satija, *Biophys. J.* 86 (2004) 1763 – 1776.
Castellana, E. T., P. S. Cremer, *Surf. Sci. Rep.* 61 (2006) 429 – 444.
Crank, J., *The Mathematics of Diffusion*, Oxford University Press, London, U.K., 1956, p. 27.
Delahay, P., J. E. Strassner, *J. Am. Chem. Soc.* 73 (1951) 5219 – 5222.
Eisenberg, M., J. E. Hall, C. A. Mead, *J. Membrane Biol.* 14 (1973) 143 – 176.
Erdelen, C., L. Häussling, R. Naumann, H. Ringdorf, H. Wolf, J. Yang, *Langmuir* 10 (1994) 1246 – 1250.
Friedrich, M. G., V. U. Kirste, J. Zhu, R. B. Gennis, W. Knoll, R. Naumann, *J. Phys. Chem. B* 112 (2008) 3193 – 3201.
Fringeli, U. P., M. Fringeli, *Proc. Natl. Acad. Sci. U.S.A.* 76 (1979) 3852 – 3856.
Groves, J. T., N. Ulman, S. G. Boxer, *Science* 275 (1997) 651 – 653.
Gufter, P. C., D. Pum, U. B. Sleytr, B. Schuster, *Biochim. Biophys. Acta* 1661 (2004) 154 – 165.

Guidelli, R., G. Aloisi, L. Becucci, A. Dolfi, M. R. Moncelli, F. Tadini Buoninsegni, *J. Electroanal. Chem.* 504 (2001) 1–28.
Guidelli, R., L. Becucci, *Soft Matter* 7 (2011a) 2195–2201.
Guidelli, R., L. Becucci, *J. Solid State Electrochem.* 15 (2011b) 1459–1470.
Guidelli, R., L. Becucci, in: *Bioelectrochemistry*, R. C. Alkire, D. M. Kolb, J. Lipkowski (Eds.); Vol. 13 *Advances in Electrochemical Science and Engineering*, Wiley-VCH, Weinheim, Germany, 2011c.
Guidelli, R., L. Becucci, *Soft Matter* 8 (2012a) 3374–3379.
Guidelli, R., L. Becucci, in: *Applications of Electrochemistry and Nanotechnology in Biology and Medicine*, N. Eliaz (Ed.); Vol 53 *Modern Aspects of Electrochemistry*, Springer, New York, 2012b.
Guidelli, R., L. Becucci, *Biochim. Biophys. Acta* 1858 (2016) 748–755; 1858 (2016) 2957.
Hall, J. E., *Biophys. J.* 15 (1975) 934–936.
Hall, J. E., I. Vodyanoy, T. M. Balasubramanian, G. R. Marshall, *Biophys. J.* 45 (1984) 233–247.
He, L., J. W. F. Robertson, J. Li, I. Kärcher, S. M. Schiller, W. Knoll, R. Naumann, *Langmuir* 21 (2005) 11666–11672.
Herrero, R., F. Tadini Buoninsegni, L. Becucci, M. R. Moncelli, *J. Electroanal. Chem.* 445 (1998) 71–80.
Herrero, R., M. R. Moncelli, R. Guidelli, M. Carla', A. Arcangeli, M. Olivotto, *Biochim. Biophys. Acta* 1466 (2000) 278–288.
Heyse, S., O. P. Ernst, Z. Dienes, K. P. Hofmann, H. Vogel, *Biochemistry* 37 (1998) 507–522.
Hillebrandt, H., G. Wiegand, M. Tanaka, E. Sackmann, *Langmuir* 15 (1999) 8451–8459.
Hodgkin, A. L., A. F. Huxley, *J. Physiol.* 117 (1952) 500–544.
Honig, B. H., W. L. Hubbell, R. F. Flewelling, *Ann. Rev. Biophys. Chem.* 15 (1986) 163–193.
Jeuken, L. J. C., S. D. Connell, P. J. F. Henderson, R. B. Gennis, S. D. Evans, R. J. Bushby, *J. Am. Chem. Soc.* 128 (2006) 1711–1716.
Jeuken, L. J. C., N. N. Daskalakis, X. Han, K. Sheikh, A. Erbe, R. J. Bushby, S. D. Evans, *Sen. Actuat. B* 124 (2007) 501–509.
Junghans, A., I. Köper, *Langmuir* 26 (2010) 11035–11040.
Kalb, E., S. Frey, L. K. Tamm, *Biochim. Biophys. Acta* 1103 (1992) 307.
Kessel, A., D. S. Cafiso, N. Ben-Tal, *Biophys. J.* 78 (2000) 571–583.
Knoll, W., I. Köper, R. Naumann, E.-K. Sinner, *Electrochim. Acta* 53 (2008a) 6680–6689.
Knoll, W., R. Naumann, M. Friedrich, J. W. F. Robertson, M. Lösche, F. Heinrich, D. J. McGillivray, B. Schuster, P. C. Gufler, D. Pum, U. B. Sleytr, *Biointerphases* 3 (2008b) FA125–FA135.
Latorre, R., O. Alvarez, *Physiol. Rev.* 61 (1981) 77–150.
Lecompte, M.-F., A.-C. Bras, N. Dousset, I. Portas, R. Salvayre, M. Ayrault-Jarrier, *Biochemistry* 37 (1998) 16165–16171.
Leitch, J., J. Kunze, J. D. Goddard, A. L. Schwan, R. J. Faragher, R. Naumann, W. Knoll, J. R. Dutcher, J. Lipkowski, *Langmuir* 25 (2009) 10354–10363.

Lingler, S., I. Rubinstein, W. Knoll, A. Offenhäusser, *Langmuir* 13 (1997) 7085–7091.
Lipowsky, R., U. Seifert, *Mol. Crystals* 202 (1991) 17–25.
Lu, X., A. Leitmannova Ottova, H. T. Tien, *Bioelectrochem. Bioenerg.* 39 (1996) 285–289.
Mathew, M. K., P. Balaram, *FEBS Lett.* 157 (1983) 1–5.
Mauro, A., R. P. Nanavati, E. Heyer, *Proc. Natl. Acad. Sci. U.S.A.* 69 (1972) 3742–3744.
McGillivray, D. J., G. Valincius, D. J. Vanderah, W. Febo-Ayala, J. T. Woodward, F. Heinrich, J. J. Kasianowicz, M. Lösche, *Biointerphases* 2 (2007) 21–33.
Menestrina, G., K.-P. Voges, G. Jung, G. Boheim, *J. Membr. Biol.* 93 (1986) 111–132.
Miller, I. R., in: *Topics in Bioelectrochemistry and Bioenergetics*, G. Milazzo (Ed.), John Wiley & Sons, Chichester, 1981.
Moncelli, M. R., L. Becucci, R. Guidelli, *Biophys. J.* 66 (1994) 1969–1980.
Moncelli, M. R., L. Becucci, A. Nelson, R. Guidelli, *Biophys. J.* 70 (1996) 2716–2726.
Moncelli, M. R., L. Becucci, *J. Electroanal. Chem.* 433 (1997) 91–96.
Moncelli, M. R., L. Becucci, F. Tadini Buoninsegni, R. Guidelli, *Biophys. J.* 74 (1998) 2388–2397.
Montal, M., P. Mueller, *Proc. Natl. Acad. Sci. U.S.A.* 69 (1972) 3561–3566.
Mossman, K. D., G. Campi, J. T. Groves, M. L. Dustin, *Science* 310 (2005) 1191.
Mottamal, M., T. Lazaridis, *Biophys. Chem.* 122 (2006) 50–57.
Mueller, P., D. O. Rudin, H. T. Tien, W. C. Wescott, *Circulation* 26 (1962) 1167–1171.
Muller, R. U., A. Finkelstein, *J. Gen. Physiol.* 60 (1972) 263–284.
Muller, R. U., G. Orin, C. S. Peskin, *J. Gen. Physiol.* 78 (1981) 171–200.
Naumann, R., A. Jonczyk, R. Kopp, J. van Esch, H. Ringsdorf, W. Knoll, P. Gräber, *Angew. Chem. Int. Ed. Engl.* 34 (1995) 2056–2058.
Naumann, R., A. Jonczyk, C. Hampel, H. Ringsdorf, W. Knoll, N. Bunjes, P. Gräber, *Bioelectrochem. Bioenerg.* 42 (1997) 241–247.
Naumann, R., E. K. Schmidt, A. Jonczyk, K. Fendler, B. Kadenbach, T. Liebermann, A. Offenhäusser, W. Knoll, *Biosens. Bioelectr.* 14 (1999) 651–662.
Naumann, R., T. Baumgart, R. Gräber, A. Jonczyk, A. Offenhäusser, W. Knoll, *Biosens. Bioelectr.* 17 (2002a) 25–34.
Naumann, C. A., O. Prucker, T. Lehmann, J. Ruhe, W. Knoll, C. W. Frank, *Biomacromolecules* 3 (2002b) 27–35.
Naumann, R., D. Walz, S. M. Schiller, W. Knoll, *J. Electroanal. Chem.* 550–551 (2003) 241–252.
Neher, E., B. Sakmann, *Nature* 29 (1976) 799–802.
Nelson, A., A. Benton, *J. Electroanal. Chem.* 202 (1986) 253–270.
Nelson, A., N. Auffret, J. Borlakoglu, *Biochim. Biophys. Acta* 1021(1990) 205–216.
Nelson, A., *J. Electroanal. Chem.* 303 (1991a) 221–236.
Nelson, A., *J. Chem. Soc. Faraday Trans.* 87 (1991b) 1851–1856.
Nelson, A., *Langmuir* 12 (1996) 2058–2067.
Nelson, A., *Langmuir* 13 (1997) 5644–5651.
Ottova-Leitmannova, A., H .T. Tien, *Progr. Surf. Sci.* 41 (1992) 337.
Pawlak M., S. Stankowski, G. Schwarz, *Biochim. Biophys. Acta* 1062 (1991) 94–102.
Pieta, P., J. Mirza, J. Lipkowski, *Proc. Natl. Acad. Sci. USA* 109 (2012) 21223–21227.

Plant, A. L., *Langmuir* 9 (1993) 2764–2767.
Plant, A. L., M. Gueguetchkeri, W. Yap, *Biophys. J.* 67 (1994) 1126–1133.
Plant, A. L., *Langmuir* 15 (1999) 5128–5135.
Raguse, B., V. Braach-Maksvytis, B. A. Cornell, L. G. King, P. D. J. Osman, R. J. Pace, L. Wieczorek, *Langmuir* 14 (1998) 648–659.
Raistrick, I. D., J. Ross Macdonald, D. R. Franceschetti in: *Impedance Spectroscopy. Emphasizing Solid Materials and Systems*, J. Ross Macdonald (Ed.), John Wiley & Sons, New York, 1987.
Rao, N. M., A. L. Plant, V. Silin, S. Wight, S. W. Hui, *Biophys. J.* 73 (1997) 3066–3077.
Reimhult, E., F. Hook, B. Kasemo, *Langmuir* 19 (2003) 1681–1691.
Rusling, J. F., B. Wang, S.-E. Yun in: *Bioelectrochemistry. Fundamentals, Experimental Techinques and Applications*, P. N. Bartlett (Ed.), John Wiley & Sons, Chichester, England, 2008.
Sackmann, E., *Science* 271 (1996) 43–48.
Salamon, Z., Y. Wang, G. Tollin, H. A. Macleod, *Biochim. Biophys. Acta* 1195 (1994) 267–275.
Salamon, Z., Y. Wang, J. L. Soulages, M. F. Brown, G. Tollin, *Biophys. J.* 71 (1996) 283–294.
Salamon, Z., G. Tollin, *Biophys. J.* 71 (1996) 848–857.
Salamon, Z., H. A. Macleod, G. Tollin, *Biophys. J.* 73 (1997) 2791–2797.
Salamon, Z., D. Huang, W. A. Cramer, G. Tollin, *Biophys. J.* 75 (1998) 1874–1885.
Salamon, Z., G. Lindblom, G. Tollin, *Biophys. J.* 84 (2003) 1796–1807.
Salamon, Z., S. Davanathan, I. D. Alves, G. Tollin, *J. Biol. Chem.* 280 (2005) 11175–11184.
Sansom, M. S. P., *Prog. Biophys. Molec. Biol.* 55 (1991) 139–235.
Schiller, S. M., R. Naumann, K. Lovejoy, H. Kunz, W. Knoll, *Angew. Chem. Int. Ed. Engl.* 42 (2003) 208–211.
Schmickler, W., R. Guidelli, *Electrochim. Acta* 127 (2014) 489–505.
Schultz, S. G., *Basic Principles of Membrane Transport*, Cambridge University Press, London, 1980.
Schwarz, G., S. Stankowski, V. Rizzo, *Biochim. Biophys. Acta* 861 (1986) 141–151.
Sek, S., T. Laredo, J. R. Dutcher, J. Lipkowski, *J. Am. Chem. Soc.* 131 (2009) 6439–6444.
Shai, Y., *Biochim. Biophys. Acta* 1462 (1999) 55–70.
Stankowski, S., M. Pawlak, E. Kaisheva, C.H. Robert, G. Schwarz, *Biochim. Biophys. Acta* 1069 (1991) 77–86.
Stoodley, R., J. Shepherd, K. M. Wasan, D. Bizzotto, *Biochim. Biophys. Acta* 1564 (2002) 289–297.
Stora, T., J. H. Lakey, H. Vogel, *Angew. Chem. Int. Ed.* 38 (1999) 389–392.
Tadini Buoninsegni, F., R. Herrero, M. R. Moncelli, *J. Electroanal. Chem.* 452 (1998) 33–42.
Tanaka, M., E. Sackmann, *Phys. Stat. Sol. (a)* 203 (2006) 3452–3462.
Toby, A., A. Jenkins, R. J. Bushby, N. Boden, S. D. Evans, P. F. Knowles, Q. Liu, R. E. Miles, S. D. Ogier, *Langmuir* 14 (1998) 4675–4678.
Tosteson, M. T., D. C. Tosteson, *Biophys. J.* 45 (1984) 112–114.

Tosteson, M. T., J. J. Levy, L. H. Caporale, M. Rosenblatt, D. C. Tosteson, *Biochemistry* 26 (1987) 6627–6631.
Tosteson, M. T., O. Alvarez, W. Hubbell, R. M. Bieganski, C. Attenbach, L. H. Caporales, J. J. Levy, R. F. Nutt, M. Rosenblatt, D. C. Tosteson, *Biophys. J.* 58 (1990) 1367–1375.
Vockenroth, I. K., C. Ohm, J. W. F. Robertson, D. J. McGillivray, M. Lösche, I. Köper, *Biointerphases* 3 (2008) FA68–FA73.
Vodyanoy, I., J. E. Hall, T. M. Balasubramanian, *Biophys. J.* 42 (1983) 71–82.
Vodyanoy, I., J. E. Hall, V. Vodyanoy, *Biophys. J.* 53 (1988) 649–658.
Wang, L.-G., Y-H. Li, H. T. Tien, *Bioelectrochem. Bioenerg.* 36 (1995) 145–147.
Williams, L. M., S. D. Evans, T. M. Flynn, A. Marsh, P. F. Knowles, R. J. Bushby, N. Boden, *Langmuir* 13 (1997a) 751–757.
Williams, L. M., S. D. Evans, T. M. Flynn, A. Marsh, P. F. Knowles, R. J. Bushby, N. Boden, *Supramol. Sci.* 4 (1997b) 513–517.
Yang, L., T. A. Harroun, T. M. Weiss, L. Ding, H. W. Huang, *Biophys. J.* 81 (2001) 1475–1485.
Yantorno, R., S. Takashima, P. Mueller, *Biophys. J.* 38 (1982) 105–110.
Zawisza, I., A. Lachenwitzer, V. Zamlynny, S. L. Horswell, J. P. Goddard, J. Lipkowski, *Biophys. J.* 85 (2003) 4055–4075.
Zawisza, I., X. Bin, J. Lipkowski, *Langmuir* 23 (2007) 5180–5194.

7

Auxiliary Techniques

Nowadays, many powerful techniques allowing the investigation of electrode/electrolyte interfaces at the molecular level or their characterization are available. However, only a limited number of them are suitable for investigations of metal-supported biomimetic membranes, since they must fulfill the requirement of not deteriorating the biological material. The techniques most commonly employed for this purpose are surface plasmon resonance (SPR), infrared reflection–absorption spectroscopy, neutron reflectivity, fluorescence microscopy, scanning probe microscopy (SPM), Langmuir–Blodgett deposition, and quartz-crystal microbalance. In this chapter, we briefly describe their features, complemented by a few illustrative applications.

7.1 Physical Properties of Electromagnetic Waves

All types of electromagnetic radiation can be described by an electromagnetic wave propagating through a given medium. A *planar electromagnetic wave* (i.e., a wave with a flat wave front) traveling in a transparent, isotropic medium along its propagation direction, with the electric field vector $\boldsymbol{E}$ exclusively located in a single plane, is said to be ***linearly polarized***. In particular, if the wave travels in vacuum, the oscillating electric field is expressed by the following equation:

$$E(x,t) = E_0 \cos\left(k_0 x - \omega t + \varphi\right) = \mathrm{Re}\left\{E_0 \exp\left[\mathrm{i}\left(k_0 x - \omega t + \varphi\right)\right]\right\} \quad \text{with } k_0 = 2\pi/\lambda_0. \tag{7.1}$$

E_0 is the ***amplitude*** of $\boldsymbol{E}$, k_0 is the magnitude of the ***wave vector*** $\boldsymbol{k_0}$ in vacuum, ω is the angular frequency, φ is the phase shift, and λ_0 is the ***wavelength*** in vacuum. In Eq. 7.1, the propagation direction is provisionally identified with the x-axis. To simplify notations, in what follows, we use the exponential expression, disregarding the fact that we are dealing with its real part. The direction of $\boldsymbol{k}$ of an electromagnetic wave is that along which the wave propagates, that is, the x-axis. The electric field $\boldsymbol{E}$ is perpendicular to the propagation direction, whereas the corresponding magnetic field, $\boldsymbol{H}$, is perpendicular to both $\boldsymbol{E}$ and the propagation direction. The speed of a wave in vacuum is a universal constant denoted by c.

A wave moving in a condensed, transparent, nonabsorbing isotropic medium travels at a speed v_p less than c, called *phase velocity*. The $n = c/v_p$ ratio, referred to as the *index of refraction* (or *refractive index*) of the medium, describes how the radiation propagates through the medium and is a measure of its optical density. The phase velocity v_p of a wave is given by the ratio of its wavelength λ to the *period* T, defined as the amount of time in seconds that the wave takes to travel one wavelength. The *frequency* ν is the number of cycles per second, and hence the reciprocal of T. Consequently, we have $v_p = \lambda\nu$. Note that the speed c of an electromagnetic wave in vacuum does not depend upon frequency. For nonmagnetic materials, n is equal to the square root of the optical dielectric constant ε: $n = \sqrt{\varepsilon}$. We then have $\lambda = v_p/\nu = (c/n)/\nu = \lambda_0/n$, where λ_0 is the wavelength in vacuum. The wave vector $\boldsymbol{k}$ in a medium can be written as

$$\boldsymbol{k} = \frac{2\pi}{\lambda}\mathbf{u} = n\frac{2\pi}{\lambda_0}\mathbf{u} = n\boldsymbol{k}_0, \tag{7.2}$$

where $\mathbf{u}$ is the unit vector along the propagation direction.

If the medium is absorbing, the radiation undergoes an exponential decay along its pathway, and its electric field is modified by introducing an exponential damping factor as follows:

$$\begin{aligned} E &= E_0 \exp\left[\mathrm{i}\,(kx - \omega t + \varphi)\right] \exp\left(-\kappa k_0 x\right) \\ &= E_0 \exp\left[\mathrm{i}\left(nk_0 x - \omega t + \varphi\right)\right] \exp\left(-\kappa k_0 x\right), \end{aligned} \tag{7.3}$$

with $k_0 = 2\pi/\lambda_0$, where use was made of Eq. (7.2). The positive parameter κ is called the *attenuation coefficient* or *extinction coefficient*. Recalling that $\mathrm{i}^2 = -1$, Eq. (7.3) can be written in a more compact form:

$$E = E_0 \exp\left[\mathrm{i}\left(\hat{n}k_0 x - \omega t + \varphi\right)\right] \quad \text{with } \hat{n} = n + \mathrm{i}\kappa. \tag{7.4}$$

Here, $\hat{n}$ is a complex refractive index that accounts for both the n and κ values of the medium; depending on whether κ is equal to zero or greater than zero, the medium is nonabsorbing or absorbing.

When the electromagnetic wave is absorbing, the time-averaged value of the energy dissipated per unit volume per unit time is called the *intensity* I. It is related to the amplitude E_0 of the electric field of the radiation via the following equation:

$$I = n\kappa\nu\left\langle E_0^2\right\rangle, \tag{7.5}$$

where $\langle E_0^2\rangle = E_0^2/2$ is the *mean squared electric field strength* (MSEFS). From Eqs. 7.3 and 7.5, it follows that the light intensity in the propagation direction is given by

$$I = I_0 \exp\left(-2\kappa k_0 x\right) = I_0 \exp\left(-4\pi\kappa x/\lambda_0\right) = I_0 \exp\left(-\alpha x\right), \tag{7.6}$$

where I_0 is the light intensity at the origin, $x = 0$ (Beden and Lamy, 1988). The last member of Eq. 7.3 is just an expression of *Lambert–Beer law*, in which the *absorption coefficient* α is related to the attenuation coefficient k.

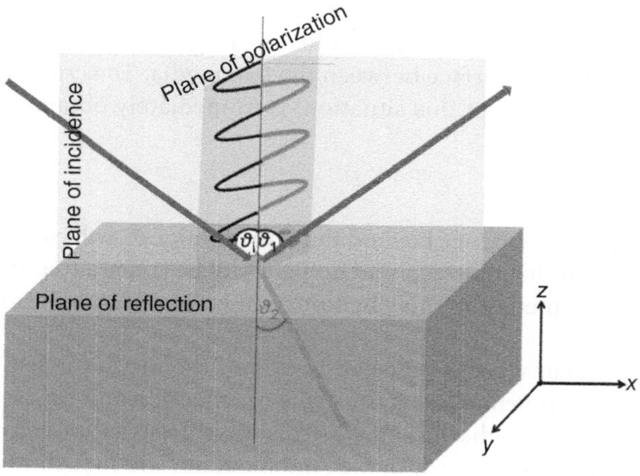

Figure 7.1 Incident, reflected, and transmitted radiations at the interface between two phases of different refractive index.

When a linearly polarized electromagnetic wave impinges on a planar surface separating two homogeneous, isotropic, semi-infinite phases (called the *plane of reflection*), each one characterized by its own index of refraction n and extinction coefficient κ, both the intensity and the state of polarization are changed. Thus, a certain percentage of the incident radiation is *transmitted* (*refracted*), while the remaining fraction is *reflected*, as shown in Fig. 7.1. Here and in the following, the plane of reflection is identified with the $x-y$ plane of an orthogonal coordinate system. The plane containing the propagation direction of the incident radiation and the z-axis, normal to the plane of reflection, is called the *plane of incidence.* The intersection line between the plane of reflection and that of incidence is identified with the x-axis. The incident light, the reflected light, and the light transmitted into the second phase, called the *refracted light,* all lie in the plane of incidence. The angles formed by the incident and the reflected light with the surface normal (the z-axis) are called the *angle of incidence* (θ_i) and the *angle of reflection* (θ_1) respectively, and assume equal values. The angle formed by the refracted light with the surface normal is called the *angle of refraction* (θ_2).

If the two semi-infinite phases separated by the plane of reflection are both nonabsorbing, the relationship between the angle of reflection and the angle of refraction is expressed by *Snell's law* (Bordo and Rubahn, 2005):

$$n_1 \sin\theta_1 = n_2 \sin\theta_2, \tag{7.7}$$

where n_1 is the refractive index of the phase from which the light originates and n_2 is that of the phase where it enters. In the trivial and unusual case of $n_1 = n_2$, the angle of refraction coincides with that of incidence ($\theta_1 = \theta_2 = \theta_i$).

If n_1 is greater than n_2, from Eq. 7.7 it follows that θ_2 is greater than θ_1, and the refracted light deviates from the normal to the plane of reflection more than the incident light. In this case, a progressive increase in the angle of incidence, $\theta_i = \theta_1$, will ultimately determine a critical situation in which θ_2 equals 90° and the refracted light travels along the interface between the two media. The *critical angle* of incidence, θ_c, corresponding to this situation, is immediately obtained from Eq. 7.7 by setting $\sin\theta_2 = 1$:

$$\sin\theta_c = n_2/n_1. \tag{7.8}$$

A further increase in the angle of incidence beyond the critical value θ_c will cause the light to be totally reflected at the interface, and no light will be transmitted to the second, less optically dense medium. This phenomenon is referred to as the *total internal reflection.*

Even though the entire incident wave is reflected back into the medium from which it originates, a small wave is generated at the plane of reflection and travels along the interface between the two adjacent media. This wave decays exponentially when proceeding toward the interior of the second medium, which justifies its name of *evanescent wave.* The penetration depth d of this evanescent wave into the second medium depends upon the refractive indices of the two media and the wavelength, according to the equation (Binder, 2003):

$$d = \lambda/\left[2\pi n_1\sqrt{\sin^2\theta_1 - (n_2/n_1)^2}\right]. \tag{7.9}$$

Finally, if the incident light moves from an optically dense medium to a more optically dense one, that is, if n_1 is less than n_2, a fraction of it will be transmitted to the second medium for any value of the angle of incidence, because the refracted radiation bends toward the normal to the interface with respect to the incident radiation in view of Snell's equation 7.7.

The geometrical considerations expressed by Snell's equation state nothing about the fraction of the incident intensity of the radiation reflected from the interface, called the *reflectivity coefficient* R, and the fraction transmitted, called the *transmission coefficient* T. This is due to the fact that these quantities depend on the polarization state of the electromagnetic wave, not considered by Snell's equation. Since the intensity of a linearly polarized wave is proportional to the MSEFS, $\langle E^2\rangle$, the reflectivity coefficient can also be written as $R = I_1/I_i = \langle E_1{}^2\rangle/\langle E_i{}^2\rangle$, where I_1 and I_i are the intensities of the reflected and incident light and E_1 and E_i are the corresponding electric field strengths.

So far, we have considered a linearly polarized wave, namely a wave whose electric field vector $\boldsymbol{E}$ oscillates in a single plane containing the direction of propagation of the wave. We refer to this plane as the *plane of polarization,* although in other texts it is called the plane of vibration. When a linearly polarized wave impinges on a planar interface, the direction of propagation of the wave and the normal to the interface determine the plane of incidence. It is only at this point that one must account for the relative position of the plane of polarization with respect to the plane of incidence, both containing the direction of propagation of the wave, as shown in Fig. 7.1. If the plane of polarization coincides with the plane of incidence, the light is said to be *p-polarized* and the magnetic field $\boldsymbol{H}$

oscillates in the plane normal to the plane of incidence that contains the direction of propagation. Conversely, if the plane of polarization is perpendicular to the plane of incidence, the light is said to the *s-polarized* and ***H*** oscillates in the plane of incidence. For any other value of the dihedral angle between the plane of polarization and the plane of incidence, we can regard the electric field ***E*** as consisting of a component $\boldsymbol{E}_p$ lying in the plane of incidence (the p-polarized component) and a component $\boldsymbol{E}_s$ lying in the plane normal to the plane of incidence (the s-polarized component).

Before proceeding further, it is convenient to digress briefly into the procedures adopted to obtain polarized waves. Most sources of radiation are *incoherent* and *unpolarized* because they consist of a random mixture of electromagnetic waves having different spatial characteristics, wavelengths, phases, and polarization states. To obtain sinusoidal waves with one particular direction, frequency, phase, and polarization state, an unpolarized radiation is passed through a *linear polarizer*, that is, an optical filter that transmits a wave with a specific polarization, while blocking waves with all other polarizations. Besides linear polarizers, *circular polarizers* may also be employed; these exploit the property of certain special anisotropic materials, whose refractive index depends on their crystallographic orientation relative to the propagation direction of the radiation, called *birefringent materials*.

In this connection, it should be noted that a refracted beam varies not only its direction with respect to the incident beam, but also its phase shift φ. This is apparent from Eq. 7.4, where, at constant t and x, a change in the refractive index $\hat{n}$ requires a change in φ in order to maintain the electric field E constant. Thus, if a linearly polarized beam is incident on a birefringent material, such as calcite, at an oblique angle with respect to its main optical axis, the two components of the electric field of the refracted beam along the main optical axis and along the axis normal to it will undergo different phase shifts. Isotropic materials may also exhibit birefringence if their symmetry is distorted by a mechanic stress. This induced birefringence is referred to as *photoelastic effect*.

This is the principle on which a widely employed circular polarizer, called *photoelastic modulator* (PEM), is based. It consists of a transparent material capable of becoming birefringent if stressed by compression or stretching (e.g., a bar of ZnSe). A piezoelectric transducer glued to the photoelastic bar converts a periodic voltage into a periodic expansion and compression of the bar. The PEM is operated at its resonant frequency, which depends on the size of the device. The beam is linearly polarized at 45° with respect to the main axis of the PEM before impinging on it, as shown in Fig. 7.2. The "quarter-wave plate" in the figure is the PEM, and its main optical axis is marked by the horizontal line. The 45° angle of the linearly polarized light causes the two orthogonal components of the electric field along the main optical axis and along the axis normal to it (the vertical line on the quarter-wave plate) to be equal in magnitude.

The electric field component of the transmitted beam that is oriented perpendicular to the main axis of the PEM (the vertical component in the figure) remains unchanged as the optical element of the PEM is alternatively compressed and

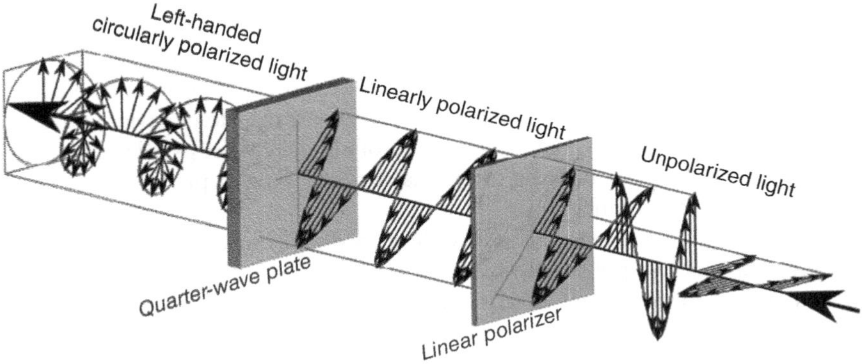

Figure 7.2 Unpolarized, linearly polarized, and left-handed circularly polarized light. Source: https://commons.wikimedia.org/wiki/File:Circular.Polarization.Circularly.Polarized.Light_And.Linearly.Polarized.Light.Comparison.svg.

stretched by the piezoelectric transducer. Conversely, the component oriented along the main axis of the device undergoes a positive phase shift during PEM compression relative to other component, and a negative phase shift during PEM elongation. The phase difference between the two components at any instant of time is called *retardation*. If the retardation is set equal to one-quarter of the wavelength, as shown in Fig. 7.2, the PEM rotates the plane of polarization by 90° during each quarter, generating a circularly polarized wave. During one cycle of the voltage applied to the PEM, the polarization oscillates between right circular and left circular, with linear polarization states in between.

The spiral-shaped curve in Fig. 7.3 is the helical trace described by the tip of the electric field vector of a left circularly polarized wave, that is, a wave whose electric field vector rotates counterclockwise when the wave vector points toward the observer. If the vertical plane in this figure represents the plane of incidence of the wave, the sinusoidal curve lying on this plane is the p-polarized component, E_p, of the electric field along the propagation direction. On the other hand, the sinusoidal curve lying on the horizontal plane, perpendicular to the plane of incidence, is the corresponding s-polarized component. Note that E_s is shifted by one-quarter of a wavelength with respect to E_p. Thus, the electric field vector in the plane of incidence attains a maximum one-quarter of a wavelength before attaining it in the plane normal to the plane of incidence. In this respect, a circularly polarized wave allows the effect of both the p- and s-polarized wave to be monitored practically simultaneously during a single measurement. This advantageous feature is exploited in polarization modulation infrared reflection absorption spectroscopy (PM IRRAS) (cf. Section 7.3).

The reason why the knowledge of the polarization of a radiation is crucial for the determination of the reflectivity coefficient R is to be found in the molecular nature of reflection and refraction. The oscillating electric field of a radiation tends to move the oppositely charged poles of the dipolar molecules at the interface

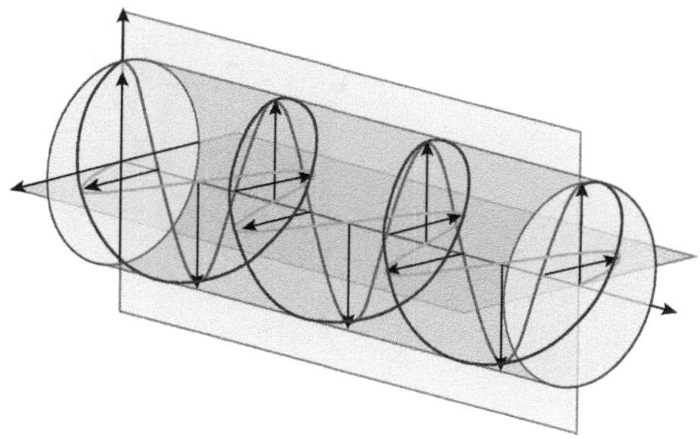

Figure 7.3 The p- and s-polarized components of circularly polarized light. Source: https://commons.wikimedia.org/wiki/File:Circular.Polarization.Circularly.Polarized.Light_And.Linearly.Polarized.Light.Comparison.svg.

in opposite directions and, by continuously inverting its direction, alternatively decreases and increases their dipole length. In turn, the resulting oscillating electric dipoles reradiate light, mainly perpendicularly to their axis, but not along it. It is, therefore, evident that, if a p-polarized refracted beam turns out to be perpendicular to the reflected one, the dipolar molecules oscillating under the influence of the refracted beam will reradiate light mainly in the direction of the latter, but not in the direction perpendicular to it, namely along the reflected beam. Since these same oscillating dipoles also generate the reflected light, this will be suppressed under the aforementioned geometrical condition, expressed by the equation $\theta_1 + \theta_2 = 90°$ (Meeten, 1997; Zamlynny, 2002). In view of Snell's law in Eq. 7.7, this condition can be written as

$$n_1 \sin\theta_1 = n_2 \sin\left(90° - \theta_1\right) = n_2 \cos\theta_1 \quad \rightarrow \quad \theta_1 = \arctan\left(n_2/n_1\right). \quad (7.10)$$

The angle of reflection that fulfills this condition is called the *Brewster's angle* and is denoted by θ_B. This name also applies to the corresponding angle of incidence. For a glass medium ($n_2 \approx 1.5$) in air ($n_1 \approx 1$), the Brewster's angle with visible light is approximately 56°, while for the air/water interface ($n_2 \approx 1.33$), it is approximately 53°. The Brewster's angle is also referred to as the *polarizing angle*, because an unpolarized light that is reflected from a surface at this angle is entirely s-polarized. Hence, a glass plate placed at Brewster's angle in a light beam can be used as a polarizer.

The aforementioned qualitative considerations serve to understand that the polarization of light must be necessarily taken into account to estimate the reflectivity coefficient R. This problem was solved by the French physicist Augustin-Jean Fresnel (1821). At a planar interface between two homogeneous transparent media, the reflectivity coefficient R_s for s-polarized light and that, R_p, for the p-polarized one are expressed by the following *Fresnel equations*:

$$R_s = \left|\frac{\hat{n}_1 \cos\theta_i - \hat{n}_2 \cos\theta_2}{\hat{n}_1 \cos\theta_i + \hat{n}_2 \cos\theta_2}\right|^2; \quad R_p = \left|\frac{\hat{n}_1 \cos\theta_2 - \hat{n}_2 \cos\theta_i}{\hat{n}_1 \cos\theta_2 + \hat{n}_2 \cos\theta_i}\right|^2. \quad (7.11)$$

In these equations, θ_i is the angle of incidence and $\hat{n}_1$, $\hat{n}_2$ are the complex refractive indices of the two media. From these reflectivity coefficients, the corresponding

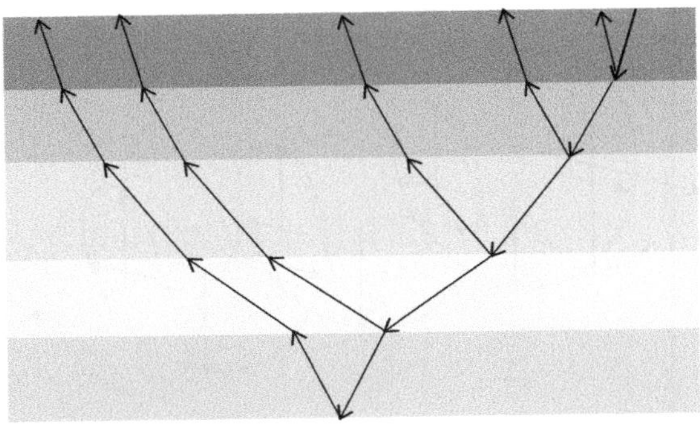

Figure 7.4 Fresnel layer model of a multilayer interface. Lighter halftones simulate lower optical densities.

transmission coefficients T_p and T_s are immediately obtained:

$$T_s = 1 - R_s; \quad T_p = 1 - R_p. \tag{7.12}$$

As a rule, practical applications deal with multiple layers of transparent materials having different complex refractive indices and different thicknesses. This requires the determination of the angles of reflection and refraction at each phase boundary, in order to determine the reflectivity coefficients. Figure 7.4 shows a Fresnel layer model of a multilayer interface. To solve this more complicated problem on the basis of Fresnel equations, an optical matrix method is conveniently employed. This method uses as inputs the complex refractive indices and thicknesses of all the layers, other than the thicknesses of the first and the last layers, which are assumed to be infinitely thick (Hansen, 1968; Hecht and Zajac, 1975). The incident beam is regarded as a planar wave of given wavelength λ and directed at a given angle with respect to the axis normal to the plane of all the phase boundaries. The matrix method provides information on the angle of refraction and the ratio of the light transmitted to that reflected at each phase boundary. With the aid of sufficiently accurate values of the complex refractive indices and thicknesses for each Fresnel layer, the optical matrix method allows the calculation of the MSEFS at each position along the whole multilayer system. Moreover, if a reasonable estimate of the different refractive indices can be made by independent means, a comparison between the experimental reflectivity coefficient and that estimated from the Fresnel equations may be employed to extract the thickness of the layers (Schmidt et al., 1998).

7.2 Surface Plasmon Resonance

SPR is an optical technique capable of generating an electromagnetic wave at the interface of a dielectric material; this wave, called *evanescent wave*, rapidly decays within the material such as to monitor the material behavior within a thickness of a few thousands of ångströms from the surface. To generate an evanescent wave, it is necessary to excite *surface plasmons*, which are collective electronic oscillations in a metal surface layer.

Over a broad frequency range, the optical properties of metals can be explained by a *plasma model*, according to which a gas of free electrons of number density n (the *conduction electrons*) moves within a lattice of metal ion cores. For alkali metals, this range extends up to the ultraviolet, while for noble metals interband transitions occur at visible frequencies, limiting the validity of this model. The plasma model ignores the details of the lattice potential and electron–electron interactions, confining itself to incorporating some aspects of the band structure into the effective optical mass m of each electron. The electrons oscillate in response to an electromagnetic field, and their motion is damped by collisions with the metal ion cores. This damping is lower, the shorter the radiation wavelength λ is. If λ is small enough with respect to the *mean free path* of electrons (i.e., the average distance traveled by an electron between two consecutive collisions), the damping is negligible. For simplicity, we limit ourselves to considering this situation. In such a case, the motion of a free electron under the action of an electric field $\boldsymbol{E}$ is expressed by Newton's second law:

$$m\frac{\mathrm{d}^2\boldsymbol{x}}{\mathrm{d}t^2} = -e\boldsymbol{E}, \tag{7.13}$$

where $\boldsymbol{x}$ is the vector displacement along the direction of the electric field $\boldsymbol{E}$ and $-e\boldsymbol{E}$ is the electric force acting on the electron. Upon ascribing a harmonic time dependence to $\boldsymbol{E}$, we have

$$\boldsymbol{E} = \boldsymbol{E}_0 \exp\left(-\mathrm{i}\omega t\right), \tag{7.14}$$

where $\omega = 2\pi\nu$ is the angular frequency. A particular solution of Eq. 7.13 describing the oscillation of the electron is

$$\boldsymbol{x}(t) = \boldsymbol{x}_0 \exp\left(-\mathrm{i}\omega t\right). \tag{7.15}$$

Replacing $\boldsymbol{x}(t)$ from Eq. 7.15 into Eq. 7.13 yields

$$m\frac{\mathrm{d}^2\boldsymbol{x}}{\mathrm{d}t^2} = m\boldsymbol{x}_0(-\mathrm{i}\omega)^2 \exp\left(-\mathrm{i}\omega t\right) = -e\boldsymbol{E}(t) \rightarrow \boldsymbol{x}(t) = \frac{e}{m\omega^2}\boldsymbol{E}(t). \tag{7.16}$$

It should be noted that the vector displacement has the same direction as the electric field $\boldsymbol{E}(t)$, due to the harmonic dependence of the latter upon time. The electron movement in this direction generates a dipole moment that, being directed from the negative to the positive pole by definition, has an opposite direction with respect to $\boldsymbol{E}(t)$. The *electric polarization vector* $\boldsymbol{P}$, whose magnitude measures the amount of dipole moment per unit volume, is therefore given by

$$\boldsymbol{P} = -ne\boldsymbol{x}(t) = -\frac{ne^2}{m\omega^2}\boldsymbol{E}, \tag{7.17}$$

where $-ne$ is the charge of free electrons per unit volume. The direction of $\boldsymbol{P}$ is opposite to that of the time-harmonic electric field that generates it.

In this respect, the situation is opposite to that encountered in a dielectric medium under the action of a constant electric field, such as that depicted in Fig. 2.5. Referring to this geometrically simple situation, the overall dipole moment of an ideal cylinder of unit cross-sectional area, with its two bases in contact with the plates of the parallel plate capacitor and its axis normal to them, is clearly given by the product of its volume by $\boldsymbol{P}$, namely $d\boldsymbol{P}$, where d is the distance between the plates. The dipole moment of our cylinder is also equal to the product of the surface polarization charge density $\sigma_{2,\mathrm{p}}$ on the right-hand side plate by the distance d that separates it from the opposite charge density $\sigma_{1,\mathrm{p}}$ on the left-hand side plate. Hence, the only nonzero component, P_x, of $\boldsymbol{P}$ in the direction of the constant electric field $\boldsymbol{E}$ and of the x-axis is given by $P_x = \sigma_{2,\mathrm{p}} = -\sigma_{1,\mathrm{p}}$. More generally, in the three-dimensional case the properties of the polarization vector $\boldsymbol{P}$ are expressed by the following equations:

$$\sigma_\mathrm{p} = \boldsymbol{P} \cdot \hat{\mathbf{n}}; \quad \rho_\mathrm{p} = -\mathrm{div}\,\boldsymbol{P}, \tag{7.18}$$

where $\hat{n}$ is the unit vector normal to the surface of the dielectric and ρ_p is the volume density of the polarization charges. The relationship between $\boldsymbol{P}$ and ρ_p is analogous that in Eq. 2.10, which relates the electric field $\boldsymbol{E}$ to the volume density ρ of the free charges.

Another equivalent way of looking at the electric field between the two plates in Fig. 2.5 consists in regarding it as created in vacuum by a *true surface charge density* equal to $(\sigma_1 + \sigma_{1,\mathrm{p}})$ on one plate and by an equal and opposite true surface charge density $(\sigma_2 + \sigma_{2,\mathrm{p}})$ on the other plate:

$$E_x = \frac{\sigma_1 + \sigma_{1,\mathrm{p}}}{\varepsilon_0} = \frac{\sigma_1 - P_x}{\varepsilon_0} \quad \rightarrow \quad \sigma_1 = \varepsilon_0 E_x + P_x. \tag{7.19}$$

In this geometrically simple situation, σ_1 coincides with the only nonzero component along the x-axis of another important vector of electrostatics, the *electric displacement* $\boldsymbol{D}$. In the three-dimensional case, Eq. 7.19 takes a more general form:

$$\boldsymbol{D} = \varepsilon_0 \boldsymbol{E} + \boldsymbol{P}. \tag{7.20}$$

In the presence of polarization charges, $\boldsymbol{D}$ obeys an equation entirely analogous to Poisson's equation for the electric field $\boldsymbol{E}$ in vacuum (cf. Eq. 2.10):

$$\mathrm{div}\ \boldsymbol{D} = \mathrm{div}\left(\varepsilon_0 \boldsymbol{E} + \boldsymbol{P}\right) = \rho, \tag{7.21}$$

where ρ is the volume density of the true charges. In a homogeneous and isotropic medium, $\boldsymbol{D}$ is proportional to $\boldsymbol{E}$. To verify this, let us again consider the parallel plate capacitor, for which the $\sigma_1/(\sigma_1 + \sigma_{1,\mathrm{p}})$ ratio was defined as the dielectric constant ε in Eq. 2.17. Recalling that σ_1 equals D_x and $\sigma_{1,\mathrm{p}}$ equals $-P_x$, and taking Eq. 7.19 into account, we obtain

$$\varepsilon \equiv \frac{\sigma_1}{\sigma_1 + \sigma_{1,\mathrm{p}}} = \frac{D_x}{\sigma_1 - P_x} = \frac{D_x}{\varepsilon_0 E_x} \quad \rightarrow \quad D_x = \varepsilon_0 \varepsilon E_x. \tag{7.22}$$

Analogously, for the three-dimensional case, we have

$$\boldsymbol{D} = \varepsilon_0 \varepsilon \boldsymbol{E}. \tag{7.23}$$

As opposed to the polarization $\boldsymbol{P}$ in a dielectric, which has the same direction as the constant electric field $\boldsymbol{E}$ that generates it, that of the free electrons in a metal has an opposite direction with respect to the corresponding time-harmonic electric field. In the latter case, the electric displacement is obtained by replacing $\boldsymbol{P}$ from Eq. 7.17 into Eq. 7.20, yielding

$$\boldsymbol{D} = \varepsilon_0 \boldsymbol{E} + \boldsymbol{P} = \varepsilon_0 \left(1 - \frac{ne^2}{\varepsilon_0 m\omega^2}\right) \boldsymbol{E} \equiv \varepsilon_0 \left(1 - \frac{\omega_{\mathrm{p}}^2}{\omega^2}\right) \boldsymbol{E}$$
$$\equiv \varepsilon_0 \varepsilon_m(\omega) \boldsymbol{E} \quad \text{with } \omega_{\mathrm{P}} \equiv \left(\frac{ne^2}{\varepsilon_0 m}\right)^{1/2}. \tag{7.24}$$

The quantity ω_{P} is called the *plasma frequency* of the free electron gas and the proportionality constant between $\boldsymbol{D}$ and $\boldsymbol{E}$, $\varepsilon_{\mathrm{m}}(\omega)$, is known as the *dielectric function*. The expression for $\varepsilon_{\mathrm{m}}(\omega)$ in a metal is clearly negative at frequencies $\omega < \omega_{\mathrm{p}}$, where metals retain their metallic character. Conversely, the dielectric function in a dielectric medium is always positive, since it is given by the dielectric constant ε (cf. Eq. 2.18, with $\sigma_1 = D_x$). The plasma frequency can be regarded as the natural frequency of a free oscillation of the electron gas. The quanta of these charge oscillations are called *volume plasmons*, to be distinguished from *surface plasmons*, which come into play in SPR. For many metals, the experimental value of the plasma frequency lies in the ultraviolet region.

Stating that at low frequencies the dielectric function of metals is real and negative involves an assumption. In fact, at low frequencies, the wavelength λ of the electromagnetic radiation cannot be regarded as negligibly small with respect to the mean free path of electrons, and their oscillatory motion is appreciably damped. Hence, the left-hand side of the equation of motion (Eq. 7.13) should include a term proportional to $m\,\mathrm{d}\mathbf{x}/\mathrm{d}t$, which accounts for the damping. This generates an imaginary contribution to $\varepsilon_{\mathrm{m}}(\omega)$, which increases with decreasing frequency and was ignored for simplicity. The simplified expression for $\varepsilon_{\mathrm{m}}(\omega)$ in Eq. 7.24 would predict a positive value for $\omega > \omega_{\mathrm{p}}$. However, for noble metals (e.g., Au, Ag, Cu) the applicability of the plasma model breaks down at these frequencies, due to the occurrence of interband transitions. In practice, a gradual increase in frequency causes a decrease in the negative value of the real component of $\varepsilon_{\mathrm{m}}(\omega)$, which approaches zero while remaining negative. Moreover, the imaginary component increases appreciably over the region of interband transitions, instead of decreasing with increasing frequency as predicted by the plasma model (Maier, 2007). In spite of the rough assumptions involved in the derivation of Eq. 7.24, what is important for our purposes is the concept that the dielectric function of a metal conductor depends strongly upon the frequency and that its real component is negative at the frequencies of the visible region of the spectrum. In what follows, the imaginary component of the dielectric function of a metal is ignored and $\varepsilon_{\mathrm{m}}(\omega)$ denotes the sole real component. This corresponds to an ideal situation in which the motion of the free electrons within the metal is undamped.

Let us now consider an electromagnetic wave that impinges on the smooth planar interface between a dielectric, nonabsorbing half-space with positive real dielectric constant ε_{d} and an adjacent metal half-space characterized by a dielectric function $\varepsilon_{\mathrm{m}}(\omega)$. Let the $x-y$ and $x-z$ planes define the planes of reflection and of incidence, respectively, as in Fig. 7.1. We already stated that the requirement of metallic character implies a negative value for the real component of $\varepsilon_m(\omega)$, which in turn requires the frequency ω to be $<\omega_p$. This simple geometry may induce the incident electromagnetic wave to generate a particular wave confined to the interfacial region and propagating along the x-direction, with an exponential decay in the perpendicular z-direction. This evanescent surface wave, which is guided along the interface in much the same way that light can be guided by an optical fiber, is called *surface plasmon polariton* (SPP).

If the incident electromagnetic wave is p-polarized, the only nonzero electric field components of the induced SPP lie in the plane of incidence, $x-z$. On the other hand, the magnetic field of the SPP, being normal to the corresponding electric field, has only one nonzero component normal to the plane of incidence, that is, along the y-axis. Upon solving the differential equations of the electromagnetic wave theory as applied to SPPs, the dependence of the two electric field components, E_x and E_z, and of the single magnetic field component, H_y, upon the z coordinate turns out to be expressed by the product $\exp(i\beta x)\exp(-k_d z)$ for the half-space ($z > 0$) occupied by the dielectric, and by the product $\exp(i\beta x)\exp(k_m z)$ for the half-space ($z < 0$) occupied by the metal (Maier, 2007). Moreover, these field components do not vary along the y-axis. The $\exp(i\beta x)$ exponential factor indicates that the field components exhibit harmonic propagation along the x-axis; β, called the *propagation constant*, is the component of the surface wave vector along this propagation direction. The signs in the z-dependent exponents are chosen in such a way as to predict an exponential decay along both positive and negative directions of the z-axis, provided the parameters k_d and k_m are both real and positive. The parameters k_d and k_m are the components of the surface wave vector perpendicular to the interface between the two media, and their reciprocals define the evanescent decay length of the electric and magnetic fields perpendicular to this interface.

Substituting the complete expressions of E_x, E_z, and H_y into the appropriate boundary conditions yields the two following expressions:

$$\frac{k_d}{k_m} = -\frac{\varepsilon_d}{\varepsilon_m}, \tag{7.25a}$$

$$\beta = \frac{\omega}{c}\sqrt{\frac{\varepsilon_d \varepsilon_m}{\varepsilon_d + \varepsilon_m}}. \tag{7.25b}$$

Here, ε_m is the real component of the dielectric function of the metal, $\omega = 2\pi\nu$ is the angular frequency, and c is the speed of light in vacuum. According to Eq. 7.25*a*, ε_d and ε_m must have opposite signs for both k_d and k_m to assume positive values. This implies that a surface wave can exist only at interfaces between a dielectric insulator and a metal conductor. Equation 7.25b expresses the dependence of the propagation constant β upon the frequency of the surface wave. Hence, as such, it represents a *dispersion relation*, namely a relationship between the wave vector of a radiation and its frequency. In passing, it should be noted that if the incident electromagnetic radiation propagating in the dielectric and impinging on the dielectric/metal interface is s-polarized, the solution of the differential equations of the electromagnetic wave theory as applied to SPPs allows positive

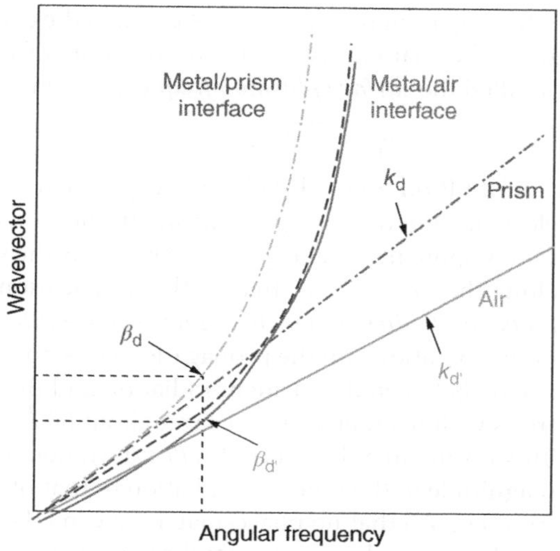

Figure 7.5 Dispersion relations at metal/prism and metal/air interfaces and prism coupling to surface plasmon polaritons.

values for both k_d and k_m only if the amplitudes of the pertinent electric and magnetic fields equal zero. In other words, SPPs can only be generated by incident p-polarized waves.

Let us now examine the properties of SPPs by taking a closer look at their dispersion relation. The dash-dotted curve in Fig. 7.5 shows a plot of the propagation constant β at a metal/prism interface against the angular frequency ω. At ω values much less than the plasma frequency ω_p, the dielectric function $\varepsilon_m(\omega)$ assumes large negative values in view of Eq. 7.24. In this case, $|\varepsilon_m|$ is $\gg \varepsilon_d$ and Eq. 7.25 for β simplifies as follows:

$$\beta \cong \frac{\omega}{c}\sqrt{\varepsilon_d}. \tag{7.26}$$

Plotting this expression in Fig.7.5 yields the dash-dotted straight line of slope $\sqrt{(\varepsilon_d)}/c$, tangent to the β versus ω plot at zero frequency. Recalling that $\sqrt{(\varepsilon_d)}$ is equal to the refractive index n of the prism, which is in turn equal to c/v_p, where v_p is the phase velocity of the electromagnetic wave in the prism, it is apparent that this straight line is also the plot of the magnitude, k_d, of the incident wave vector propagating in this transparent dielectric:

$$\frac{\omega}{c}\sqrt{\varepsilon_d} = \frac{\omega}{c}n = \frac{\omega}{v_p} = \frac{2\pi\nu}{v_p} = \frac{2\pi}{\lambda} = k_d. \tag{7.27}$$

With increasing frequency, the propagation constant β departs progressively from the k_d versus ω straight line, ultimately tending to infinity. Thus, by replacing $\varepsilon_m(\omega)$ from Eq. 7.24 into Eq. 7.25b, we obtain

$$\beta = \frac{\omega}{c}\sqrt{\frac{\varepsilon_d\varepsilon_m}{\varepsilon_d+\varepsilon_m}} = \frac{\omega}{c}\sqrt{\frac{\varepsilon_d\left(1-\omega_p^2/\omega^2\right)}{\varepsilon_d+\left(1-\omega_p^2/\omega^2\right)}} = \frac{\omega}{c}\sqrt{\frac{\varepsilon_d\left(\omega^2-\omega_p^2\right)}{\left(\varepsilon_d+1\right)\omega^2-\omega_p^2}}. \tag{7.28}$$

It is apparent that β tends to $+\infty$ when the expression $[(\varepsilon_d + 1)\omega^2 - \omega_p^2]$ tends to zero. The characteristic value of the plasma frequency at which β goes to infinity is called the *surface plasmon frequency* and is given by

$$\omega_{SP} = \omega_p / \sqrt{\varepsilon_d + 1}. \tag{7.29}$$

An SPP on a metal/dielectric $x-y$ planar interface is exited by a p-polarized electromagnetic wave propagating in the $x-z$ plane of incidence if its wave vector component, k_x, along the x-axis equals the wave vector component of the SPP along the same x-axis, that is, the propagation constant β. This situation, called *phase-matching*, determines a *destructive interference* between the reflected part of the radiation and the propagation constant. In other words, an energy transfer occurs between the incident radiation and the collective oscillations of free electrons within a metal surface layer about 50 Å thick, that is, the *surface plasmons*. This results in a decrease of *reflectivity*, defined as the square of the ratio of the magnitude of the reflected radiation to that of the incident radiation. It is evident from Fig. 7.5 that no phase-matching can take place when a wave propagating in a dielectric medium impinges directly on a planar interface between this medium and a metal. In fact, k_d is lower than β_d at any given ω value, and $k_{d,x}$ is even lower, being equal to $k_{d,x} = k_d \sin\theta_i$, where θ_i is angle of incidence. The incident radiation is totally reflected by the dielectric/metal interface (*total internal reflection*) even by approaching θ_i to its maximum $\pi/2$ value.

To achieve phase-matching, one must cause the radiation propagating in a dielectric medium d of refractive index n_d to reach the interface between the metal and a medium of dielectric constant $n_{d'} < n_d$. This necessarily requires the addition of a third, very thin layer. One configuration, called the *Otto geometry*, uses a thin air layer, whereas another one, referred to as the *Kretschmann geometry*, adopts a thin metal film. The two-layer system consisting of two semi-infinite phases that was examined so far can be regarded as a limiting case of this three-layer system, in which the thickness of the intermediate thin layer tends to zero. Hence, the conclusions drawn for the two-layer system remain substantially valid.

In the Otto geometry, a light wave propagating in a prism with a high refractive index n_d (1.5–1.6) impinges on its interface with a thin air film, whose refractive index $n_{d'}$ equals 1.000293. If the angle of incidence is higher than the critical value above which total reflection takes place, the incident wave produces a wave that propagates along the interface, while decaying very rapidly in the direction perpendicular to it. In view of its character, such a wave is again referred to as an evanescent wave, although it should not be confused with an SPP. If the thickness of the air layer is chosen properly (typically few microns), the evanescent wave can couple with a surface plasmon at the underlying metal/air interface (see Fig. 7.6A) generating an SPP. For the coupling to occur, the wave vector of the evanescent wave must equate the propagation constant $\beta_{d'}$ of the SPP. That this is now possible is shown in Fig. 7.5, where the propagation constant $\beta_{d'}$ at the metal/air interface and the magnitude $k_{d'}$ of the wave vector in air are plotted against ω and represented by solid curves. Since n_d is greater than $n_{d'}$, the slope of the k_d versus ω plot is higher than that of the $k_{d'}$ versus ω plot, in view of Eq. 7.27. Hence,

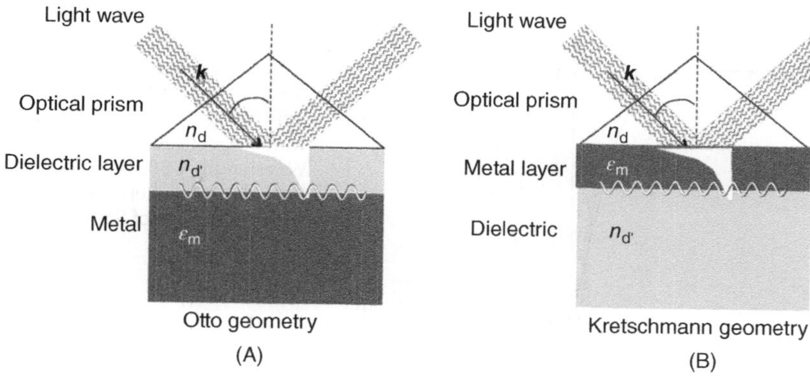

Figure 7.6 Excitation of surface plasmon polaritons in the Otto and Kretschmann geometries.

k_d in Fig. 7.5 is greater than the propagation constant $\beta_{d'}$ at the metal/air interface at usual angular frequencies (see the dashed vertical segment of constant ω). Phase-matching can then be realized by decreasing the angle of incidence until $k_{d,x}$ comes to coincide with $\beta_{d'}$.

In the Kretschmann geometry, the additional thin film consists of a thin metal layer interposed between a high refractive index prism and a phase of lower refractive index (usually air or an aqueous solution). In experiments with biomimetic membranes, the metal layer consists of a gold film evaporated on a prism and supporting the membrane in direct contact with an aqueous solution. When a light wave propagating in the prism impinges on the metal film, a part of the light is reflected back into the prism and a part propagates in the metal in the form of an inhomogeneous electromagnetic wave (Fig. 7.6B). This phenomenon, which also occurs with Otto geometry, is called *attenuated total reflection* (ATR). This inhomogeneous wave decays exponentially in the direction perpendicular to the prism/metal interface and is again referred to as an evanescent wave. If the metal film is sufficiently thin (about 50 nm for light in the visible spectrum), the evanescent wave penetrates through the metal film and couples with a surface plasmon at the outer boundary of the metal film.

As a matter of fact, the presence of the metal film alters the propagation constant of the SPP with respect to that, $\beta_{d'}$, predicted for the interface between the dielectric of lower refractive index $n_{d'}$ (air or aqueous solution) and the metal. Thus, it increases the propagation constant by an amount $\Delta\beta_{d'}$, which accounts for the finite thickness of the film and the presence of the prism (Homola, 2006). The phase-matching is, therefore, attained when the wave vector component, $k_{d,x} = k_d \sin\theta_i$, of the wave propagating in the prism is made equal to $\beta_{d'} + \Delta\beta_{d'}$ by adjusting the angle of incidence θ_i:

$$k_d \sin\theta_i = k_{d,x} = \beta_{d'} + \Delta\beta_{d'} = \frac{\omega}{c}\sqrt{\frac{\varepsilon_{d'}\varepsilon_m}{\varepsilon_{d'} + \varepsilon_m}} + \Delta\beta_{d'}. \tag{7.30}$$

The curve of the propagation constant, $\beta_{d'} + \Delta\beta'$, against ω lies between those of β_d and $\beta_{d'}$ (see the dashed curve in Fig. 7.5).

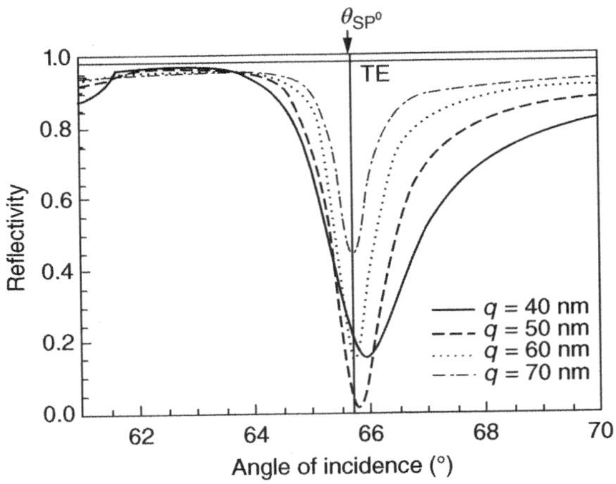

Figure 7.7 Reflectivity against the angle of incidence for four different thickness values, q, of the gold film. Prism ($n_d = 1.51$); gold film ($\varepsilon_w = -25 + 1.44i$); and water ($n_{d'} = 1.329$). Wavelength = 800 nm. Source: Homola (2006). Reproduced with permission of Springer.

Figure 7.7 exemplifies the behavior of the reflectivity R against the angle of incidence for four different thicknesses of a gold film interposed between a prism and water. At lower angles of incidence, R is only slightly lower than the unit value expected for total reflection, due to some slight radiation absorption in the prism. As θ approaches the condition for phase-matching, as expressed by Eq. 7.30, R decreases attaining a deep minimum at a θ_i value called *resonant angle of incidence*. The depth of the reflectivity minimum depends on the thickness of the metal film. The strongest excitation of a surface plasmon, yielding a zero value of reflectivity, occurs for a well-defined thickness of the metal film.

Prism, metal, solution, and film thickness being left unaltered, the value of the resonant angle of incidence depends on the presence of any molecules capable of accumulating on the surface of the metal film. The weak penetration of the evanescent wave renders the resonant angle of incidence particularly sensitive to the amount of material that accumulates on the metal film. This is the principle on which *SPR sensors with angular modulation* are based (Homola and Piliarik, 2006). These sensors use a monochromatic light wave and measure the reflectivity as a function of the angle of incidence. The sensor output is the resonant angle of incidence, which can be calibrated to the refractive index $n_{d'}$ of the dielectric d′. The sensor measures the change, $\Delta n_{d'}$, in the refractive index with varying the molecular mass accumulating on the sensor surface, starting from the beginning of the interaction to be monitored. Under these conditions, $\Delta n_{d'}$ is directly proportional to the surface concentration of the analyte. In general, as the analyte concentration in the bulk solution increases, the sensor output tends to a maximum limiting value, which corresponds to the maximum surface concentration attainable by the analyte. The latter may consist, for example, of the specific ligand of a receptor previously anchored to the surface of the sensor; in this case, the sensor response attains a limiting value as soon as all receptor molecules are bound by the ligand.

SPR is a surface-sensitive technique particularly appropriate for measuring the optical thickness of ultrathin films adsorbed on metals, such as thiolipid monolayers and lipid bilayers tethered to gold, or lipid monolayers self-assembled on gold-supported thiolipid monolayers. It is also conveniently employed for following the kinetics of vesicle fusion, with the formation of lipid monolayers on hydrophobic surfaces or lipid bilayers on hydrophilic surfaces. The angular position of the minimum of the SPR *reflectivity curves* (i.e., the curves of reflectivity versus the angle of incidence) is critically dependent on the thickness of the layer adsorbed on the support surface. When macromolecules assemble onto a dielectric layer at the metal surface, the reflectivity minimum is shifted to a higher angle.

Physical layer thicknesses can be calculated by quantitative modeling of the reflectivity curves. Modeling of the SPR response is carried out by using a computer program that calculates reflection and transmission of p-polarized light in a stratified structure (a stack of parallel slabs) sandwiched between a semi-infinite medium (the prism) and a semi-infinite ambient medium (the sample solution). All slabs must be assumed to be homogeneous and isotropic. The refractive index and the thickness of each slab are used as input data. The calculation can be based, for example, on a 2×2 scattering matrix derived by using the Fresnel equations (cf. Eq. 7.11). Thus, for instance, a gold-supported tBLM consisting of a thiolipid monolayer, with a self-assembled lipid monolayer on top of it, can be simulated by two slabs of different refractive index and thickness (Schmidt et al., 1998). The exact refractive index of the compounds forming layers on the metal surface is unknown; however, reasonable approximations for refractive indices are $n = 1.45$ for proteins and $n = 1.5$ for lipids. Once the refractive indices are established, the layer thickness can be extracted from a comparison between experimental and calculated reflectivity curves.

By recording the reflectivity at a particular fixed angle of incidence close to resonance, the kinetics of adsorption at an interface can be monitored as a function of time. The reflectivity at the fixed angle increases with time from its value prior to the adsorption of the molecules, attaining a maximum limiting value when the adsorption process terminates. An example is provided by the reflectivity versus time curve in Fig. 7.8 (Vockenroth et al., 2008b) relative to the functional characterization of vesicle fusion on a gold-supported thiolipid monolayer and of the subsequent incorporation of the channel-forming toxin α-hemolysin into the resulting tBLM. The addition of the vesicle dispersion results in a substantial increase in thickness, caused by the gradual self-assembly of a lipid monolayer and revealed by the strong increase of the reflectivity. This is accompanied by a dramatic increase in the resistance of the lipid bilayer, as measured by EIS. Upon addition of α-hemolysin, a further slight increase in thickness denotes the incorporation of the toxin into the membrane. This is confirmed by the strong decrease in the membrane resistance by more than three orders of magnitude, which could hardly be ascribed to a mere nonspecific adsorption of the protein on the polar head region of the lipid monolayer.

A variant of SPR spectroscopy was developed by Salamon et al. (1997). It involves the modification of the Kretschmann geometry by the interposition of a SiO_2 layer, from 400 to 800 nm thick, between an Ag film deposited on the prism and a lipid bilayer deposited on the outer surface of the SiO_2 layer. This variant,

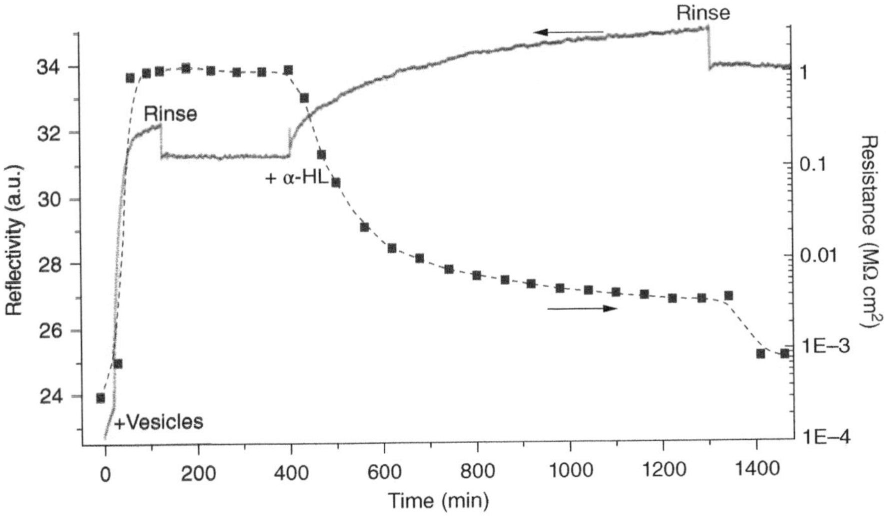

Figure 7.8 Plots of reflectivity (solid curve) and resistance (dotted curve and squares) at a gold-supported hexaethyleneoxy-based thiolipid monolayer, following addition of DPhPC vesicles, rinsing and further addition of 180 nM α-hemolysin. Source: Vockenroth et al. (2008b). Reproduced with permission of American Chemical Society.

referred to as *coupled plasmon-waveguide resonance* (CPWR) *spectroscopy*, determines the appearance of an s-polarized resonance component, which is precluded in a conventional SPR experiment. This system allows an estimate of the refractive index and the extinction coefficient of the composite film not only with p-polarized light, as in conventional SPR, but also with s-polarized light, that is, in the direction parallel to the bilayer plane.

7.3 Infrared Spectroscopy

Infrared (IR) spectroscopy is commonly employed to determine the structure of organic molecules in solution by identifying specific chemical bonds. An infrared radiation is absorbed whenever its frequency matches that of the vibration of some chemical bond, provided that such a vibration determines a change in the permanent dipole moment. The number of vibrational modes is determined by the degrees of freedom of the molecule and, hence, depends on its geometry. Nowadays, modern IR techniques are also capable of probing the behavior of organic molecules adsorbed at electrode/solution interfaces, including those forming biomimetic membranes. Two techniques that are commonly employed for this purpose are IR-attenuated total reflection (IR ATR) spectroscopy and IR reflection absorption (IRRA) spectroscopy.

IR ATR spectroscopy uses the same Kretschmann geometry employed in SPR with visible radiations (cf. Section 7.2) (Osawa et al., 1986). The cell is equipped with an optical window (usually a Si or Ge hemisphere) that works in the regime of total reflection. The electrode consists of a thin metal film (<100 nm) directly deposited onto the optical window. A film of adsorbed molecules is formed on the metal film, as it is brought in contact with an aqueous solution. The metal film must be much thinner than the wavelength of the incident IR radiation to ensure good coupling between the optical window and the aqueous medium on the opposite side of the metal film. The evanescent wave of the attenuated totally reflected IR radiation penetrates the metal film and decays exponentially with distance inside the aqueous solution, such that the species adsorbed on the metal film show more absorption compared with those in the bulk. Unfortunately, the sputter metal deposition methods yield rough polycrystalline films. Consequently, the IR absorbance at the metal surface is not uniform across the film and is not representative of the average adsorbed molecular film.

IR spectroscopy investigations on organic molecules adsorbed at single-crystal electrodes can be carried out by applying the IRRAS technique. In this case, the electrode is pressed against an optical window made from BaF_2, CaF_2, LiF, or ZnSe in contact with an electrolyte solution, until it is separated from the window by a thin layer of solution, whose thickness amounts to several micrometers. The unavoidable penetration of the IR beam across the resulting thin-layer cell, before reaching the electrode surface, causes the spectrum of the organic film on this surface to be superimposed on the spectrum of the aqueous electrolyte. It would be hard to correct for the background spectrum by acquiring a spectrum in the absence of the adsorbed organic film through the reproduction of exactly the same thin-layer cell thickness and all other experimental conditions employed in its presence. Correction for the background spectrum is, therefore, made by either polarization modulation or potential modulation. The former technique is called *polarization modulation Fourier transform infrared reflection–absorption spectroscopy* (PM FTIRRAS), or more briefly PM IRRAS, whereas the second is called *subtractively normalized interfacial Fourier transform infrared spectroscopy* (SNIFTIRS) (Zamlynny, 2002; Zamlynny and Lipkowski, 2006; Leitch, 2014). Grating spectrometers, which utilize prisms and diffraction gratings as monochromators, have now been completely replaced by *Fourier transform infrared* (FTIR) instruments. With grating spectrometers, the light is dispersed by the monochromator, and the detector receives each wavelength separately, with the result that the intensity is low. On the other hand, with FTIR instruments all wavelengths are collected and received simultaneously by the detector, and this high energy throughput determines a high signal-to-noise ratio.

In a PM IRRAS experiment, the wave polarization is modulated between the directions parallel (p-polarization) and perpendicular (s-polarization) to the plane of incidence of the IR beam. Figure 7.9 shows the incident electric field vectors of the p-polarized ($\boldsymbol{E}_{ip}$) and s-polarized waves ($\boldsymbol{E}_{is}$), as well as the electric field vectors, $\boldsymbol{E}_{rp}$ and $\boldsymbol{E}_{rs}$, of the corresponding reflected waves. The electric field of the electromagnetic wave at the surface is a vectorial sum of the fields of the incident and reflected waves. Depending on the phase shift of the reflected wave, the sum gives enhancement or attenuation of the electric field. The phase

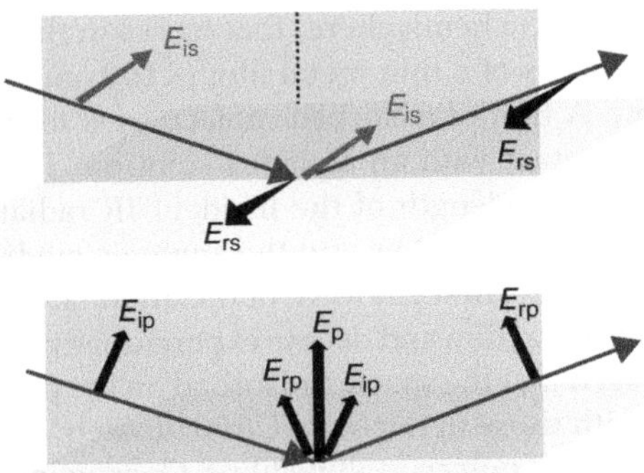

Figure 7.9 Phase shift of the electric field vector for s-polarized (top) and p-polarized (bottom) radiation at an isotropic metal surface.

shift of the electric field vector of the s-polarized wave upon reflection is nearly 180° for all the angles of incidence, and hence the resulting s-polarization, $\boldsymbol{E}_\mathrm{s} = \boldsymbol{E}_\mathrm{is} + \boldsymbol{E}_\mathrm{rs}$, is practically zero (*destructive interference*). In contrast, the phase shift of the electric field vector of the p-polarized wave strongly depends on the angle of incidence. In practice, at an angle of incidence of 80.7° (grazing angle of incidence) $\boldsymbol{E}_\mathrm{ip}$ and $\boldsymbol{E}_\mathrm{rp}$ add in phase, and the resulting p-polarization, $\boldsymbol{E}_\mathrm{p} = \boldsymbol{E}_\mathrm{ip} + \boldsymbol{E}_\mathrm{rp}$, is almost twice $\boldsymbol{E}_\mathrm{ip}$ (*constructive interference*). However, it should be noted that the x-component of $\boldsymbol{E}_\mathrm{p}$ is also canceled at the metal surface.

Only the adsorbed molecules having a nonzero component of the *transition dipole moment* oriented along the z-axis (i.e., normally to the plane of reflection) can absorb the p-polarized incident radiation. The transition dipole is proportional to the derivative of the dipole moment with respect to a change of the length of its vibrational coordinates. Its magnitude and direction can be calculated *ab initio* by calculating the change in the dipole moment when the vibrational coordinates are varied. Hence, the transition dipole is a property of the molecule and depends on its symmetry. The oscillating electric field $\boldsymbol{E}_\mathrm{p}$ continuously changes the coordinates, thus providing information on the orientation of the transition dipole moment, but only if the latter is not parallel to the plane of reflection $x-y$, because the $\boldsymbol{E}_\mathrm{p}$ component along the x-axis equals zero.

In contrast with the adsorbed molecules, the molecules dissolved in the thin-layer cell absorb both the p and s components of the radiation. Thus, the s-component of the modulated signal, while being insensitive to the adsorbed molecules, accounts for the absorption of the dissolved ones and of the solvent molecules, and can be used to obtain the background spectrum. On the other hand, the p-polarized component is enhanced at the metal surface by constructive interference and is utilized to obtain the sample spectrum. Modulation between s- and p-polarization is made at a high frequency (100 kHz) using the PEM described in Section 7.1, and ultimately yields the reflection spectrum expressed by the following equation:

$$\frac{\Delta R}{\langle R\rangle} = \frac{|R_\mathrm{s} - R_\mathrm{p}|}{(R_\mathrm{s} + R_\mathrm{p})/2}. \tag{7.31}$$

R_s and R_p are the reflectivity coefficients of the s- and p-polarized components of the modulated IR beam from the thin-layer cell. In general, R_p accounts for the reflectivity of all adsorbed and dissolved molecules in the thin-layer cell, and R_s only for that of the dissolved molecules. In Eq. 7.31, their difference is divided by the arithmetic mean of the two reflectivities. For the randomly oriented dissolved molecules to respond to s-polarization in the same way and to the same extent as they respond to p-polarization, the s- and p-polarized components of the incident IR beam penetrating the thin-layer cell must be equal; this is realized by linearly polarizing the beam impinging on the PEM at 45° with respect to the its main axis, as shown in Fig. 7.2. It should be noted that the PM IRRAS is sensitive not only to the surface species but also to the molecules that are in close vicinity of the metal surface, at a distance less than 1 μm. Consequently, the background due to the aqueous electrolyte cannot be completely removed from the $\Delta R/\langle R \rangle$ spectrum.

The experimental signal received by the detector after the passage of the beam through the PEM and the thin-layer cell alternates between the s- and p-polarizations according to the equation:

$$I_D(t) = \frac{I_s + I_p}{2} + \frac{I_s - I_p}{2} \cos\left(\frac{\lambda_0}{\lambda}\pi \cos\left(\omega_m t\right)\right). \tag{7.32}$$

Here, I_p and I_s are the intensities of the p- and s-polarized reflected radiation, ω_m is the angular frequency of the PEM, t is the time, λ is the wavelength of the polarized radiation and λ_0 is the particular wavelength value at which the amplitude of the voltage V_m applied to the PEM makes the maximum phase shift imposed by the PEM equal to π. The spectra obtained using the PM IRRAS can also be represented by the following equation:

$$\Delta S \equiv \frac{\Delta I}{\langle I \rangle} = \frac{\left|I_s - I_p\right|}{\left(I_s + I_p\right)/2}. \tag{7.33}$$

This equation, which expresses the so-called *differential reflectance spectrum* of the adsorbed surface species, is identical with Eq. 7.31. Thus, it can be obtained from the latter equation by multiplying both its numerator and its denominator by the common value I_i of the s- and p-polarized components, $I_{i,s}$ and $I_{i,p}$, of the incident radiation, in view of the definition of reflectivity coefficient. The equality $I_{i,s} = I_{i,p}$ represents a constant requirement in PM IRRAS measurements.

To obtain the differential reflectance spectrum, the signal from the detector must be decoded. One demodulation technique uses a lock-in amplifier and a low-pass filter. A more convenient demodulation technique uses real-time sampling electronics to detect not only the second but also higher harmonics of the PEM frequency to obtain the intensity difference and the average signals. After demodulation, the spectra must be further processed to remove the artifacts introduced by the PEM, in order to be able to extract quantitative information.

As a rule, ΔS is plotted against the *wavenumber* $\overline{\nu} = 1/\lambda$, defined as the number of wavelengths per unit length. For example, Fig. 7.10 shows a ΔS versus $\overline{\nu}$ plot for the C=O stretching band at a Au(111)-supported DOPC bilayer in a D_2O

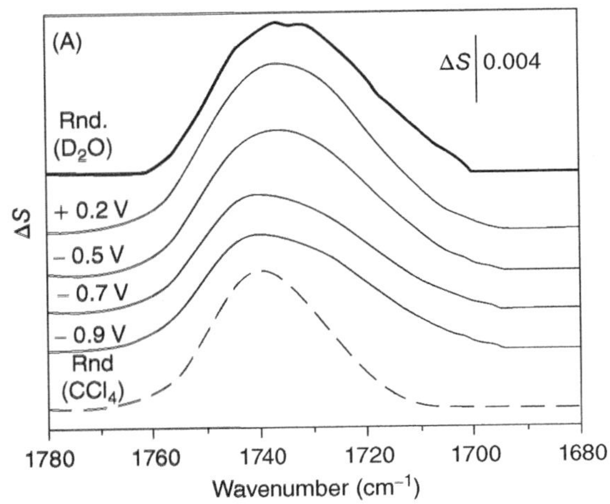

Figure 7.10 PM FTIRRA spectra in the ν(C=O) stretching region of a DOPC bilayer at an Au(111) electrode in 0.1 M NaF/D_2O solution at potentials indicated in the figure. Thick lines are spectra calculated for a 4.5 nm thick DOPC bilayer in different environments: (solid line) D_2O; (dashed line) CCl_4 solution. Source: Zawisza et al. (2003). Reproduced with permission of Elsevier.

solution of 0.1 M NaF, at different potentials versus Ag/AgCl (3 M KCl) (Zawisza et al., 2003). The top curve (thick solid line) was obtained from DOPC vesicles dispersed in D_2O and the bottom one (dashed line) from a DOPC solution in CCl_4. The composite band for the dispersion of DOPC vesicles in D_2O may be deconvoluted into two bands with maxima at 1741 and 1727 cm^{-1}, which are assigned to the nonhydrated and hydrated carbonyl ester groups, respectively. This assignment is consistent with the band of DOPC dissolved in the aprotic solvent CCl_4 being dominated by the peak at about 1740 cm^{-1}. The shift in the C=O stretching band center for the adsorbed DOPC bilayer toward increasing wavenumber values with a negative shift in the applied potential denotes a progressive decrease in the hydration of the ester groups, which accompanies the gradual detachment of the bilayer from the electrode surface.

The integrated intensity of a band in an absorption spectrum is proportional to the square of the scalar product of the corresponding transition dipole moment and the electric field $\boldsymbol{E}$ of the photon. At the metal surface, the direction of the electric field $\boldsymbol{E}_p$ of the p-polarized radiation is normal to the surface. This allows an estimate of the angle between the transition dipole moment and the surface normal. If the direction of the transition dipole with respect to the coordinates of the adsorbed molecule is known, the integrated intensity of a band may, therefore, be used to determine the orientation of the molecule at the metal surface. Since the transition dipole moment of the carbonyl stretch is known to be parallel to the C=O bond direction, which in turn is normal to the initial fragments of the β- and γ-acyl chains (see Fig. 7.11), the value of such an angle is a measure of the average orientation of the ester group. The mean value of this tilt angle with respect to the surface normal passes from about 53° to about 61° with a negative shift of the applied potential from −0.3 to −1.0 V, which determines the gradual detachment of the bilayer from the electrode surface. This corresponds to an increased alignment of the glycerol fragments of the acyl chains along the direction of the surface normal. Figure 7.11 shows a schematic diagram of the DOPC

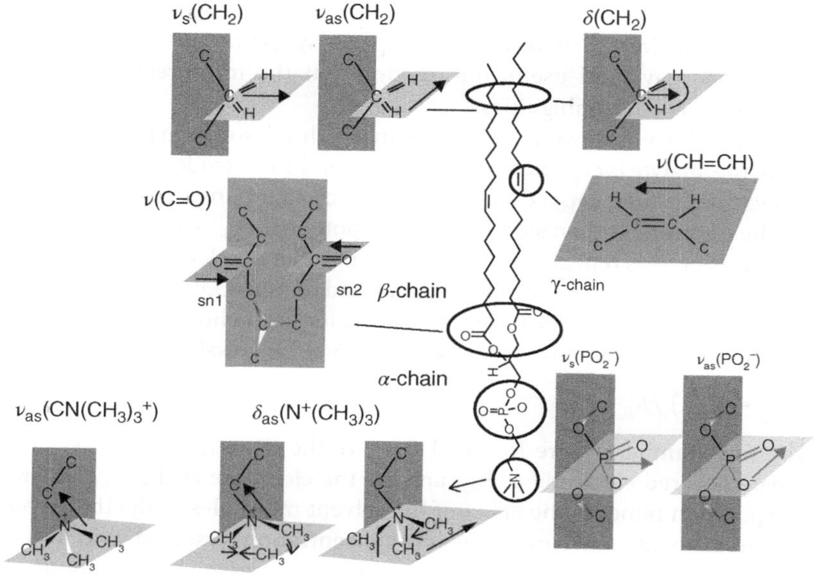

Figure 7.11 Schematic diagram of the DOPC molecule and the directions of the transition dipole moments estimated from PM IRRAS. Source: Zawisza et al. (2003). Reproduced with permission of Elsevier.

molecules and the directions of the transition dipole moments estimated from the integrated intensities of the different absorption bands of the PM IRRAS of an Au(111)-supported DOPC bilayer. Symmetric and asymmetric stretch bands are denoted by ν_s and ν_{as}, respectively, whereas bending bands are denoted by δ. Insertion of a tethered 1-thio-β-D-glucose monolayer between the Au(111) surface and the DMPC bilayer creates a hydrophilic cushion that decreases the tilt of the acyl chains and strengthens the chain–chain interactions, as verified by PM IRRAS (Matyszewska et al., 2016).

When IR reflection absorption spectroscopy is used to investigate thin films adsorbed at electrode surfaces *in situ*, the goal of subtracting the signal due to the solvent and of differentiating between the molecules adsorbed on the electrode surface and those present in the solution can be achieved by modulating the applied electric potential with SNIFTIRS, rather than the photon polarization with PM IRRAS. As a rule, PM IRRAS is to be preferred to SNIFTIRS for the investigation of biological material, since the potential modulation intrinsic in the SNIFTIRS technique may easily damage biomolecules adsorbed on the electrode surface, whereas PM IRRAS does not require the electrode potential to be changed. The SNIFTIRS technique involves recording the difference spectra between the upper and lower limits of the electric potential modulation (Pettinger et al., 2001), using only p-polarized radiation. The technique requires the IR spectrum of the adsorbed species of interest (the analyte) to be distinctly different between the upper and lower potential limits. If at one potential limit there is no absorption spectrum of the analyte, this requirement may be accurately satisfied. Even if there is an absorption spectrum at both potential limits, SNIFTIRS may yield useful information, but the interpretation of the results may conceal some ambiguities.

Let us consider the most favorable situation in which a desorption potential E_{des} exists, where the analyte is completely removed from the electrode surface. In this case, the electrode is first stepped to E_{des}, where a certain number of spectra are collected. The electrode is then stepped to some potential E_{ads}, where the analyte is adsorbed. This cycle is repeated for a sufficient number of selected potentials at which the analyte is adsorbed to various extents, and the spectra of all the selected potentials are collected. The SNIFTIRS spectrum for each chosen potential can be plotted as follows (Zamlynny, 2002; Zamlynny and Lipkowski, 2006):

$$\left(R_{E_{ads}} - R_{E_{des}}\right) / R_{E_{des}} \tag{7.34}$$

against the wavenumber, where $R_{E_{ads}}$ and $R_{E_{des}}$ are the IR reflectivity curves for the adsorbate covered electrode at E_{ads} and for the electrode at the E_{des} potential. This expression removes the effect of the solvent molecules in the thin-layer cell, but ignores the difference in the analyte concentration dissolved in it, which is clearly a maximum at E_{des} and decreases with an increase in the amount of adsorbed analyte at the other selected potentials. Different procedures have been adopted to correct for this difference. This requires prior knowledge about the adsorptivity of the analyte molecules on the electrode as a function of the applied potential. A theoretical SNIFTIRS spectrum of the analyte desorbed into the electrolyte solution of the thin-layer cell is then simulated. This requires knowledge of the angle of incidence, the thin-layer cell thickness and the isotropic optical constants of analyte, window material and electrode. The simulation yields the following equation:

$$\left(R_0 - R_{\Gamma_{des}}\right) / R_0 \cong \left(R_0 - R_{\Gamma_{des}}\right) / R_{\Gamma_{des}}. \tag{7.35}$$

Here, $R_{\Gamma_{des}}$ denotes the reflectivity of the stratified medium (window/electrolyte/metal) that contains a hypothetical homogenous solution generated by the dissolution in the thin-layer cell of the surface concentration Γ of the analyte at the E_{ads} potential in Eq. 7.34; R_0 is the reflectivity of the same stratified medium in the absence of the analyte. These reflectivities are calculated using the Fresnel equations in the matrix form. Since the difference between $R_{\Gamma_{des}}$ and R_0 is small, the latter quantity can be replaced by the former in the denominator of Eq. 7.35. It is readily seen that subtracting the calculated spectrum of Eq. 7.35 from the experimental SNIFTIRS spectrum of Eq. 7.34

$$\left(R_{E_{ads}} - R_{E_{des}}\right) / R_{E_{des}} - \left(R_0 - R_{\Gamma_{des}}\right) / R_{\Gamma_{des}} \cong \left[R_{E_{ads}} - \left(R_{E_{des}} - R_{\Gamma_{des}}\right) - R_0\right] / R_{E_{des}} \tag{7.36}$$

yields a SNIFTIRS spectrum in which the subtraction of the quantity $(R_{E_{des}} - R_{\Gamma_{des}})$ corrects for the analyte molecules dissolved in the thin-layer cell at the potential E_{ads}.

7.4 Neutron Reflectivity

Having zero charge and negligible electric dipole, neutrons interact with matter via nuclear forces, which have a very short range, of the order of 10^{-6} nm. Since the sizes of nuclei are typically 10^5 times smaller than the distances between them, neutrons can travel long distances in materials without being scattered or absorbed, covering distances up to 0.01–0.1 m. In view of their low mass, neutrons share the properties of waves, and have amplitude and phase. When a neutron impinges on a nucleus, its wave vector may change direction while maintaining its magnitude (*elastic scattering*), or it may change both direction and magnitude (*inelastic scattering*). The energy of the neutron is often too small to change the internal energy of the scattering nucleus; in this case, the scattering does not change the neutron energy and is elastic. However, if the nucleus is moving while the neutron is arriving and recoils during the collision, it can impart energy to the neutron or absorb energy from it, giving rise to inelastic scattering (Pynn, 2009).

A beam of neutrons can be represented by a plane wave $\Psi_i = \exp(ikx)$, where x is the propagation direction and $k = 2\pi/\lambda$ is the wave vector. The square of the amplitude of the neutron wave at any point gives the probability that the neutron will be found at that point. The range of the potential created by a nucleus is short compared with the wavelength λ of a neutron, which amounts to a few ångströms in usual scattering experiments; hence, the nucleus is effectively a *point scatterer* for neutrons. The *scattered neutron wave* is isotropic, and its wave function can be written as $\Psi_r = (-b/r)\exp(ikr)$, if the scattering nucleus is set at the origin, $r = 0$, of the coordinate system. The factor $(1/r)$ accounts for the inverse square law that applies to all wave motions; the intensity of the scattered neutron beam, expressed by the square of the amplitude of the wave function, is inversely proportional to the square of the distance from the source.

The quantity b, referred to as the *scattering length* of the nucleus, measures the strength of the interaction between the neutron and the scattering nucleus. The minus sign is arbitrary and is used so that a positive value for b denotes a repulsive interaction potential. The scattering length is a complex number, but the imaginary component is only important for nuclei that have a high absorption coefficient, such as boron and cadmium; in all other cases, it can be treated as a real quantity. Different isotopes of the same element have different b values. A remarkable example is provided by ^1H and ^2H, namely hydrogen and deuterium, with the latter often labeled D. Hydrogen has a scattering length of -3.74×10^{-5} Å and deuterium of 6.67×10^{-5} Å. The differences in scattering lengths from one isotope to another are used in various isotopic-labeling techniques to increase the amount of information available from a particular neutron-scattering experiment. The *scattering cross section*, denoted by σ, is an area related to the scattering length b by the simple relation $\sigma = 4\pi b^2$. It can be regarded as the cross-sectional area of the nucleus, as "seen" by the neutron.

During a collision between two particles, their total momentum is conserved. Noting that the momentum of a particle whose wave vector is $\boldsymbol{k}$ equals $h\boldsymbol{k}/(2\pi)$,

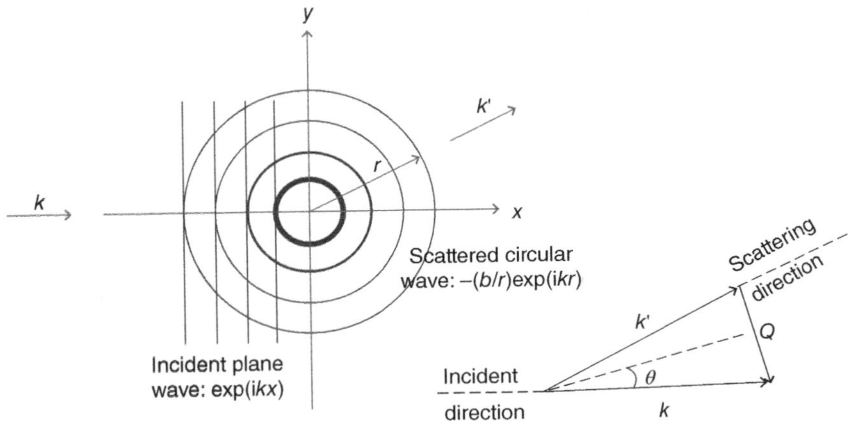

Figure 7.12 Incident plane wave, circular wave scattered by a scattering center at $r=0$, and a scattering triangle formed by the wave vectors of an incident and a scattered neutron and by the scattering vector $\boldsymbol{Q}$.

the amount of momentum given up by a neutron during its collision with a nucleus, called *momentum transfer*, is given by $h\boldsymbol{Q}/(2\pi) = h(\boldsymbol{k} - \boldsymbol{k}')/(2\pi)$, where $\boldsymbol{k}$ is the wave vector of the incident neutron and $\boldsymbol{k}'$ is that of the scattered neutron. The quantity $\boldsymbol{Q} = \boldsymbol{k} - \boldsymbol{k}'$ is known as the *scattering vector*. In the case of elastic scattering, no energy is lost and $\boldsymbol{k}$ and $\boldsymbol{k}'$ have the same magnitude. Hence, the triangle formed by the two vectors and by their difference $\boldsymbol{Q}$, called *scattering triangle*, is an isosceles triangle, as shown in Fig. 7.12. Denoting by 2θ the *scattering angle* formed by the two wave vectors $\boldsymbol{k}$ and $\boldsymbol{k}'$, simple trigonometric considerations, which are evident from the figure, yield the following relation:

$$\sin\theta = \frac{Q/2}{k} \quad \rightarrow \quad Q = 2k\sin\theta = \frac{4\pi\sin\theta}{\lambda}. \tag{7.37}$$

When an incident neutron impinges on a regular array of atoms, all atomic sites become the centers of scattered spherical waves, which overlap each other at all points in space. For a three-dimensional assembly of nuclei, the resultant scattered wave is

$$\Psi_s = -\sum_i (b_i/r)\exp(ikr)\exp(i\boldsymbol{Q}\cdot\boldsymbol{r}), \tag{7.38}$$

which includes the scattering vector $\boldsymbol{Q} = \boldsymbol{k} - \boldsymbol{k}'$, where $\boldsymbol{k}$ and $\boldsymbol{k}'$ are the wave vectors of the incoming and scattered neutrons, respectively (Jackson, 2008).

There will be some points where the disturbances from different waves reinforce one another and other points where they cancel out. These two phenomena are called *constructive* and *destructive interferences*, respectively. In practice, the waves of the scattered neutrons moving away from the array will only travel in those directions along which they interfere constructively. Constructive interference occurs when the waves *reflected* from adjacent scattering planes of atoms

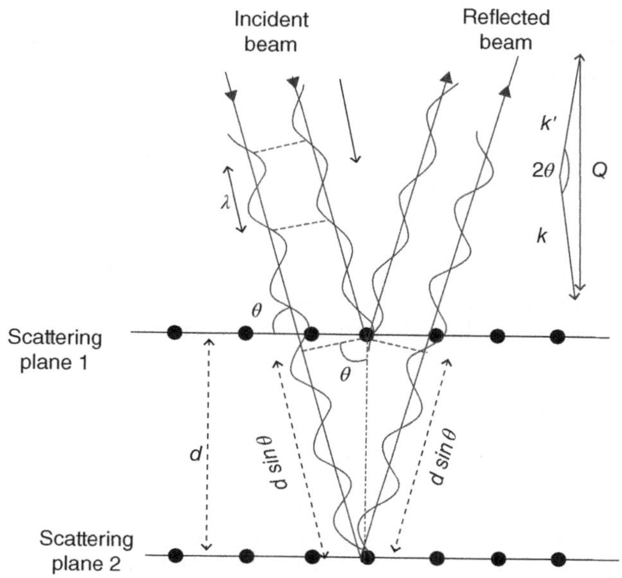

Figure 7.13 Constructive interference between waves reflected from two adjacent scattering planes.

remain in phase. This happens when the difference in distance traveled by the waves reflected from adjacent planes is an integral multiple of their wavelength. This situation is depicted in Fig. 7.13, where the extra distance traveled by the wave reflected from scattering plane 2 contains five wavelengths and is enclosed between the intersection points of the incident and reflected beams with the two dashed segments drawn normally to them. Denoting the complement of the angle of incidence by θ, since it equals one half of the scattering angle 2θ, the angle enclosed between any of the aforementioned two dashed segments and the dash-dotted segment expressing the distance d between to two adjacent scattering planes equals θ because its two sides are perpendicular to those of the angle θ. The requirement for constructive interference can, therefore, be expressed by the following equation:

$$2d \sin\theta = n\lambda, \tag{7.39}$$

where n is an integer. Not unexpectedly, this relationship is identical with the well-known *Bragg's law* for coherent scattering of X-rays from a crystal lattice.

Substituting $\sin\theta$ from Eq. 7.39 into Eq. 7.37, we obtain a further expression, $Q = 2n\pi/d$, for the magnitude of the scattering vector $\boldsymbol{Q}$, which is normal to the scattering planes. The quantity d provides a measure of the size of the structure that one wishes to probe by neutron reflectivity. Hence, large structures require the adoption of small Q values. In view of Eq. 7.37, this condition can be satisfied by increasing λ or by decreasing θ, and by modulating one of these two quantities. A common practice involves choosing neutrons with a constant and relatively high neutron wavelength (e.g., ≥ 5 Å) and in varying the scattering angle

2θ, while maintaining it at a low level (e.g., less than 1°). This technique, which allows the examination of large structures, such as colloids or viruses, is known as *small-angle neutron scattering* (SANS). The directly measured quantity in a scattering experiment is the *differential cross section*, which is defined by

$$\frac{d\sigma}{d\Omega} = \frac{\text{number of neutrons scattered per second into } d\Omega \text{ in a given direction}}{\text{number of incident neutrons per unit area per second } \times d\Omega}, \tag{7.40}$$

where $d\Omega$ is an infinitesimal solid angle. Upon considering the velocity of the scattered neutrons within the solid angle, from the equation of the scattered wave in Eq. 7.38, the following expression for the differential cross section is obtained:

$$\frac{d\sigma}{d\Omega}(\mathbf{Q}) = \frac{1}{N}\left|\sum_i^N b_i \exp\left(\mathrm{i}\mathbf{Q}_i \cdot \mathbf{r}\right)\right|^2. \tag{7.41}$$

This is a function of the scattering vector $\mathbf{Q}$.

So far, only scattering nuclei were considered. However, the length scales involved in many relevant problems are much larger than atomic dimensions and one must think in terms of material properties. This can be done by introducing a quantity called the *scattering length density* (SLD):

$$\rho = \frac{\sum_i^n b_i}{\overline{V}}, \tag{7.42}$$

which is obtained by summing the coherent scattering lengths b_i of all the n atoms over a small volume $\overline{V}$. This volume must be sufficiently small, because extending it beyond a certain limit would cause the details of the atomic structure to be lost. The sum of Eq. 7.41 is then replaced by an integral over the whole sample volume, in which the scattering lengths b_i of the single nuclei are now replaced by the SLD, $\rho(\mathbf{r})$, and the integral is normalized by the sample volume:

$$\frac{d\Sigma}{d\Omega}(\mathbf{Q}) \equiv \frac{N}{V}\frac{d\sigma}{d\Omega}(\mathbf{Q}) = \frac{1}{V}\left|\int_V \rho(\mathbf{r}) \exp(\mathrm{i}\mathbf{Q}\cdot\mathbf{r})\, d\mathbf{r}\right|^2. \tag{7.43}$$

The differential cross section is clearly affected by the inhomogeneities of the SLD $\rho(\mathbf{r})$.

In many cases, samples measured by SANS consist of particles with a uniform SLD ρ_1 that are dispersed in a uniform matrix with a SLD ρ_2. Examples are represented by colloidal dispersions and biological macromolecules in water. In this case, the integral in Eq. 7.43 can be separated into an integral over the volume, V_1, occupied by the particles and a term expressing the difference, $(\rho_1 - \rho_2)$, between the scattering length densities of the particle and of the matrix:

$$\frac{d\Sigma}{d\Omega}(\mathbf{Q}) \equiv \frac{1}{V}(\rho_1 - \rho_2)^2\left|\int_{V_1} \exp(\mathrm{i}\mathbf{Q}\cdot\mathbf{r})\, d\mathbf{r}_1\right|^2. \tag{7.44}$$

The difference $(\rho_1 - \rho_2)$, called the *contrast factor*, encapsulates the material properties (density, composition) and scattering lengths for both the particles and the matrix, whereas the integral term describes the spatial arrangement of the particles. Equation 7.44 is the basis of an important technique used in SANS known as *contrast-matching*. If particles are embedded in a matrix whose SLD is made equal to that of the particles, the latter will be invisible. In practice, the SLD of the matrix is often varied by choosing a matrix that contains hydrogen, such as water. By replacing different fractions of ^1H atoms with ^2H atoms, a large range of scattering length densities can be achieved for the matrix, thanks to the significantly different scattering length densities of these two isotopes. This approach finds successful applications to biological problems.

So far, we have only considered experiments probing the structure of bulk matter. In discussing SANS, we assumed that we could calculate the differential cross section by adding the scattering from each nucleus in the sample. This involves the assumption that a neutron is only scattered once on passing through the sample. This assumption, called *Born approximation*, relies on the observation that multiple scattering has weak effects, such as to be neglected. In spite of this assumption, neutrons become sensitive to surface structure when they impinge on the surface at sufficiently high angles. In fact, neutrons are totally reflected from a smooth surface provided the angle of incidence (i.e., the angle between the incident beam and the normal to the surface plane) is higher than a particular value θ_c, known as the *critical angle* (cf. Eq. 7.8). In general, a wave is said to undergo total reflection on a surface when none of it is refracted by passing through it. Under these conditions, the Born approximation does not hold and neutron scattering can no longer be regarded as weak. Specular reflection is defined as reflection in which the angle of reflection equals the angle of incidence. This property is satisfied by total reflection, but specular reflection also occurs at incident angles lower than the critical value. Naturally, as the angle of incidence decreases with respect to the critical angle, less and less of the incident neutrons are reflected by the surface. In both cases, the scattering vector $\boldsymbol{Q}$ is necessarily perpendicular to the surface (cf. Fig. 7.13, even though it refers to a bulk situation). Denoting the axis normal to the smooth surface by z, Q_z equals $2k_z$, where k_z is the component of the incident neutron wave along the z-axis, and the neutron wave vector can only change in the z-direction. As such, *reflectometry* measurements provide information about the structure perpendicular to the interface. The critical angle θ_c decreases with an increase in the coherent SLD of the material and the neutron wavelength.

The *reflectivity* R measures the fraction of neutrons reflected from the surface. Reflection and refraction of neutrons at smooth surfaces satisfy the same laws as those for s-polarized electromagnetic waves. Thus, R equals unity for $\theta \geq \theta_c$, whereas is expressed by Fresnel's law for $\theta < \theta_c$. The Fresnel calculation is also extended to the case of a thin film deposited on a bulk substrate. A beam incident on such a system will be multiply reflected and refracted at the two interfaces of the film. The same method of matching wave functions can be used to obtain an expression for the reflectivity R of multiple layers. These calculations are only strictly valid for smooth interfaces between layers of homogenous material. As a matter of fact, real samples are inhomogeneous and there is a variation of the SLD perpendicular to the surface.

Neutron reflectometry involves measuring the reflectivity R as a function of the scattering vector Q_z. A SLD profile is then generated, from which a calculated reflectivity profile is extracted, to be compared with that obtained from experiment. As a rule, an SLD profile is determined by subdividing the interfacial layer into different slabs (*box model*), one for each chemically distinct component, and ascribing a uniform SLD to each of them. Roughness of the interfaces between adjacent slabs may be implemented in the model by smoothing the steps between the different SLD values. An experimental reflectivity profile cannot be uniquely described by a single SLD profile, although some help may come from independent knowledge of the system investigated. This deficiency can be overcome by the use of multiple contrasts. Thus, if one can measure the same system while changing only the SLD of one or the other bulk side of the interfacial layer, then the phase information can be recovered and the reflectivity data can be inverted to get the SLD profile. This can be achieved, for example, by changing the solvent (e.g., by exchanging H_2O for D_2O), if there is no solvent penetration into the interfacial layer.

Figure 7.14A shows plots of RQ_z^4 versus Q_z at different applied potentials for a DPTL monolayer anchored to a gold layer deposited on a Si/SiO_x substrate via a thin Cr adhesion layer (Vockenroth et al., 2008a). The model used for generating the SLD profiles shown in Fig. 7.14B consisted of the following slabs: Si/SiO_x/Cr/Au/tetraethyleneoxy spacer/lipid monolayer/electrolyte in D_2O. The vertical distance plotted along the abscissa, starting from the Si substrate, runs in parallel with the elongated structure of the DPTL molecule reported in the inset of the figure. In view of the relatively high SLD of D_2O, the level of the SLD plateau along the spacer region is a rough measure of the water content in this region; it increases gradually as the applied potential is shifted from 0 to −600 mV. Irreversible changes are only observed at −900 mV, with an appreciable increase in the water content in the spacer, apparently due to partial desorption of DPTL molecules from the gold surface. At this point, a return to 0 mV does not restore the initial situation, and a large amount of water remains trapped in the spacer region.

An example of SLD profile along a gold-supported tBLM incorporating a protein is shown in Fig. 7.15A. The tBLM consists of a monolayer of a thiolipid denoted by the WC14 acronym and characterized by a hexaethyleneoxy spacer, which is diluted with β-mercaptoethanol molecules to increase the hydrophilicity of the spacer region and to favor the incorporation of the bulky α-hemolysin channel (McGillivray et al., 2009). A DPhPC monolayer on top of the WC14 monolayer completes the lipid bilayer moiety of the tBLM. A schematic picture of the tBLM incorporating an α-hemolysin molecule with its X-ray structure is shown in Fig. 7.15B, side by side with the corresponding SLD profile. This was generated by a box model with layers of uniform SLD for each of the following chemically distinct components of the interfacial region: Si, SiO_x, Cr, Au, hexaethyleneoxy spacer, inner lipid leaflet, outer lipid leaflet, polar heads, and solvent. Three different solvents were used: D_2O, H_2O, and a D_2O/H_2O mixture with an SLD of 4×10^{-6} Å^{-2}, denoted by CM4. Isostructural samples

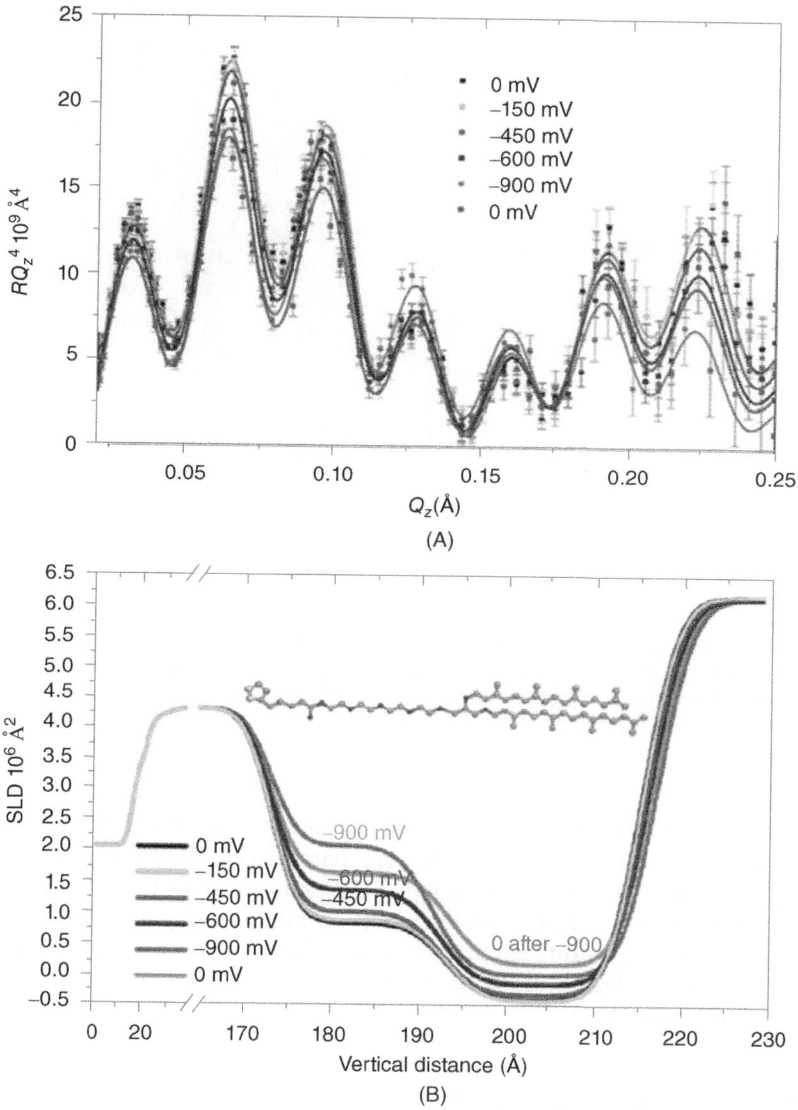

Figure 7.14 (A) RQ_z^4 versus Q_z plots for a DPTL SAM at various potentials against a reference electrode having a potential of 288 mV versus the NHE. The solid lines are the reflectivity fits of the experimental data by a box model. (B) SLD profiles of the models that are best fits to the data shown in (A). The box model consisted of the following layers: Si, SiO_x, Cr, Au, tetraethyleneoxy spacer, lipid monolayer, electrolyte. Source: Vockenroth et al. (2008a). Reproduced with permission of American Vacuum Society.

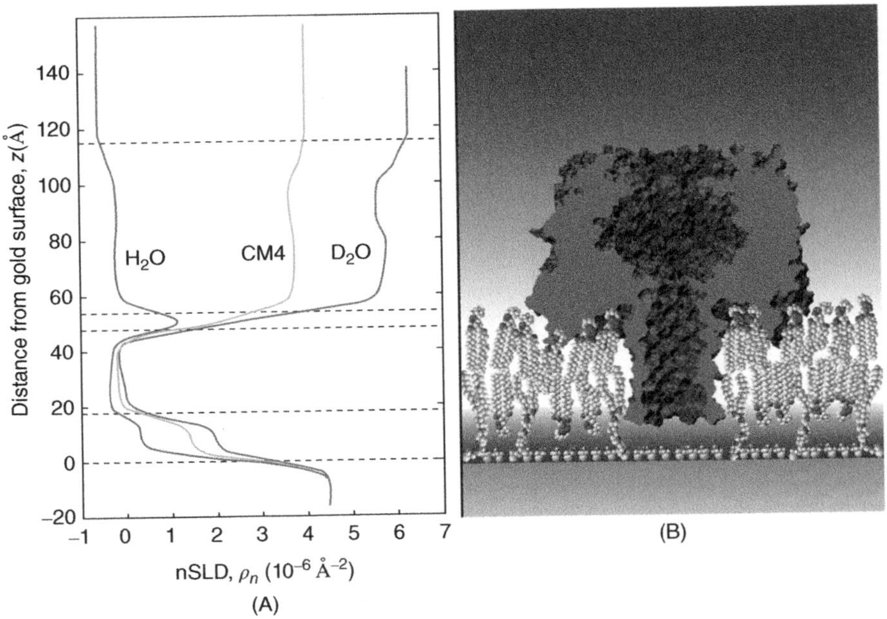

Figure 7.15 Schematic picture of a WC14/DPhPC tBLM reconstituted with α-hemolysin (B) and of the corresponding SLD profiles (A), resulting from the simultaneous fitting of a molecular model to five neutron reflectivity data sets. The SLD profiles contain the calculated contribution of the α-hemolysin X-ray crystal structure at a lateral density and insertion depth derived from the model fit. Source: McGillivray et al. (2009). Reproduced with permission of Elsevier.

with distinct *isotopic contrast* were fitted simultaneously by consistently refining the corresponding SLD profiles, which were parameterized in terms of the underlying molecular structures. From Fig. 7.15A, it is apparent that D_2O has a much higher SLD than H_2O, and that the spacer region is satisfactorily hydrated. A comparison of the SLD profile in the presence of the α-hemolysin channel with respect to that in its absence indicates a strong dehydration of the polar head region induced by the channel.

7.5 Fluorescence Microscopy

Fluorescence is the outcome of a three-stage spontaneous process that takes place in certain molecules called *fluorophores* or *fluorescent dyes* (generally polyaromatic hydrocarbons or heterocycles). At room temperature, these molecules occupy the lowest vibrational level, S_0, of the ground electronic state. The process responsible for the fluorescence of fluorophores is illustrated by the simple electronic-state diagram in Fig. 7.16.

After absorbing a photon of energy $h\nu_{ex}$, the fluorophore is excited from the ground state S_0 to one of the vibrational levels of a higher singlet state, S_1; this

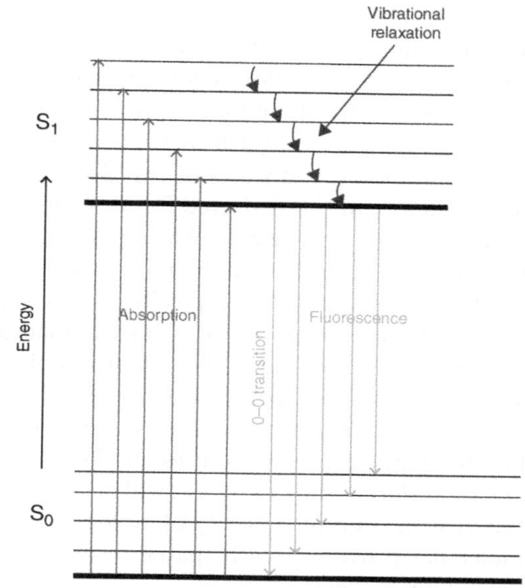

Figure 7.16 Simplified electronic-state diagram, where the energies of some vibrational levels of the two electronic states S_0 and S_1 are sketched.

represents the first stage. Naturally, excitation can result in the molecule reaching any of the vibrational levels associated with each electronic state. Since the energy is absorbed as discrete quanta, this gives rise to a series of absorption bands, which constitute the ***absorption spectrum***. The excited state only exists for a few nanoseconds. During this period, the excited molecule interacts with its environment in many different ways, rapidly losing its excess of vibrational energy by collision and falling to the lowest vibrational level of the lowest excited state, namely the singlet excited state S_1, from which the fluorescent emission occurs. From this level, the molecule can return to any of the vibrational levels of the ground state, emitting a photon of energy $h\nu_{em}$ in the form of fluorescence. This gives rise to a ***fluorescence emission spectrum*** (Lakowicz, 2006).

If this process takes place for all the molecules that absorb light, then the ***fluorescence quantum yield*** **Φ**, defined as the ratio of the number of fluorescence photons emitted to the number of photons absorbed, will be a maximum, that is, unity. Often, however, other parallel processes take place, such as fluorescence resonance energy transfer (FRET), intersystem crossing, internal conversion, or collisional quenching, which can depopulate molecules from S1. Hence, the ***fluorescence quantum yield*** **Φ** turns out to be less than unity, and may even be almost zero. In view of energy loss during the excited state lifetime, $h\nu_{em}$ is smaller than $h\nu_{ex}$. Therefore, the wavelength of the emitted fluorescence photons is longer than that of the photons responsible for the excitation. This wavelength difference is termed the *Stokes shift*, which allows the emitted fluorescent photons to be easily distinguished from the excitation photons, leading to the possibility of a very low background in fluorescent studies. Only the transition from the lowest vibrational level in the ground electronic state to the lowest vibrational level in the first excited state, called 0–0 transition, is common to both the absorption and emission spectra. All other absorption transitions require more energy than any transition in the fluorescence emission. Hence, the emission spectrum lies at wavelengths longer than that of the 0–0 transition.

FRET, or *Förster energy transfer*, is a distance-dependent interaction between the electronic excited states of two dye molecules, a donor and an acceptor, in which excitation is transferred from the donor to the acceptor without emission of a photon (Valeur, 2002). An important consequence of this transfer is that the emission of light by the donor is greatly reduced. The acceptor may or may not be fluorescent. FRET typically occurs over a distance of 1–10 nm, comparable with the dimensions of biological macromolecules. The rate of the energy transfer from the donor to the acceptor molecule is inversely proportional to the sixth power of the separation R between them. More precisely, the *efficiency* of FRET, that is, the fraction of the energy transfer event occurring per donor excitation event, is expressed by the equation $\Phi_{FRET} = 1/[1 + (R/R_0)^6]$, where R_0, called *Förster radius,* is the distance at which energy transfer is 50% efficient. This allows the investigation of a variety of biological phenomena that produce changes in molecular proximity, with spatial resolution beyond the limits of conventional optical microscopy.

The interpretation of FRET is based on dipole–dipole interactions between the transition dipoles of the donor and acceptor. In fact, studying the fluorescence characteristics of a molecule practically amounts to studying the electrodynamics of a nearly point-size dipole from an optical viewpoint. For FRET to take place, the transition dipoles of donor and acceptor must be approximately parallel. Moreover, the absorption spectrum of the acceptor must overlap the fluorescence emission spectrum of the donor. Important applications of FRET are investigations on spatial distribution and assembly of protein complexes, receptor/ligand interactions, distribution and transport of lipids, and membrane fusion assays. In most applications, the donor and acceptor dyes are different, in which case FRET can be detected by the appearance of sensitized fluorescence of the acceptor (if it is fluorescent) or by the quenching of the donor fluorescence.

Quenching refers to any process that causes a reduction in the quantum yield of a given fluorescence process; it can be either collisional or static. Collisional quenching occurs if the *quencher* comes into contact with the excited fluorophore during the fluorescence lifetime and causes the dye to return to the ground state without emitting a photon. In this respect, FRET can be regarded as a particular collisional quenching mechanism. On the other hand, static quenching is due to the formation of a ground-state complex between the fluorescent molecule and the quencher, that is, before excitation occurs. The complex is not fluorescent and has a unique absorption spectrum.

When a fluorophore is optically excited at a distance from a metal surface less than about 100 nm, its lifetime decreases by nonradiative energy transfer into electron–hole pairs (*excitons*) and into propagating surface plasmons in the metal, and the fluorescence is essentially quenched (Helen and Axelrod, 1987). Unlike the Förster transfer, the rate of nonradiative energy transfer from the excited fluorophore to a planar gold surface is inversely proportional to the third power of their distance. In practice, the fluorescence intensity of the fluorophore increases gradually with increasing distance of the fluorophore from a gold layer, until it remains completely unquenched at a distance of about 30 nm (Tawa and Morigaki, 2005).

Fluorescence measurements are carried out with a *fluorescence microscope* equipped with a high-intensity light source (usually a mercury arc lamp), which emits light in a broad spectrum from visible through ultraviolet. Most fluorescence microscopes use incident illumination to illuminate the sample from above (*epifluorescence illumination*). In this way, the objective lens is used for both illuminating the sample and collecting the fluorescent light. The wavelength of the excitation illumination is selected by placing after the light source a filter that limits light transmission to a narrow range of wavelengths. The light then impinges on a beam splitter and is reflected down through the objective lens and onto the sample. Fluorescent molecules within the specimen (either endogenous or exogenous) absorb the light and reirradiate the fluorescent light of longer wavelength. The objective lens collects this emitted fluorescent light, which then passes through the beam splitter. Any excitation light reflected by the sample is blocked by a third filter (barrier filter). Thus, only the light emitted from fluorescent molecules within the specimen is observed or photographed.

This procedure is used to determine the detachment of self-assembled films from an electrode as the applied potential is made sufficiently negative. To this end, a small amount of a fluorescent lipophilic dye is added to the organic material before self-assembling it on the metal surface, under the reasonable assumption that the dye remains incorporated in the resulting film. Fluorescence can only be observed for dye molecules that are separated from the electrode surface by a certain distance. At an octadecanol monolayer deposited on Au(111), little fluorescence is observed not only over its potential range of stability, but also at more negative potentials, where the electric capacitance increases due to film restructuring (Bizzotto et al., 2004). Only at potentials negative enough to displace the organic molecules from the electrode surface does the fluorescence of the accompanying dye molecules increase appreciably. The fluorescence images show bright spots much larger than typical micelles, indicative of aggregates or small crystallites of the solid surfactant. Scanning back the applied potential causes readsorption of the organic material, with a resulting decrease in the fluorescence signal and gradual disappearance of the aggregates from the images.

On lipid-coated mercury, as distinct from lipid-coated gold, fluorescence can be detected even if the fluorophore is at a small distance from the electrode surface. In fact, the fluorescence quenching efficiency is by two to three orders of magnitude lower on mercury than on gold, also thanks to the perfectly smooth surface of mercury, as compared with the more or less rough surface of solid electrodes. Upon scanning the potential progressively in the negative direction up to film desorption, the fluorescence intensity maintains a constant low level along the flat capacitance minimum (cf. curve *a* in Fig. 6.21); a large increase in fluorescence is observed at far negative potentials, indicating a separation of the lipid monolayer from the electrode surface. By scanning back the potential, the fluorescence decreases slowly down to the positive potential limit; however, neither the fluorescence nor the capacitance recovers the pristine value, denoting a defective reformed monolayer. In contrast to a similar system on gold, the fluorescent particles or aggregates on Hg are freely mobile, yielding fluorescence images free from spots (Bizzotto et al., 2004).

Fluorescence microscopy is also utilized to estimate the long-range lateral mobility of lipid molecules in a lipid mono- or bilayer, a fundamental property of biological membranes. The fluidity of plasma membranes should be preserved in supported biomimetic membranes. Free movement of lipid molecules enables the biomimetic membrane to react to the presence of proteins, charges, and physical forces in a dynamic and responsive manner. A satisfactory fluidity allows biomimetic membranes to reorganize upon interaction with external perturbations, mimicking the functionality of living cell membranes. In particular, lateral mobility enables a biomimetic membrane to incorporate large membrane proteins from their detergent solutions by making space for them; it also determines the spontaneous separation of the components of a lipid mixture (demixing), giving rise to the formation of important lipid microdomains, called *lipid rafts*.

The fluidity and lateral mobility of biomimetic membranes can be characterized quantitatively by a technique called *fluorescence recovery after photobleaching* (FRAP), one of the most popular ways of measuring molecular diffusion in membranes. It relies on introducing a small amount of fluorescent probe molecules, usually covalently bound to lipids or a membrane protein, into the membrane. All fluorescent dyes emit light of one wavelength (e.g., green), after they have absorbed light of shorter wavelength (e.g., blue). However, if a light of very high intensity and relatively short wavelength is delivered to the dye, this will undergo *photobleaching*, a process whereby the high-intensity light renders the dye unable to fluoresce. A typical fluorophore can undergo a finite number of excitation–relaxation cycles prior to photobleaching. For a photostable fluorophore, such as tetramethylrhodamine, photobleaching occurs after about 10^5 cycles. In contrast, fluorescein photobleaches very easily. The rate of the photobleaching is often proportional to the intensity of illumination.

Using an epifluorescence microscope, an excitation light of appropriate wavelength is focused on a spot of the biomimetic membrane (a few tens of micrometers in diameter), to determine the starting level of fluorescence. A short burst of intense excitation light is then projected onto the chosen spot. This may be the whole spectrum of a mercury lamp for a few minutes, or a strong laser beam for a few tens of milliseconds. As soon as the fluorescent molecules are monitored again by the microscope, the spot will appear as a black area filled with photobleached molecules and surrounded by fluorescently tagged molecules that have not been photobleached. If these molecules are able to diffuse, they will, thus confirming the membrane fluidity. As they diffuse, the photobleached molecules and the fluorescent molecules will begin to mix, and the latter will tend to migrate into the black area, as shown in Fig. 7.17C. This will cause the blackened area to gradually increase in brightness, while the fluorescent area will become a little

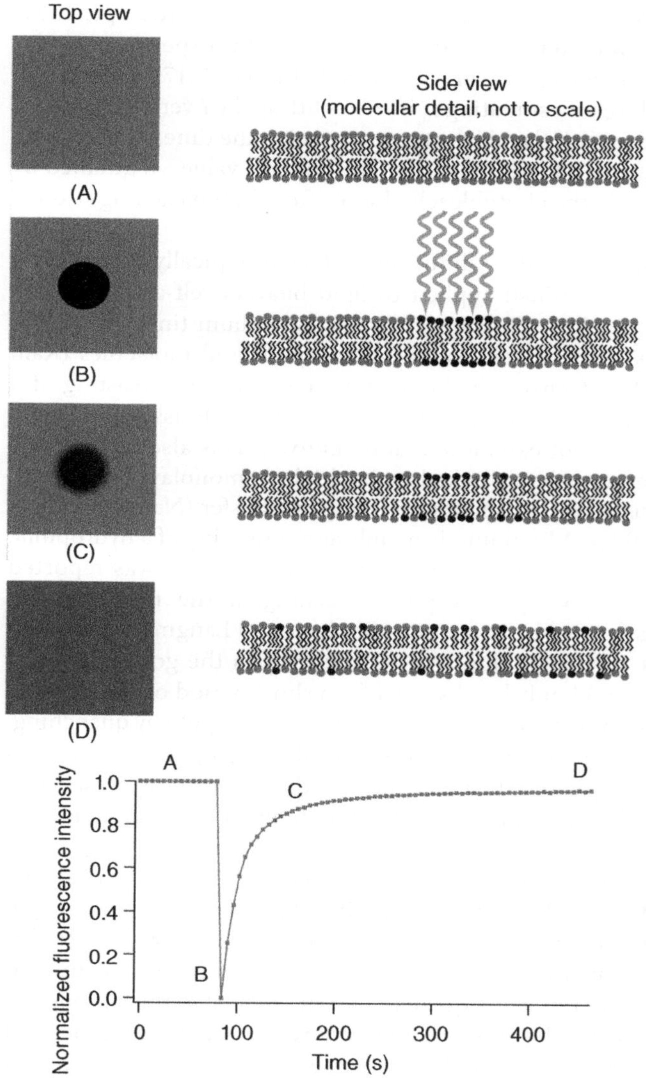

Figure 7.17 Lipid bilayer uniformly labeled with a fluorescent tag (A), selectively photobleached by a small and intense light pulse (B), monitored as the bleached dye diffuses out of the beached area and new dye diffuses into it (C), and monitored after uniform intensity is ultimately restored (D). Source: https://commons.wikimedia.org/wiki/File:Frap_diagram.svg.

less bright. The gradual fluorescence recovery within the given spot is followed as a function of time, thus permitting an estimate of the lipid diffusion coefficient (Fig. 7.17). Assuming a Gaussian profile for the bleaching beam, the diffusion constant D can be calculated from the simple equation $D = r^2/(4t_d)$, where r is the beam radius and t_d is the *characteristic diffusion time*. To this end, raw data are collected by a light sensor that measures the light through the microscope. These data are displayed on a computer screen and, by the end of the experiment, a plot of the fluorescence intensity I versus time, such as that in Fig. 7.17, is produced. An approximate procedure for estimating t_d involves fitting the I versus t curve by the equation $I = A[1 - \exp(-t/\tau)]$ and in setting t_d equal to the time, $t_{1/2}$, at which I equals $A/2$, namely $t_{1/2} = -\tau \ln 0.5$. As a rule, the limiting value, A, attained by the fluorescence intensity in the photobleached spot after photobleaching is lower than the pristine value.

In supported membranes, the lipid diffusion coefficient typically ranges from 1×10^{-8} to 1×10^{-7} cm^2 s^{-1}. When applied to lipid bilayers self-assembled on smooth supports such as silica, glass, quartz, mica, or indium tin oxide (ITO), FRAP usually confirms a satisfactory lateral mobility of lipid molecules (Kalb et al., 1992; Parikh, 2008). Conversely, biomimetic membranes consisting of a thiolipid monolayer tethered to a gold electrode, with a self-assembled lipid monolayer on top of it, do not exhibit lateral mobility. This is also true for the distal lipid monolayer noncovalently linked to the thiolipid monolayer, no matter if obtained by vesicle fusion or by Langmuir–Blodgett transfer (Naumann et al., 2002; Baumgart et al., 2003). A biomimetic membrane consisting of a hydrophilic spacer tethered to gold, with a lipid bilayer formed on top of it, was reported to exhibit fluorescence recovery after photobleaching, if the noncovalently bound lipid bilayer was formed by Langmuir–Blodgett and Langmuir–Schaefer transfers (Baumgart et al., 2003). Fluorescence recovery in the gold-supported biomimetic membrane could only be observed for a short period of time, due to the gradual energy transfer from the fluorophore molecules to gold by quenching. No fluorescence recovery could be observed by forming the lipid bilayer on top of the hydrophilic spacer by vesicle fusion. Evidently, the unavoidable presence of adsorbed and hemifused vesicles prevents lateral mobility of the lipid molecules of the distal monolayer.

The separation distance of a fluorescent probe from a film-coated metal surface within a range of about 100 nm by exploiting the quenching process can also be estimated by utilizing SPR, instead of epifluorescence microscopy. In this case, rather than estimating the distance from the metal at which a photoexcited fluorophore can escape fluorescence quenching, one estimates the distance at which the fluorophore can still be photoexcited by the evanescent wave generated at the dielectric/metal interface of an SPR apparatus. This technique is referred to as *surface plasmon fluorescence spectroscopy* (SPFS). A fluorophore close to a back-illuminated dielectric/metal interface can be excited by the evanescent wave. For this purpose, in order to be excited, the fluorophore should be located within the decay length of the evanescent field, which is typically several tens to hundreds of nanometers for applicable wavelengths (Tawa and Morigaki, 2005).

Total internal reflection fluorescence microscopy (TIRFM) is a further technique belonging to the same category as SPFS. To selectively illuminate and excite fluorophores, TIRFM uses the evanescent wave generated at a glass/water interface (Axelrod, 1981). This evanescent wave is generated only when the incident light is totally internally reflected at such as interface. The evanescent electromagnetic field decays exponentially from the interface, and thus penetrates to a depth of only approximately 100 nm into the sample medium. Thus, TIRFM allows a selective visualization of surface regions, such as a lipid bilayer interposed between the glass and an aqueous solution. SPFS has some advantages over TIRFM. Thus, SPFS generates a more intense evanescent field than TIRFM and reduces the background contribution from the excitation light source, thanks to the blocking effect exerted by the thin metal film. On the other hand, the presence of the metal layer decreases the radiative quantum yield of the fluorophore by quenching at distances from the metal surface shorter than about 10 nm.

Fluorescence laser scanning microscopy is used to obtain images of the lateral heterogeneity of biomimetic membranes, and also of cell membranes (Bagatolli, 2006). Gel-phase and liquid-ordered microdomains and the liquid-disordered matrix appear as differently colored areas in these images. Giant unilamellar vesicles (GUVs) are the most commonly employed model membranes for this purpose, since their size, of the order of a few tens of micrometers, is close to that of living cells. Moreover, they can also be formed from natural lipid extracts and native membranes, and their molecular composition can be accurately controlled.

Confocal laser scanning microscopy (CLSM or LSCM) and *fluorescence lifetime imaging microscopy* (FLIM) are usually employed for obtaining high-resolution optical images. In a confocal laser scanning microscope, a laser beam passes through an aperture and is then focused by an objective lens into a small focal volume within a fluorescent sample. A mixture of the emitted fluorescence light and the reflected laser light is then recollected by the objective lens. A beam splitter separates the light mixture, by allowing only the laser light to pass through, while reflecting and deviating by 90° the fluorescence light into a pinhole; this blocks the out-of-focus light and directs the fluorescence light to a detection apparatus, which converts the light signal into an electrical signal recorded by a computer. The detected light originating from the illuminated volume element within the sample represents one pixel in the resulting image. As the laser scans the surface of interest, a whole image is gradually obtained pixel-by-pixel and line-by-line.

To impart florescence to the sample, a small amount of one or two lipophilic fluorophores must be added. A single fluorophore is sufficient if it partitions unevenly between the different domains of the GUV vesicular membrane, imparting a markedly different fluorescence intensity to them. It should be noted that the partition of a fluorescent probe into particular membrane microdomains does not depend on their phase state. Thus, the same fluorophore may well show preferential partitioning into the gel phase in a gel/fluid phase mixture and into the fluid phase in a different gel/fluid phase mixture. As a rule, the addition of two different fluorophores partitioning into different membrane domains and emitting light of different color is preferred (Margineanu et al., 2007).

FLIM presents a number of advantages over CLSM. This is particularly true for *two photon excitation fluorescence microscopy*, one of the most promising techniques in biological and medical imaging (Bagatolli, 2006). Two-photon excitation is a nonlinear process whereby a fluorophore absorbs two photons simultaneously. Each photon provides half the energy needed for excitation. The high photon densities required for two-photon absorption are realized by focusing a high-peak laser light source on a diffraction-limited spot through a high numerical aperture objective. In the areas above and below the focal plane, two-photon absorption does not take place, due to insufficient photon flux. This phenomenon permits a sectioning of the image of a three-dimensional object, such as a giant vesicle, into different parallel planes, without using emission pinholes as in confocal microscopy. Among the advantages of two-photon excitation, we may mention improved background discrimination, minimal photodamage to living cell specimens, and reduced photobleaching of fluorophores. In addition, the characteristics of two-photon excitation fluorescence microscopy allow the use of UV light, thus permitting the photoexcitation of UV-sensitive fluorescent probes, such as LAURDAN. This hydrophobic probe belongs to the family of polarity-sensitive fluorophores and is ideal for studying the lateral structure of membranes. Thus, it is widely used among fluorescence investigators (Bagatolli, 2003; 2006). It distributes equally well into the ordered and disordered lipid phases. Nonetheless, it shows a phase-dependent emission spectral shift, namely, bluish in the liquid-ordered phase and greenish in the liquid-disordered one. Its transition dipole moment is aligned parallel to the hydrocarbon tails of the membrane.

Two-photon FLIM (2P-FLIM) combines the advantages of two-photon excitation with those of fluorescence lifetime imaging. Its description requires a few considerations on *time-resolved fluorescence*. In traditional, steady-state fluorescence spectroscopy, a light source provides continuous illumination of a large population of fluorophores. The resulting fluorescence intensity I is an average, over all fluorophores of the sample and over time, of the intensity decays of the single fluorophores. If the decay is a single exponential, it is expressed by the equation $I = \alpha \exp(-t/\tau)$, where α is a pre-exponential factor and τ is the *fluorescence lifetime*. It follows that the steady-state fluorescence intensity is determined by several decay laws, which cannot be extracted from it (de Almeida et al., 2009). To determine the decay law of a fluorophore, time-resolved experiments are performed through direct measurements in the time domain. Since the various molecules of a sample, once simultaneously excited by a short laser pulse, do not emit a photon at exactly the same time, their decay must be thought of as obeying a given rate law rather than occurring at a specific time after excitation. By monitoring the time required by individual molecules to emit their photons and by then combining the resulting data, one can generate an intensity versus time graph displaying an exponential decay curve typical of these processes.

This goal is generally achieved using *time-correlated single photon counting* (TCSPC). With this technique, one measures the time elapsed between a sample excitation by a pulsed laser and the arrival of the emitted photon at the detector. TCSPC requires a defined "start," provided by the electronics steering the laser pulse, and a defined "stop" signal, realized by detection with single-photon-sensitive detectors (Wahl et al., 2013). The measurement of this time delay is repeated many times in consecutive cycles, to account for the statistical nature of the fluorophore emission. The delay times are inserted into a histogram that plots the occurrence of photon emission over time after the excitation pulse. The probability of registering more than one photon per cycle must be reduced to a minimum, to guarantee that the histogram of photon arrivals represents the actual time decay law. It is, therefore, crucial to keep the probability of cycles with more than one photon low. To this end, the average count rate of the single-photon detector should be at most 1–5% of the excitation rate. As expected, the time delays between start signals and corresponding stop signals (*start–stop times*) become less frequent the longer they are. Hence, a histogram of start–stop times against the number of the corresponding counts yields the decay curve of the fluorophore. This situation is schematically depicted in Fig. 7.18, which shows the stop signals due to photon emission against the time elapsed from the start signal for a series of different cycles, as well as the histogram resulting from a much higher number of cycles. Note that some cycles lack stop signals, while the others report only one stop signal.

2P-FLIM is extensively used for imaging lateral inhomogeneity in vesicular membranes, although it has also been applied at lipid mono- and bilayers deposited on glass, after ascertaining by FRAP a lipid lateral mobility high enough to guarantee a spontaneous and rapid phase separation. 2P-FLIM was also employed at a mercury-supported tBLM formed by electrodepositing mercury at the tip of a platinum microelectrode, about 20 μm in diameter, from a pH 1 mercurous nitrate aqueous solution (Becucci et al., 2010). This tBLM consists of a DPTL thiolipid monolayer anchored to mercury, with a monolayer of a sphingomyelin/dioleoylphosphatidylcholine/cholesterol (47:47:6) mixture self-assembled on top of it; addition of 1 mol% LAURDAN bestows fluorescence

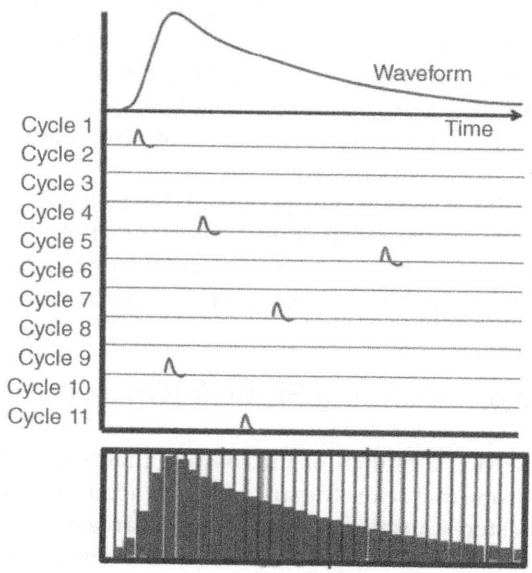

Figure 7.18 TCSPC measurement principle.

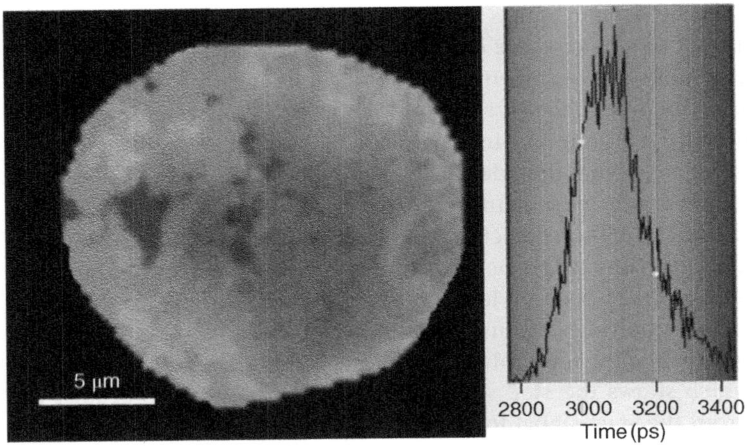

Figure 7.19 Two-photon (λ excitation = 760 nm) fluorescence lifetime 21 × 21 μm image (left panel) and fluorescence lifetime distribution histogram (right panel) of a distal monolayer of DOPC/PSM/Chol (47:47:6) mixture, labeled with 1 mol% LAURDAN, at room temperature. The mixture is the distal monolayer of a micro tBLM. The color code in the FLIM image is that indicated in the histogram on the right. Source: Becucci et al. (2010). Reproduced with permission of Royal Society of Chemistry.

to the lipid mixture. The high lipid lateral mobility imparted to the tBLM by the liquid state and perfectly smooth surface of mercury ensures a rapid phase demixing. The fluorescence lifetime image in Fig. 7.19 reveals a high degree of heterogeneity of the distal lipid mixture, characterized by the coexistence of gel, liquid-ordered, and liquid-disordered phases. The microdomains with an irregular percolative-like shape have lifetimes of 3450 ± 50 ps and are ascribed to the gel phase. The small roundish microdomains immersed in the surrounding matrix have lifetimes of 2830 ± 50 ps and are attributed to the liquid-ordered phase (the rafts). Finally, the matrix, with intermediate fluorescence lifetimes, is ascribed to the coexistence of the liquid-disordered phase and of liquid-ordered and gel-phase microdomains of size below the resolution of the microscope (about 0.4 μm radial).

FLIM can also be operated in the frequency domain. In this case, the laser used as the excitation light source is modulated (1 – 200 MHz) and the fluorescence lifetime is calculated by measuring the phase shift of fluorescence and the decrease in its amplitude using a gain-modulated detector.

7.6 Scanning Probe Microscopy

SPM is a branch of microscopy that forms images of surfaces using a physical probe that scans the specimen. To acquire an image, a scanning probe microscope raster-scans the probe over a small area of the sample, measuring some local property, such as height, friction, and magnetism. Of these techniques, atomic force microscopy (AFM) and scanning tunneling microscopy (STM) are the most commonly employed, finding application in numerous branches of science.

7.6.1 Atomic Force Microscopy

AFM operates by measuring the force between a probe and the sample. A piezoelectric scanner, consisting of a "tripod" of three piezo crystals responsible for scanning in the x, y, and z directions, moves the probe with respect to the sample. It may be attached either to the stage holding the sample or to the probe. Normally, the probe is a sharp tip consisting of a 3–6 μm tall pyramid with 15–40 nm end radius. The tip is attached to a flexible microcantilever, which bends under the influence of the force, acting like a microscopic diving board. The bending determines a change in the angle of inclination of the cantilever, which is measured by reflecting a laser beam off of the cantilever and into a position-sensitive detector (a split photodiode). The detector gauges the small movements of the laser spot and converts them into an electrical signal, thus measuring the cantilever deflection (see Fig. 7.20). The piezoelectric scanner moves the tip over a two-dimensional grid of locations across the surface of the sample (the $x-y$ plane). Via a feedback loop activated by the position-sensitive detector, the scanner also moves the tip up and down along the vertical direction (the z-axis). In fact, by raster-scanning the asperities of the sample surface in the absence of a vertical movement, the tip or the sample might mechanically abrade atop the highest hills, whereas tip and sample might separate or disconnect down in the deepest valleys. Moreover, there is always some unavoidable small tilt between the sample surface and the $x-y$ plane of the scanning device, such that the force would continually grow while scanning in one direction and decrease while scanning in the opposite one.

The force between the tip and a sample varies with the tip–surface distance. When the tip is sufficiently far from the surface, this force equals zero and the

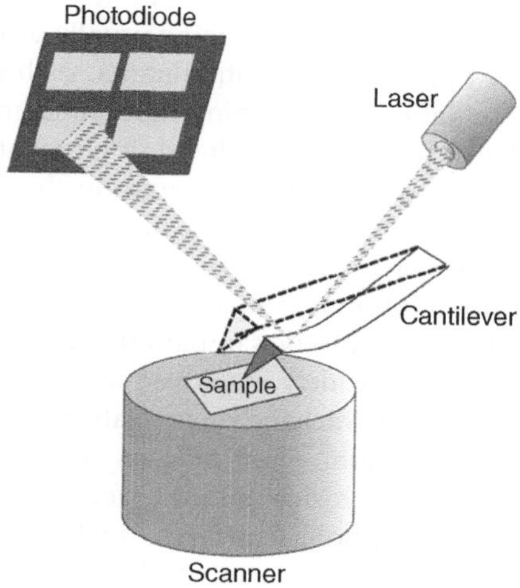

Figure 7.20 Scheme of a system for converting the deflection of an AFM cantilever into an electrical signal.

device is said to be "out of feedback." As the tip gradually approaches the surface, van der Waals attractive forces, arising from interactions between transient electric dipole moments, may first come into play. Then, at very short tip–surface distances (a few ångströms), very strong repulsive forces arise between the tip and sample atoms, due to the overlap of their electronic orbitals. When these repulsive forces become predominant, tip and sample are considered to be in contact. An atomic force microscope may operate in three different imaging modes: *contact*, *noncontact*, and *tapping*. Moreover, the contact mode may be realized by keeping constant either the scanner height or the force.

The *constant-height mode* is realized by disconnecting the feedback circuit for scanner movements along the z-axis. In this mode, the image is generated from the variation of the cantilever deflection. On the other hand, in the *constant-force mode* the cantilever deflection is kept constant via an input to the feedback circuit that controls the scanner movement along the z-axis, in such a way that the force applied to the sample remains constant. In this mode, the image is generated from the z-motion of the scanner. The constant-force mode is preferred to the constant-height mode for most applications.

In the noncontact mode, the tip does not contact the sample surface. The cantilever is made to oscillate just above its resonant frequency and the amplitude of the oscillation is set constant in the proximity of the free amplitude. The van der Waals forces and any other long-range forces extending above the sample surface would tend to decrease the amplitude of the cantilever oscillation. The feedback loop system maintains a constant oscillation amplitude by adjusting the average tip-to-sample distance. In this case, the image of the sample surface is provided by the tip-to-sample distance at each (x, y) data point. The tapping mode is something between the contact and the noncontact mode, in that its amplitude is set between 50% and 60% of the free amplitude. In most applications, the tapping mode is preferred to both the contact and the noncontact mode. In fact, the contact mode imaging is heavily influenced by frictional and adhesive forces, and can damage samples and distort image data. On the other hand, the noncontact mode imaging usually provides low resolution and can be hampered by thin layers of contaminants, such as water, which can interfere with its wide oscillation. The tapping mode imaging reduces the aforementioned drawbacks because, on the one hand, it eliminates frictional forces by intermittently contacting the surface whereas, on the other hand, its amplitude is just sufficient to prevent the tip from being trapped by adhesive meniscus forces from any contaminant layers. Hence, tapping mode normally provides higher resolution with minimum sample damage, by imaging the force of the intermittent contacts of the tip with the sample surface.

AFM is widely employed for monitoring vesicle fusion on solid metals (e.g., gold and silver) and on insulating materials (e.g., silica and mica) with formation of lipid mono- and bilayers. As a rule, vesicle fusion yields supported lipid bilayers on hydrophilic surfaces, such as mica, and supported lipid monolayers on hydrophobic surfaces, such as alkanethiol monolayers tethered to gold. However, when forming planar lipid monolayers and bilayers by vesicle fusion, the problem of vesicle mere adsorption or partial fusion on the substrate cannot be disregarded. Thus, fusion of small unilamellar vesicles (SUVs) onto a binary mixture of EO3-cholesterol thiolipid molecules and hydrophilic 6-mercaptohexanol molecules tethered to gold, such as that depicted in Fig. 6.29, yields AFM images showing heightened areas; their diameter being close to that of the SUVs denotes the presence of adsorbed vesicles (Jeuken et al., 2006). The presence of a membrane protein, such as *cytochrome bo3*, in these vesicles increases the number density and the size of the heightened structures ascribable to adsorbed vesicles. Analogously, fusion of large unilamellar vesicles (LUVs) on a hydrophobic support exposing dipalmitoylphosphatidylethanolamine (DPPE) alkyl chains to the aqueous phase yields tapping-mode AFM images with a number of dome-shaped structures (Zebrowska and Krysinski, 2004). Some of these structures, whose diameter is close to that of the LUVs, tend to disappear after 1 h, due to vesicle complete fusion. However, if the vesicles contain the membrane protein Na^+,K^+-ATPase, all the dome-like structures are stable even after 3 h. In general, adsorption of proteoliposomes (namely, vesicles incorporating membrane proteins) on hydrophobic surfaces prevents their complete spreading and fusion, due to the presence of protein molecules with extramembrane domains and to the hydrophilicity of the outer polar heads of the proteoliposomes (Jass et al., 2000).

The presence of adsorbed vesicles becomes predominant if a biomimetic membrane is brought into contact with vesicles directly formed from membrane extracts still containing proteins and coenzymes (Jeuken et al., 2005). These vesicles do not desorb after removing their dispersion from the solution, as shown by the AFM images in Fig. 7.21 (E) and (F). On the average, the AFM images of the adsorbed vesicles exhibit diameters larger and heights smaller than the diameters of the corresponding free vesicles in solution, indicating that they are partially semi-fused and flattened upon adsorption, and also possibly compressed by the force applied by the AFM tip. In fact, lipid structures, such as surface bound vesicles or bilayer patches resulting from their complete fusion, are known to be easily modified by interactions with the AFM tip (Richter and Brisson, 2005). Imaging in tapping mode provides least disturbance to the lipid material. Nonetheless, images in contact mode may also be obtained with minor artifacts, such as occasional tip-induced vesicle rupture, by adjusting to a minimum the forces exerted by the AFM tip.

Plots of the tip–sample interaction against distance, commonly known as *force curves*, may provide useful information on the strength of interfacial attraction or repulsion related to van der Waals, electrostatic, or steric forces. These quantitative and fundamental measurements can be collected within *x,y*-mapping routines, and analyzed within subregions of the explored AFM image. Examples of force curves are offered by panels (A)–(D) in Fig. 7.21, obtained at a binary mixture of EO3-cholesterol thiolipid molecules and hydrophilic 6-mercaptohexanol molecules tethered to gold (cf. Fig. 6.29), both before (A) and after incubation with inside-out vesicles formed from membrane extracts of *Bacillus subtilis* still containing proteins. The force curves (B)–(D)

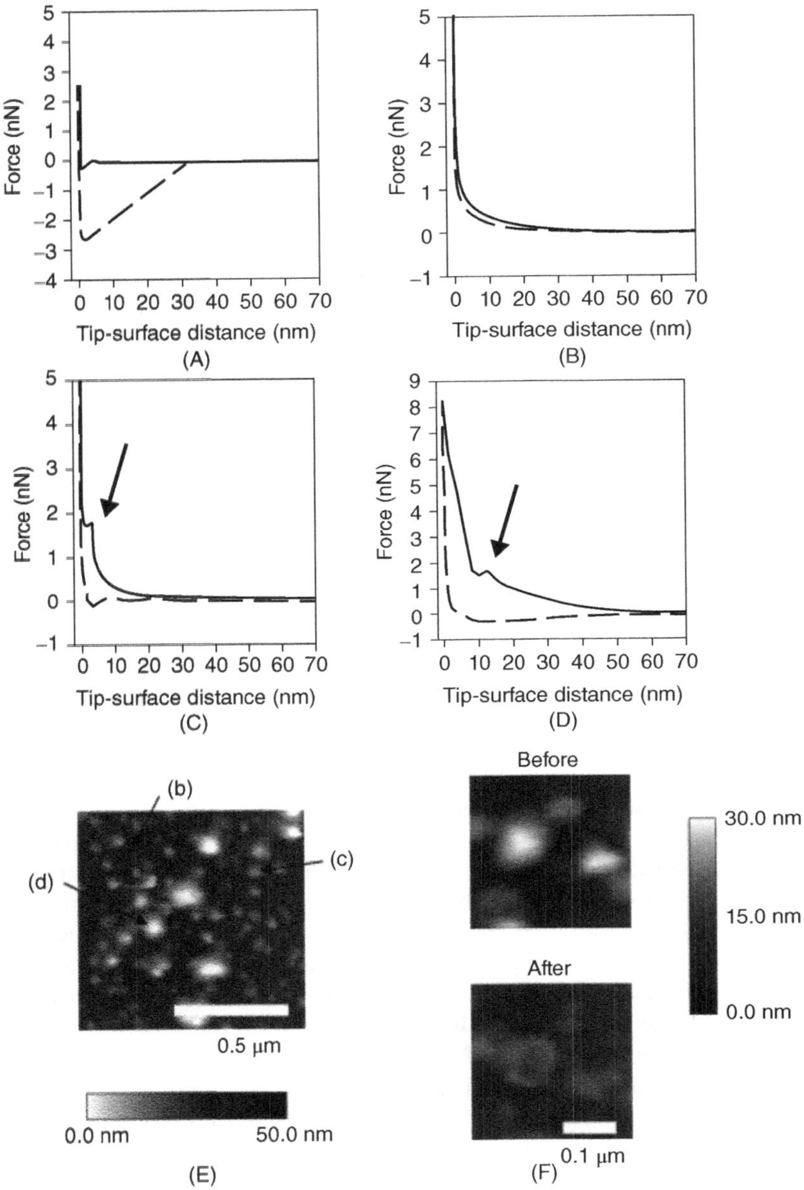

Figure 7.21 (A–D) Force measurements of template stripped gold, modified with EO3-cholesterol/6-mercaptohexanol (30:70) before (A) and after (B–D) incubation with *B. subtilis* membrane vesicles. Panel B is recorded in a region free of vesicles, and panels C and D are force curves at vesicles. The arrow indicates a typical signal obtained when the AFM tip protrudes through the phospholipid bilayer. Solid lines are for the tips approaching the surface, dashed lines for receding tips. (E) Tapping mode AFM image, prior to force measurements, indicating the positions where force curves were acquired. (F) AFM pictures before and after the force measurements. Source: Jeuken et al. (2005). Reproduced with permission of American Chemical Society.

were acquired at the positions indicated in the AFM image of panel (E), namely on a region free from vesicles, and on top of a small and a large vesicle, respectively. The force curve (A) at the Au-supported thiol mixture first decreases below the zero level at distances of a few nanometers from the surface, denoting an attractive van der Waals interaction, and then shows an abrupt increase due to very strong repulsive forces. The force curve (D), recorded on top of the large vesicle, starts increasing with respect to the zero level as soon as the tip touches the vesicle and increases smoothly with decreasing distance as long as the tip confines itself to compressing and flattening the vesicle; then, at the distance at which the tip pierces the vesicular membrane, the force curve shows a dip due to contact with the aqueous content of the vesicle, and then increases more rapidly with a further decrease in distance. The force curve (C) recorded on top of the small vesicle is similar, but obviously more compressed along the tip–surface distance axis. Finally, the force curve (B) recorded on a vesicle-free region shows no dip but differs from that, (A), obtained before the addition of vesicles, probably due to protein and lipid adsorption on the tip. Piercing of the larger vesicles causes their collapse, leaving lipid doughnuts of toroidal shape, where water remains trapped in the outer ring, as shown by panel (F).

AFM has provided direct evidence that silica can be covered by isolated vesicles that remain stable for days (Richter et al., 2003). Conversely, when adsorbing vesicles at low surface density on mica, they initially remain intact but collapse individually over a time interval ranging from minutes to hours (Richter and Brisson, 2005). The particularly smooth surface of mica imparts to vesicles and bilayer patches a certain lateral mobility, not to be confused with that of single lipid molecules. Thus, laterally mobile bilayer patches on mica tend to reshape into circular patches to minimize their line tension; increased forces on the tip can also be employed to induce the coalescence of bilayer patches. A satisfactory lateral mobility accelerates the formation of a complete lipid bilayer coating. In fact, the edges of lipid bilayer patches are thermodynamically unstable, and tend to interact with adjacent lipid material, for example, by rupturing surface-bound vesicles or by coalescing with other bilayer patches (Richter et al., 2006).

The high lateral mobility of lipid molecules on mica allows monolayers and bilayers of lipid mixtures deposited on freshly cleaved, hydrophilic mica to separate into different phases (demixing). The deposition is carried out by Langmuir–Blodgett or Langmuir–Schaefer techniques and also, less frequently, by vesicle fusion. The AFM images of these mica-supported lipid films (either monolayers in contact with air or bilayers in contact with water) show microdomains that are readily distinguished from the surrounding lipid matrix by their greater height. Sphingomyelin (SPM)/dioleoylphosphatidylcholine (DOPC) (1:1) monolayers yield images with irregularly shaped microdomains, whose size and height above the surrounding matrix depends upon the surface pressure used for their Langmuir–Schaefer deposition. Thus, the height of the microdomains decreases from 1 to 0.5 nm with increasing surface pressure from 10 to 30 mN, while their size increases. The microdomains are ascribed to an SPM-rich gel phase, while the low surrounding matrix is assigned to a

DOPC-rich liquid-disordered (l_d) phase (Yuan et al., 2002). Addition of 20% cholesterol to the SPM/DOPC (1:1) monolayer converts the gel phase into an (SPM/cholesterol)-rich liquid-ordered (l_o) phase, about 1 nm higher than the l_d matrix. The further addition of 1% of the GM1 ganglioside changes again the imaging scenario, by determining the appearance of randomly distributed dots, ranging from 40 to 150 nm in diameter and approximately 2.0 nm higher than the surrounding phase. The dots are predominantly localized in the higher, l_o phase, and are attributed to GM1-rich submicrometric islands.

Mica-supported lipid bilayers in water, obtained by depositing a DPPE monolayer and then an SPM/DOPC/cholesterol (1:1:1) monolayer on top of it, are characterized by two phases of equal height but different compactness, which are still to be ascribed to a l_o and a l_d phase. The difference with respect to a mica-supported SPM/DOPC/cholesterol (1:1:1) monolayer in air, where the l_o phase is definitely higher than the l_d phase, is probably to be ascribed to a certain "damping" effect by the underlying DPPE monolayer. Nonetheless, incorporation of 1% GM1 in the SPM/DOPC/cholesterol monolayer determines again the appearance of irregularly shaped microdomains about 2 nm higher than the surrounding bilayer, from 100 to 400 nm in size, and preferentially located in one of the two phases of equal height. That the preferred phase is the l_o one is confirmed by two-photon fluorescence-lifetime imaging microscopy (FLIM) images obtained on GUVs of identical composition by Dietrich et al. (2001) by the use of two different fluorophores. These images reveal a high GM1 concentration in SPM-rich areas, even though the lateral resolution of FLIM (about 0.5 μm radial) could not detect the presence of GM1 submicrometric islands in the l_o phase. The detection of microdomains about 100 nm in size by AFM imaging in biomimetic membranes seems to support the existence of lipid rafts smaller than 100 nm in cell membranes, as suggested from fluorescence experiments (Varma and Mayor, 1998), even though their small size is beyond the resolution of optical microscopy.

7.6.2 Scanning Tunneling Microscopy

STM is a SPM that exploits a quantum mechanical phenomenon called *tunneling effect*. In STM, a sharp metallic tip is placed very close to the surface of a metallic sample and a small *bias voltage* is applied between the tip and the sample. Under these conditions, an electron transfer (*tunneling current*) takes place between one or very few atoms of the tip apex and those on the sample surface closest to them. In general, electrons, by obeying the law of quantum mechanics, can move from a region of space to another even though the two regions are separated by a region in which they have no possibility of existing. In other words, no reference is made to possible intermediate states through which the electron might pass. However, the distance between the two regions (the STM tip and the underlying metal surface) must be very small for tunneling to occur, since the tunneling current decays exponentially with such a distance. By raster-scanning the tip over a small area of the sample, STM provides information not only about the topography of the sample, but also about the spectroscopic properties and local variations of the work function.

Similar to AFM, which can operate in a constant force mode and a constant height mode, STM can operate in a *constant current mode* and a *constant height mode*. In the constant current mode, the tip is scanned over the surface, while the vertical position of the tip varies continuously under control of a feedback loop to maintain the tunneling current constant at some preset value between 0.1 and 10 nÅ. The spatial variation of the tip height is converted into the real space image. In the constant height mode, the tip moves above the surface at a constant height fixed at some value between 3 and 10 Å, without feedback control, and it is the spatial variation of the tunneling current that is converted into the real space image. This mode requires a very smooth surface and, hence, it is not applied with biomimetic membranes. STM images in air of molecularly resolved phosphoglycerol headgroup structures at a dipalmitoylphosphatidylglycerol/1-octadecanthiol bilayer tethered to gold were reported by Gregory et al. (1994).

Most frequently, STM images of lipid films are acquired in solution. At present, high-resolution images are only obtained with metal-supported phospholipid monolayers. This requires the traditional STM technique in air to be extended to electrode/electrolyte interfaces. The resulting technique, which makes use of a potentiostatic system, is referred to as *electrochemical STM* (EC-STM). The electrochemical cell uses the conducting sample as the working electrode, and includes a reference electrode (e.g., an Ag/AgCl electrode) and a counter electrode consisting of an Au or Pt wire (Yagati et al., 2014). With respect to conventional potentiostatic three-electrode systems, it also includes a sharp tip in close proximity to the sample surface; this is not an active electrode in the electrochemical cell, although it must be kept under potential control in order to maintain a constant and suitable potential difference (*bias potential*) between the tip and the sample. This goal is achieved by utilizing the bipotentiostatic setup shown in Fig. 7.22.

A positive feedback circuitry (cf. Section 4.8) keeps the electric potential of the sample at a preset value, E_S, and that of the tip at a different preset value, E_T, with

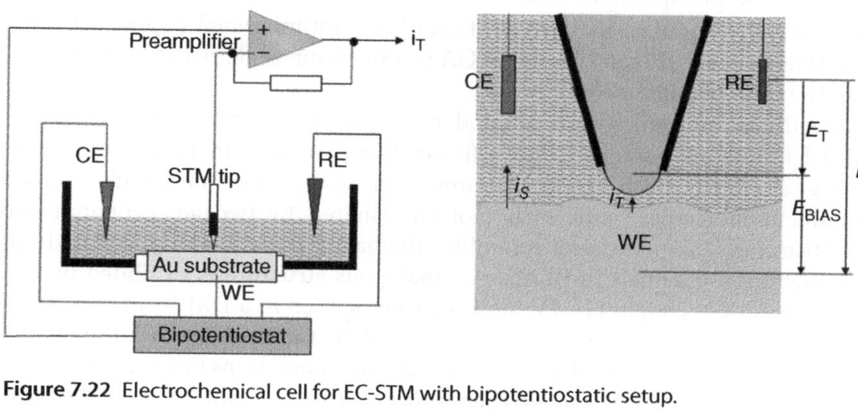

Figure 7.22 Electrochemical cell for EC-STM with bipotentiostatic setup.

respect to the same reference electrode (RE) by drawing a negligibly small current through the latter, at the expense of a finite current (i_S) between the sample and the counter electrode, CE. The difference between E_S and E_T is clearly equal to the potential difference between the tip and the sample, that is, the bias potential denoted by E_{BIAS} in Fig. 7.22. Besides the tip current, i_T, the current flowing through the tip includes other contributions, whose magnitude must be reduced by more than one order of magnitude with respect to the tip current for accurate measurements. These additional contributions are represented by double-layer charging, which introduces noise, and by faradaic currents due to electroactive trace impurities. Coating the tip, except for its extreme apex, with Apiezon wax, melted glass or copolymers reduces these contributions, in view of the fact that both faradaic and capacitive currents are directly proportional to the exposed area of the tip.

Lipkowski and his group obtained atomic-resolution STM images of dimyristoylphosphatidylcholine (DMPC) monolayers deposited on the Au(111) facets of a small bead, either by vesicle fusion or by Langmuir–Blodgett transfer (Xu et al., 2004). The brightness of the spots composing the image increases with an increase in tunneling probability. The resulting contrast may be explained by a weak coupling between electronic states in the adsorbate and in the metal substrate near the Fermi level, which gives the adsorbate a property of an antenna capable of receiving tunneling electrons (Giancarlo and Flynn, 1998). Fusion of DMPC vesicles on Au(111) causes the lipid molecules to adsorb flat with the acyl chains oriented parallel to the surface and to assemble into an ordered monolayer. After about 30 min, the molecules reorient and the monolayer is converted into a hemimicellar film. Transfer of a DMPC Langmuir monolayer onto Au(111) by the Langmuir–Blodgett technique yields a hexagonal lattice of DMPC molecules with the polar heads oriented toward the gold and the acyl chains exposed to the aqueous solution, both in the absence and presence of coadsorbed gramicidin (Sek et al., 2009). The molecules appear as bright spots formed by well-ordered and nearly vertically oriented acyl chains. A slight increase in temperature yields STM images characterized by molecular stripes ascribed to a tilted and less ordered packing of the phospholipid molecules (Smetanin et al., 2014). Similar striped images are also induced by coadsorption of the antimicrobial peptides alamethicin (Pieta et al., 2012) and trichogin GA IV (Smetanin et al., 2014), both belonging to the class of peptaibols.

Alamethicin molecules incorporated in a matrix of DMPC and egg phosphatidylcholine are arranged into a flower-like pattern with hexagonal lattice (Pieta et al., 2012). Each flower-like unit has a central pore surrounded by six alamethicin molecules, with each molecule shared by two adjacent units of the lattice (Fig. 7.23). The unit resembles the barrel-stave model postulated for alamethicin ion channels in BLMs. An analogous structure is exhibited by the STM images of trichogin GA IV molecules embedded in a DMPC matrix, even though the peptide aggregates are much smaller and less ordered (Smetanin et al., 2014). This is explained by peptide–peptide interactions between trichogin GA IV molecules being weaker than those between alamethicin molecules,

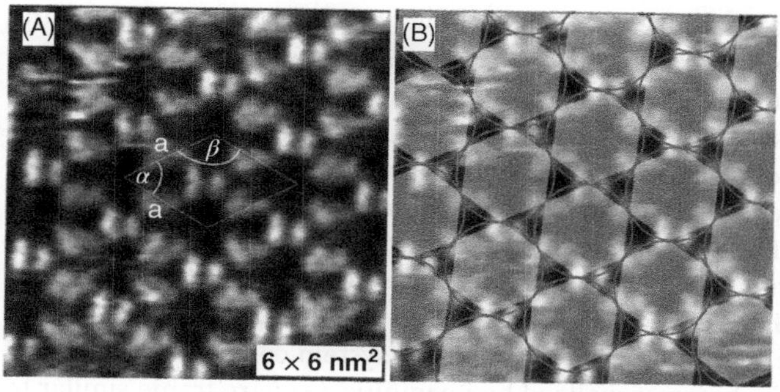

Figure 7.23 AC-STM image of flower-like structures at a Au(111)-supported DMPC monolayer incorporating alamethicin, as acquired using a constant tunneling current of 1.16 nA (A). Schematic arrangement of pores formed by alamethicin molecules (B). Source: Pieta et al. (2012). Reproduced with permission of National Academy of Sciences.

due to their shorter length. Even though alamethicin molecules span the whole hydrophobic region of BLMs and the shorter trichogin molecules are also assumed to do the same by inducing local membrane thinning, the STM images point out the tendency of these peptide molecules to aggregate into pores even within Au-supported lipid monolayers. The fact that the number of monomers forming channel-like aggregates is constantly equal to six may be induced by the underlying structure of the Au(111) support.

The images of a DMPC monolayer with a 10% molar content of gramicidin reveal a hexagonal lattice of DMPC molecules with randomly distributed, triangular-shaped dark features corresponding to cavities in the film (Sek et al., 2009). The area of the cavity agrees with that estimated for a perpendicularly oriented $\beta^{6.3}$-form of the gramicidin helix. These dark features can be explained by the strong tendency of the tryptophan groups in the C-terminus of gramicidin to form H-bonds with the polar heads of phosphatidylcholine molecules, which are turned toward the Au surface. Since the gramicidin molecules are shorter than the DMPC ones, they are expected to be buried inside the DMPC monolayer and to appear in the STM image as cavities.

7.7 Langmuir–Blodgett and Langmuir–Schaefer Transfers

A popular procedure for forming a lipid monolayer on a hydrophobic substrate or a lipid bilayer on a hydrophilic substrate makes use of vesicle fusion (cf. Section 6.4.2). In some cases, the kinetics of vesicle fusion is followed by SPR reflectivity (cf. Fig. 7.8), whereas the quality of the resulting lipid films is verified by AFM imaging. A more accurate procedure for preparing these lipid films involves transferring one or two lipid monolayers formed at the air/water interface, called *Langmuir monolayers*, onto a solid support. This procedure, referred to as *Langmuir–Blodgett* (LB) *technique*, allows an accurate control of the film thickness and compactness and of its homogeneous deposition. It takes advantage of the spontaneous tendency of amphiphilic molecules, such as lipids, to spread spontaneously on the surface of an aqueous solution upon adding a few drops of their dilute solution in a volatile and water-immiscible solvent, such as chloroform. This causes a decrease in the *surface tension*, γ, of the solution, which can be operatively defined as the force acting on the unit length of its surface. From a thermodynamic point of view, it is expressed by the partial derivative, $(\partial G/\partial A)_{T,P,n_k}$, of the Gibbs energy G with respect to the surface area A at constant temperature, pressure, and number of moles of the system components.

Its significance can be better visualized by imagining a square frame similar to a bubble wand used for creating soap bubbles, equipped with a movable, sliding side. If a soap film is stretched across the frame, it exerts on the movable side a force that tends to decrease its surface area and must be compensated for by an equal and opposite force directed outward. Since the soap film is actually a water layer of macroscopic thickness, with two surfaces containing a high surface concentration of an amphiphilic substance (the *surfactant*), the force applied by the surroundings to maintain an equilibrium situation is equal to $2l\gamma$, where l is the length of the movable side of the frame.

The origin of this force, which acts on the thermodynamic system "water layer," can be justified on the basis of the following qualitative considerations. Within the bulk water phase of the soap film, the water molecules interact strongly between themselves via H-bonds. If, *ab absurdo*, the water layer did not contain surfactant molecules (this hypothetical situation is practically unrealizable, because a pure water layer spanning the frame would not be stable), then the surface water molecules would experience an unbalanced attraction toward the bulk phase, due to the lack of interactions toward the gas phase. A force per unit length of the edge of the water surface would, therefore, be required for expanding this surface reversibly. This force is the surface tension of pure water, γ_0, which equals 72.8 mN m^{-1} at 20 °C and atmospheric pressure, where N stands for Newton, the unit of force in the MKS unit system. A force per unit length of the edge of the surface is also required in the real situation of a soap film, although it is less than γ_0 and will be simply denoted by γ. The decrease in surface tension is due to the amphiphilic molecules of the surfactant replacing the water molecules on the surface of the soap film. These molecules have no tendency to escape from the surface, because they are practically insoluble in the bulk water phase. Moreover, their hydrophilic portions are turned toward the bulk phase, where they interact attractively with the underlying water molecules of the *subphase*. In addition, their hydrophobic portions are directed toward the gas phase, and interact attractively between themselves via van der Waals forces.

The positive potential difference, $\gamma_0 - \gamma$, is termed *surface pressure* and is denoted by π. The reason for this appellation can be understood by considering the operation of the *Langmuir balance* (or *Langmuir trough*), which is used to measure this quantity (see Fig. 7.24A). It consists of a long, narrow, and shallow open container filled with water, on whose surface a movable barrier

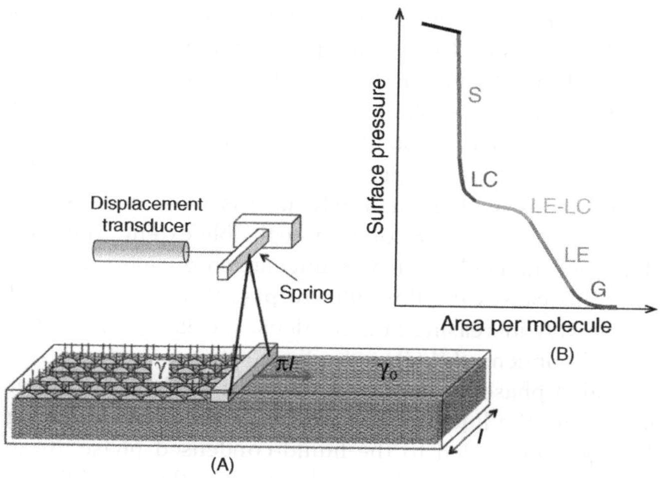

Figure 7.24 (A) Scheme of the Langmuir balance method. (B) Schematic π–A isotherm of a long-chain phospholipid.

(float) separates a clean portion of the water surface from a surface covered with a surfactant monolayer (Girard-Ergot and Blum, 2007). The amphiphilic molecules of the surfactants exert a force F on the float in order to cover a wider portion of the surface and to decrease the Gibbs energy. At equilibrium, an infinitesimal positive shift dx of the float is opposed by the force exerted by the water molecules on the other side of the float. Hence, the infinitesimal work done by F is the resultant of the works done by the advancing lipid molecules and by the retreating surface water molecules on the float:

$$F\ dx = \gamma\ dA_s + \gamma_0\ dA_w = (\gamma - \gamma_0)\ dA_s = -\pi l dx. \tag{7.45}$$

Here, l is the float length, $dA_s = l\,dx$ is the area covered by the surfactant molecules during their infinitesimal advancement, and $dA_w = -dA_s$ is the area released by the retreating water molecules. The surface pressure is given by $-F/l$. It should be noted that the force F in Eq. 7.45 is negative, because it is exerted by the system "lipid monolayer+surface water," and hence it is opposite to that exerted by the "surroundings" (the float) in the thermodynamic reasoning. In Fig. 7.24A, F is measured using a displacement transducer. Nowadays, the *Wilhelmy plate* method, which measures the absolute value of the surface tension γ, is more commonly adopted. A Wilhelmy plate consists of a platinum or filter paper immersed in the subphase. Two measurements are required to determine π, one in the subphase covered by pure water, and the other in the subphase underneath the surfactant monolayer. The force experienced by the plate is the resultant of two downward components, gravity and surface tension, and of the upward component due to the buoyancy determined by water displacement.

The Langmuir trough is commonly employed to measure the curve of the surface pressure π against the average area A occupied by lipid molecules spread on the surface of pure water or an aqueous solution (Fig. 7.24B). This curve is referred to as a $\pi-A$ *isotherm* and provides useful information on the properties of the lipid material, such as monolayer stability and occurrence of phase transitions and conformational transformations. It is obtained by continuously compressing the surface area covered by the lipid at constant temperature, while monitoring the surface pressure. The area A is usually expressed in Å^2 or nm^2/molecule. The amount of lipids initially spread is small enough to behave as a two-dimensional gas that affects the initial water surface tension to a negligible extent. The total number of the insoluble lipid molecules remains unchanged during the gradual surface compression. This causes the lipid film to pass through a series of different phases of progressively increasing number density, called gaseous (G), liquid-expanded (LE), liquid-condensed (LC), and solid (S) phase.

During the liquid-expanded phase, the curve of the $\pi-A$ isotherm is rather steep, and even steeper during the liquid-condensed and solid phases. During the transition from the liquid-expanded to the liquid-condensed phase, these two phases may coexist for a while, in which case the $\pi-A$ isotherm exhibits a plateau interposed between the rising sections ascribed to the liquid-expanded and liquid-condensed phases (see Fig. 7.24b). Ultimately, as the compression of the lipid monolayer exceeds a certain limit, the lipid film becomes mechanically unstable and patches of lipid monolayers start moving on top of each other, generating disordered multilayers. This phenomenon is termed *collapse*. This occurs when the surface pressure exceeds the so-called *equilibrium spreading pressure*, which corresponds to the equilibrium pressure between the two-dimensional state of the lipid monolayer and the three-dimensional state of the corresponding crystal. It can be regarded as spontaneously generated when a crystalline sample of the lipid material is left in contact with a pure water surface, allowing the lipid molecules to detach from the crystal and spread over the water subphase until equilibrium is reached.

A lipid monolayer floating in a Langmuir trough can be transferred, similar to a carpet, from the water surface to a flat solid support, provided the surface pressure is high enough to ensure lateral cohesion in the monolayer. For this reason, monolayer transfer is often carried out in the condensed state, while carefully avoiding the attainment of the equilibrium spreading pressure. In fact, exceeding this pressure would endanger the monolayer stability. Lipid monolayer transfer is achieved by the *Langmuir–Blodgett* technique, which makes use of a Langmuir trough with two movable barriers enclosing the lipid monolayer between two pure water surfaces (Fig. 7.25A). A Wilhelmy plate monitors the surface pressure, which is kept constant at a preset value by the two movable barriers via a feedback system during the whole transfer operation.

This technique involves lowering a substrate (e.g., a gold, mica, or silica plate), held by a dipping arm in a perpendicular orientation, into the subphase or in raising it from the subphase, under computer control; the former operation is called immersion or *downstroke*, the latter one emersion or *upstroke*. If the substrate is hydrophilic (e.g., freshly cleaved mica or polished and flame-annealed gold), an upstroke is required, since it brings the polar heads of the lipid monolayer,

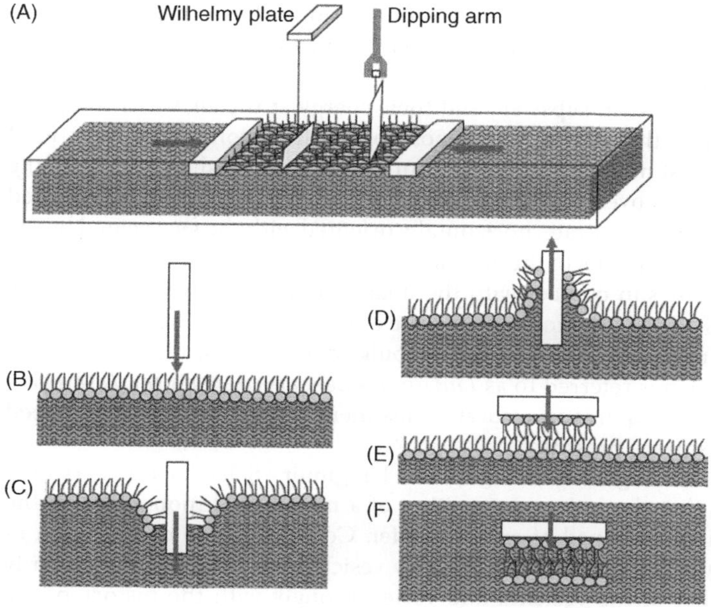

Figure 7.25 Langmuir trough for vertical transfer onto a solid substrate (A); Langmuir–Blodgett deposition onto a hydrophobic (B, C) and a hydrophilic (D) substrate; Langmuir–Schaefer deposition (E, F).

turned toward the water phase, in direct contact with the hydrophilic surface, which becomes coated by the monolayer (Fig. 7.25D). In this case, the substrate may be conveniently immersed into water before spreading the lipid on the water surface. Conversely, if the substrate is hydrophobic (e.g., a gold plate with an alkanethiol monolayer tethered to its surface), a downstroke is required, so as to bring the hydrocarbon tails of the monolayer, turned toward the gas phase, in direct contact with the substrate surface (Fig. 7.25B and C). The subsequent emersion of the resulting lipid-coated substrate is not accompanied by a change in the surface pressure monitored by the Wilhelmy plate, and hence by an adjustment of the two barriers.

After the transfer of a lipid monolayer by the Langmuir–Blodgett technique, a hydrophilic substrate becomes hydrophobic, since the acyl chains are turned outward. Hence, the downstroke of a lipid-coated substrate may cause the deposition of a further lipid monolayer, yielding a supported lipid bilayer. However, this bilayer deposition by the Langmuir–Blodgett technique is not always satisfactory, in that the second monolayer may turn out to be incomplete. This is verified by calculating the so-called *deposition ratio* (also named *transfer ratio*), defined as the decrease in the area occupied by the lipid monolayer on the Langmuir trough surface divided by the concomitant increase in the immersed area of the solid substrate. The transfer is entirely satisfactory if the deposition ratio equals unity, whereas values less than 0.8 suggest a poor homogeneity of the transferred monolayer. On gold substrates, the deposition ratio may even be equal to −1, indicating complete stripping of the first monolayer from the substrate (Zawisza et al., 2007). In fact, a delicate balance exists between the adsorption energy of the molecules on the solid substrate and their energy at the air/water interface. It may then happen that, during the deposition of the second monolayer, lipid molecules of the first monolayer desorb from the substrate and move over to the air/water interface. To overcome this problem, the second monolayer is deposited by slowly lowering the substrate, horizontally oriented with the face coated by the first monolayer lying parallel to the air/water interface and turned toward it, until the whole face comes in contact with the floating lipid monolayer (Fig. 7.25E). The resulting tail-to-tail interactions generate a substrate-supported lipid bilayer that is pushed through the interface into the bulk aqueous phase (Fig. 7.25F). This alternative approach is referred to as *Langmuir–Schaefer* deposition. A satisfactory lipid deposition requires an accurate adjustment of both the deposition speed and the transfer surface pressure.

An advantage of Langmuir–Blodgett and Langmuir–Schaefer transfers over vesicle fusion is that the strict composition of a mixed lipid monolayer on the trough surface is maintained after its transfer. Conversely, the composition of a mixed lipid layer formed on a substrate by vesicle fusion does not necessarily correspond to that of the vesicles, and varies strongly with the history of the sample preparation. On the other hand, an advantage of vesicle fusion over Langmuir–Blodgett and Langmuir–Schaefer transfers is that vesicles may easily incorporate membrane proteins, forming proteoliposomes. To this end, membrane proteins are usually stabilized in detergent micelles and then incorporated into the lipid vesicles upon removing the detergent molecules by dialysis. Fusion of proteoliposomes on a metal-supported thiolipid monolayer may then induce the insertion of the membrane proteins into the resulting lipid bilayer.

7.8 Quartz-Crystal Microbalance

The quartz-crystal microbalance (QCM) is a nanogram-sensitive technique that utilizes acoustic waves generated by oscillating a piezoelectric crystal-quartz plate to measure a mass. *Piezoelectricity* is the ability of certain materials to generate an AC voltage when subjected to mechanical stress or vibration, or to vibrate when subjected to an AC voltage, or both. In a QCM, an alternating electric field is applied to a quartz crystal sandwiched between two metal electrodes, in order to induce an alternating expansion and contraction of its crystal lattice. Typically for QCM applications, quartz crystals a few tenths of a millimeter in thickness are cut at a 35°10′ angle from their *z*-axis (Marx, 2003). This geometry provides a stable oscillation with almost no temperature fluctuation in frequency around room temperature. Au or Pt metals are usually deposited upon the upper and lower quartz surfaces. Wire leads are attached to these electrodes and connected to an oscillator circuit (Fig. 7.26A). Resonance is excited when a sufficient AC voltage is applied with a frequency close to the *resonant frequency*, f_0, of the quartz

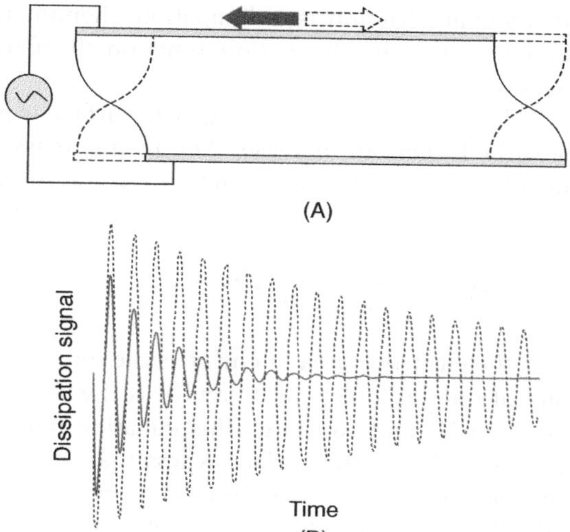

Figure 7.26 (A) Quartz crystal with alternating current applied across two gold electrodes, schematically depicted under shear stress in opposite directions. (B) Oscillatory decay of the QCM-D as it comes to rest, both before (dashed curve) and after deposition of the adsorbed material (solid curve).

crystal. The resonant condition of the QCM is achieved when the amplitude of the acoustic standing wave generated by its alternating expansion and contraction is an odd number of the thickness of the quartz plate. The resonant frequency increases with a decrease in the quartz-crystal thickness. Typically, a resonant frequency f_0 of 5 MHz is realized with a quartz plate about 330 μm thick. A mass attached to the quartz plate surface causes the crystal oscillation frequency f to decrease with respect to the intrinsic resonant frequency f_0 of the free plate surface ($\Delta f = (f_0 - f) > 0$).

The lateral amplitude of a vibrating crystal is 1–2 nm. Any mass attached to the quartz surface will tend to oscillate with the same lateral displacement and the same frequency, but not all adsorbed materials respond to this deformation (*strain*) in the same way. Let us imagine subdividing the adsorbed material, assumed to be distributed uniformly on the quartz plate, into many thin slabs parallel to the plate surface. If each individual slab is thin enough, its motion can be regarded as that of a rigid body. The strain of each slab is just its response to an applied *stress*, that is, the force acting per unit surface of the slab and generated by the AC voltage. The components of the stress vector parallel and normal to the slab are called *shear stress* and *normal stress*, respectively. If all the thin slabs move at the same velocity during the deformation, the shear stress acts to restore the material to its rest state, and the stress is said to be *elastic*. Conversely, if the slabs move at different velocities, the shear stress depends on the rate of change of the deformation over time and tends to oppose such a change; it is called *viscous stress*. More precisely, each slab moves slightly faster than the immediately underlying slab and slightly slower that the immediately overlying one. This implies that each slab will experience a forward drag from the layer above and a backward drag from the layer below. This amounts to an internal friction, called *viscosity*, which tends to counteract this velocity gradient.

Incidentally, friction is a dissipation phenomenon and, as such, it contributes to increasing the dissipation function Φ of irreversible thermodynamics (cf. Section 3.3).

If the mass attached to the quartz surface is purely elastic, that is, it is rigidly adsorbed, then an increase, Δm, in this mass is proportional to the resulting decrease in oscillation frequency, Δf, according to the so-called *Sauerbrey equation*:

$$\Delta f = -2\frac{f^2}{A\left(\mu\rho_q\right)^{1/2}}\Delta m. \tag{7.46}$$

A is the electrode area in cm^2, f is the intrinsic crystal frequency, ρ_q is the density of quartz (2.65 g cm^{-3}), and μ is the shear modulus (2.95×10^{11} dyn cm^{-2}); Δf is in Hz and Δm in g. This equation holds strictly only if the elastic mass added to the crystal surface is less than about 2% of the crystal mass and the mass is evenly distributed over the active area of the crystal. Unfortunately, many adsorbed materials incorporate both elastic and viscous portions, thus violating the requirement of a rigidly adsorbed mass.

This prompted a new approach for characterizing mass deposits with frictional dissipative losses, which led to the development of a *QCM with dissipation monitoring* (QCM-D). This microbalance switches off the AC voltage imparting an excitation close to the resonance frequency f_0 to the crystal, and monitors the resulting decay rate of the crystal oscillation, which is proportional to the energy dissipation of the oscillator (Fig. 7.26B). The decay voltage, namely the output voltage amplitude as a function of time, is mixed with a reference frequency, f_R, and filtered with a low-pass band filter. This gives an output frequency f based on the difference between f_R and f_0. This difference frequency is fit to the following exponentially decaying sinusoidal function (Dixon, 2008):

$$A(t) = A_0 \exp(-t/\tau)\sin\left(2\pi ft + \phi\right). \tag{7.47}$$

The computer repeatedly excites the crystal and records the changes in f and decay time τ, and the damping data are converted from analog to digital. The *dissipation parameter* is defined as $D = W_{\text{dissipated}}/(2\pi W_{\text{stored}})$, where $W_{\text{dissipated}}$ is the energy dissipated during an oscillation cycle and W_{stored} is that stored in the oscillating system. It is provided by the expression $D = (\pi f\tau)^{-1}$, where f and τ are obtained from the fitting with Eq. 7.47. The QCM-D technology also allows a quantitative analysis of thickness, shear elastic modulus, and viscosity of adsorbed films, well beyond the Sauerbrey regime. This result is achieved by combining the frequency and dissipation measurements from multiple harmonics and applying simulations based on viscoelastic models.

Under shear stress, well-formed supported lipid layers behave more elastically (or, in other words, less viscously) than adsorbed vesicles, coupling strongly to the motion of the crystal surface. Hence, their mass satisfies the requirements of Sauerbrey equation. Conversely, adsorbed or partially fused vesicles are substantially larger and less compact structures filled with water; moreover, water is also trapped between the vesicles and the surface. Consequently, vesicles are subject to larger deformations under shear stress. These viscoelastic features are invisible by simple resonant frequency determination. The dissipation parameter D allows viscoelasticity to be visualized as a function of time, making a distinction between intact, adsorbed vesicles (high dissipation) and lipid bilayer patches (low dissipation).

Figure 7.27 shows the changes in resonance frequency (Δf) and in dissipation (ΔD) following fusion of vesicles of different composition on silica (Richter et al., 2006). If vesicles do not adsorb, no change in f or D takes place (A). If vesicles adsorb and remain intact, forming a supported vesicular layer, their high mass causes an appreciable frequency decrease, and also a high dissipation (B). If vesicles are initially adsorbed, but subsequently fuse forming a water-free lipid bilayer of lower mass, Δf first shows a minimum and then attains a maximum constant value, while ΔD shows a maximum and then drops to an almost zero value (C). Figure 7.27D exemplifies the case in which the vesicles adsorb and rupture instantaneously, forming a lipid bilayer. Combined measurements by QCM-D and SPR have shown that isolated vesicles of egg-phosphatidylcholine (egg-PC) remain intact when bound to a silica support; a minimum critical vesicle coverage is required to initiate the conversion of surface-bound vesicles into lipid bilayer patches (Keller et al., 2000). When the critical vesicle coverage is reached, the support-induced stress of the vesicles becomes sufficient to induce their rupture. Use of the QCM-D permitted to confirm the formation of lipid bilayers on silica and of lipid monolayers on gold-supported alkanethiol monolayers by vesicle fusion, and the adsorption of intact vesicles on oxidized gold (Keller and Kasemo, 1998); silica deposition on quartz crystal was carried out by evaporation or spin coating, whereas thiol monolayers on gold were directly tethered to the gold electrode of the QCM-D by self-assembly.

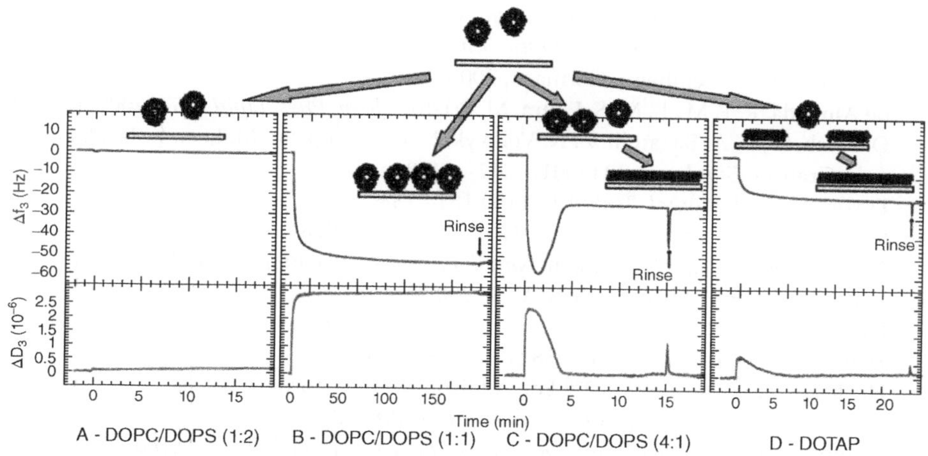

Figure 7.27 Lipid deposition pathways measured by QCM-D on silica. The legends indicate the lipids used: dioleoyltrimethylammonium-propane (DOTAP), dioleoylphosphatidylcholine (DOPC), and dioleoylphosphatidylserine (DOPS), with their molar mixing ratios. Source: Richter et al. (2006). Reproduced with permission of American Chemical Society.

It should be borne in mind that the mass obtained from QCM-D measurements corresponds to the total mass coupled to the motion of the sensor crystal, including the mass of the adsorbed biomolecules and of the solvent bound or dynamically coupled to them. This feature distinguishes mass measurements by QCM-D from those by SPR. In fact, the measured SPR signal originates from altered conditions for resonant surface plasmon excitation due to changes in the interfacial refractive index. For a simple dielectric material there exists, as a first approximation, a nearly linear relationship between the change in refractive index caused by biomolecule adsorption and the number of biomolecules at the interface; on the other hand, SPR is practically insensitive to the presence of water molecules. Hence, the mass sensed by a QCM exceeds that estimated by SPR by the amount of water molecules bound or dynamically coupled to the adsorbed biomolecules.

References

Axelrod, D., *J. Cell Biol.* 89 (1981) 141–145.

Bagatolli, L. A., *Chem. Phys. Lipids* 122 (2003) 137–145.

Bagatolli, L. A., *Biochim. Biophys. Acta* 1758 (2006) 1541–1556.

Baumgart, T., M. Kreiter, H. Lauer, R. Naumann, G. Jung, A. Jonczyk, A. Offenhäusser, W. Knoll, *J. Colloid Interface Sci.* 258 (2003) 298–309.

Becucci, L., S. Martinuzzi, E. Monetti, R. Mercatelli, F. Quercioli, D. Battistel, R. Guidelli, *Soft Matter* 6 (2010) 2733–2741.

Beden, B., C. Lamy, in: *Spectroelectrochemistry. Theory and Practice*, R. J. Gale (Ed.); Plenum Press, New York, 1988.

Binder, H., *Appl. Spectrosc. Rev.* 38 (2003) 15–69.

Bizzotto, D., Y. Yang, J. L. Shepherd, R. Stoodley, J. Agak, V. Stauffer, M. Lathuillière, A. S. Akhtar, E. Chung, *J. Electroanal. Chem.* 574 (2004) 167–184.

Bordo, V. G., H.-G. Rubahn, in: *Optics and Spectroscopy at Surfaces and Interfaces*; Wiley-VCH, Weinheim, Germany, 2005.

de Almeida, R. F. M., L. M. S. Loura, M. Prieto, *Chem. Phys. Lipids* 157 (2009) 61–77.

Dietrich, C., L. A. Bagatolli, Z. N. Volovyk, N. L. Thompson, M. Levi, K. Jacobson, E. Gratton, *Biophys. J.* 80 (2001) 1417–1428.

Dixon, M. C., *J. Biomol. Tech.* 19 (2008) 151–158.

Giancarlo, L., G. Flynn, *Annu. Rev. Phys. Chem.* 49 (1998) 297–336.

Girard-Ergot, A. P., L. J. Blum, in: *Nanobiotechnology of Biomimetic Membranes*, D. Martin (Ed.), Springer, New York, 2007.

Gregory, B. W., R. A. Dluhy, L. A. Bottomley, *J. Phys. Chem.* 98 (1994) 1010–1021.

Hansen, W. N., *J. Opt. Soc. Am.* 58 (1968) 380–390.

Hecht, A., A. Zajac, *Optics*, 2nd Ed., Addison-Wesley, Toronto, 1975.

Helen, E. H., D. Axelrod, *J. Opt. Soc. Am. B* 4 (1987) 337–350.

Homola, J., in: *Surface Plasmon Resonance Based Sensors*, J. Homola (Ed.), Springer, Berlin, 2006.

Homola, J., M. Piliarik in: *Surface Plasmon Resonance Based Sensors*, J. Homola (Ed.), Springer, Berlin, 2006.

Jackson, A. J., http://www.ncnr.nist.gov/summerschool/ss10/pdf/SANS_NR_Intro.pdf, 2008.
Jass, J., T. Tjärnhage, G. Puu, *Biophys. J.* 79 (2000) 3153–3163.
Jeuken, L. J. C., S. D. Connell, M. Nurnabi, J. O'Reilly, P. J. F. Henderson, S. D. Evans, R. J. Bushby, *Langmuir* 21 (2005) 1481–1488.
Jeuken, L. J. C., S. D. Connell, P. J. F. Henderson, R. B. Gennis, S. D. Evans, R. J. Bushby, *J. Am. Chem. Soc.* 128 (2006) 1711–1716.
Kalb, E., S. Frey, L. K. Tamm, *Biochim. Biophys. Acta* 1103 (1992) 307.
Keller, C. A., B. Kasemo, *Biophys. J.* 75 (1998) 1397–1402.
Keller, C. A., K. Glasmästar, V. P. Zhdanov, B. Kasemo, *Phys. Rev. Letters* 84 (2000) 5443–5446.
Lakowicz, J. R., *Principles of Fluorescence Spectroscopy*, 3rd Ed., Springer, New York, U.S.A., 2006.
Leitch, J. J., PhD Thesis, 2014, https://atrium.lib.uoguelph.ca/xmlui/.../Leitch_JJay_201411_Phd.pdf
Maier, S. A., *Plasmonics: Fundamentals and Applications*, Springer, New York, U.S.A., 2007.
Margineanu, A., J.-I. Hotta, M. Van der Auweraer, M. Ameloot, A. Stefan, D. Belijonne, Y. Engelborghs, A. Herrmann, K. Müller, F.C. De Schryver, J. Hofkens, *Biophys. J.* 93 (2007) 2877–2891.
Marx, K. A., *Biomacromolecules* 4 (2003) 1099–1120.
Matyszewska, D., R. Bilewicz, Z.-F. Su, F. Abbasi, J. J. Leitch, J. Lipkowski, *Langmuir* 32 (2016)1791–1798.
McGillivray, D. J., G. Valincius, F. Heinrich, J. W. F. Robertson, D. J. Vanderah, W. Febo-Ayala, I. Ignatjev, M. Lösche, J. J. Kasianowicz, *Biophys. J.* 96 (2009) 1547–1553.
Meeten, G. H., *Meas. Sci. Technol.* 8 (1997) 728–733.
Naumann, R., T. Baumgart, R. Gräber, A. Jonczyk, A. Offenhäusser, W. Knoll, *Biosens. Bioelectr.* 17 (2002) 25–34.
Osawa, M., M. Kuramitsu, A. Hatta, W. Suëtaka, H. Seki, *Surf. Sci.* 175 (1986) L787–L793.
Parikh, A. N., *Biointerphases* 3 (2008) FA22.
Pettinger, B., J. Lipkowski, M. Hoon-Khosla, *J. Electroanal. Chem.* 500 (2001) 471–478.
Pieta, P., J. Mirza, J. Lipkowski, *Proc. Natl. Acad. Sci. USA* 109 (2012) 21223–21227.
Pynn, R., in: *Neutron Applications in Earth, Energy and Environmental Sciences*, L. Liyuan, R. Rinaldi, H. Schober (Eds.), Springer, New York, 2009.
Richter, R. P., A. Mukhopadhyay, A. Brisson, *Biophys. J.* 85 (2003) 3035.
Richter, R. P., A. R. Brisson, *Biophys. J.* 88 (2005) 3422–3433.
Richter, R. P., R. Bérat, A. R. Brisson, *Langmuir* 22 (2006) 3497–3505.
Salamon, Z., H. A. Macleod, G. Tollin, *Biophys. J.* 73 (1997) 2791–2797.
Schmidt, E. K., T. Liebermann, M. Kreiter, A. Jonczyk, R. Naumann, A. Offenhäusser, E. Neumann, A. Kukol, A. Maelicke, W. Knoll, *Bionsens. Bioelectron.* 13 (1998) 585–591.

Sek, S., T. Laredo, J. R. Dutcher, J. Lipkowski, *J. Am. Chem. Soc.* 131 (2009) 6439–6444.

Smetanin, M., S. Sek, F. Maran, J. Lipkowski, *Biochim. Biophys. Acta* 1838 (2014) 3130–3136.

Tawa, K., K. Morigaki, *Biophys. J.* 89 (2005) 2750–2758.

Valeur, B., *Molecular Fluorescence*, Wiley-VCH, New York, 2002.

Varma, R., S. Mayor, *Nature* 394 (1998) 798–801.

Vockenroth, I. K., C. Ohm, J. W. F. Robertson, D. J. McGillivray, M. Lösche, I. Köper, *Biointerphases* 3 (2008a) FA68–FA73.

Vockenroth, I. K., P. P. Atanasova, A. T. A. Jenkins, I. Köper, *Langmuir* 24 (2008b) 496–502.

Wahl, M., T. Röhlicke, H.-J. Rahn, R. Erdmann, G. Kell, A. Ahlrichs, M. Kernbach, A. W. Schell, O. Benson, *Rev. Sci. Instrum.* 84 (2013) 043102.

Xu, S., G. Szymanski, J. Lipkowski, *J. Am. Chem. Soc.* 126 (2004) 12276–12277.

Yagati, A. K., J. Min, J.-W. Choi, in: *Modern Electrochemical Methods in Nano, Surface and Corrosion Science*, M. Aliofkhazraei (Ed.), InThec, Rijeka, Croatia, 2014, Chap. 3.

Yuan, C., J. Furlong, P. Burgos, L. J. Johnston, *Biophys. J.* 82 (2002) 2526–2535.

Zamlynny, V., PhD Thesis, 2002, http://www.acadiau.ca/~vzamlynn/VZ_Pdf/VZ_PhD_Thesis.pdf.

Zamlynny, V., J. Lipkowski, in: *Diffraction and Spectroscopic Methods in Electrochemistry*, Vol. 9 *Advances in Electrochemical Science and Engineering*, R. C. Alkire, D. M. Kolb, J. Lipkowski, P. N. Ross (Eds.), Wiley-VCH, Weinheim, 2006.

Zawisza, I., A. Lachenwitzer, V. Zamlynny, S. L. Horswell, J. P. Goddard, J. Lipkowski, *Biophys. J.* 85 (2003) 4055–4075.

Zawisza, I., X. Bin, J. Lipkowski, *Langmuir* 23 (2007) 5180–5194.

Zebrowska, A., P. Krysinski, *Langmuir* 20 (2004) 11127–11133.

Index

d

m

n

q

Printed and bound by CPI Group (UK) Ltd, Croydon, CR0 4YY
17/09/2026
14964322-0001